THEORY OF PLASMA INSTABILITIES

Volume 2: Instabilities of an Inhomogeneous Plasma

STUDIES IN SOVIET SCIENCE

PHYSICAL SCIENCES

1973

Densification of Metal Powders during Sintering
V. A. Ivensen

The Transuranium Elements
V. I. Goldanskii and S. M. Polikanov

Gas-Chromatographic Analysis of Trace Impurities
V. G. Berezkin and V. S. Tatarinskii

A Configurational Model of Matter
G. V. Samsonov, I. F. Pryadko, and L. F. Pryadko

Complex Thermodynamic Systems
V. V. Sychev

Crystallization Processes under Hydrothermal Conditions
A. N. Lobachev

Migration of Macroscopic Inclusions in Solids
Ya. E. Geguzin and M. A. Krivoglaz

1974

Theory of Plasma Instabilities
Volume 1: Instabilities of a Homogeneous Plasma
A. B. Mikhailovskii

Theory of Plasma Instabilities
Volume 2: Instabilities of an Inhomogeneous Plasma
A. B. Mikhailovskii

LIFE SCIENCES

1973

Motile Muscle and Cell Models
N. I. Arronet

Pathological Effects of Radio Waves
M. S. Tolgskaya and Z. V. Gordon

Central Regulation of the Pituitary-Adrenal Complex
E. V. Naumenko

STUDIES IN SOVIET SCIENCE

THEORY OF PLASMA INSTABILITIES

Volume 2: Instabilities of an Inhomogeneous Plasma

A. B. Mikhailovskii

I. V. Kurchatov Institute of Atomic Energy
Moscow, USSR

Translated from Russian by
Julian B. Barbour

SPRINGER SCIENCE+BUSINESS MEDIA, LLC

Library of Congress Cataloging in Publication Data

Mikhaĭlovskiĭ, Anatoliĭ Borisovich.
Theory of plasma instabilities.

Translation of Teoriia plazmennykh neustoĭchivosteĭ.
Includes bibliographies.
CONTENTS: v. 1. Instabilities of a homogeneous plasma.–v. 2. Instabilities of an inhomogeneous plasma.
1. Plasma instabilities. I. Title. [DNLM: 1. Nuclear physics. QC718 M636t 1974]
QC718.5.S7M5413 530.4'4 73-83899

ISBN 978-1-4899-4787-1 ISBN 978-1-4899-4785-7 (eBook)
DOI 10.1007/978-1-4899-4785-7

Dr. Anatolii Borisovich Mikhailovskii is one of the world's leading plasma theoreticians. He works at the renowned I. V. Kurchatov Institute of Atomic Energy in Moscow in the theoretical group led by Academician M. A. Leontovich. He has specialized in the field of plasma instabilities, and many of the more than 70 papers he has published in Soviet and Western journals are already regarded as classics.

The original Russian text, published by Atomizdat in Moscow in 1971, has been corrected by the author for the present edition. The translation is published under an agreement with Mezhdunarodnaya Kniga, the Soviet book export agency.

ТЕОРИЯ ПЛАЗМЕННЫХ НЕУСТОЙЧИВОСТЕЙ
2. Неустойчивости неоднородной плазмы
А. Б Михайловский

TEORIYA PLAZMENNYKH NEUSTOICHIVOSTEI
2. Neustoichivosti Neodnorodnoi Plazmy
A. B. Mikhailovskii

Originally published by Consultants Bureau, New York in 1974
Softcover reprint of the hardcover 1st edition 1974

Preface

The work of many investigators has shown that different types of instability can arise in a plasma. These instabilities arise because of a thermodynamic nonuniformity of the plasma – either the particles have a non-Maxwellian velocity distribution or the plasma is spatially inhomogeneous. An instability may be due to either or both of these factors. The instabilities due to non-Maxwellian velocity distributions are called the *instabilities of a homogeneous plasma*. If an instability can develop only when there is spatial inhomogeneity of the plasma, it belongs to the class of *instabilities of an inhomogeneous plasma*, also known as *gradient instabilities*.

Our first volume was devoted to the instabilities of a homogeneous plasma. In the present volume we shall study the main results of the theory of gradient instabilities. These instabilities play an important role in plasma physics. Interest in these instabilities derives, on the one hand, from their striking physical properties and, on the other hand, from the recognition that they are one of the principal hindrances in the way of a successful resolution of the problem of controlled thermonuclear fusion.

The gradient instabilities discussed in this book occur in a plasma in a magnetic field. It is not a plasma gradient as such that is responsible for the instability but only the component of the gradient at right angles to the magnetic field.

The gradient instabilities are strongly affected by many factors that characterize the steady state of the plasma and magnetic field. Among the most important factors we may mention the following.

1. Ratio of the plasma pressure to the magnetic field pressure. This is characterized by the number $\beta \equiv 8\pi p/B^2$. As a rule, we shall assume that the parameter β is small.

2. Degree of inhomogeneity of the plasma. This is characterized by the ratio of the particle Larmor radius ρ to the length scale a of the plasma inhomogeneity. In actual situations, the dimensionless parameter ρ/a is usually small. This will, as a rule, be assumed in this book.

3. Nature of the particle velocity distribution. An inhomogeneous plasma may consist of several streams of charged particles or contain components with an anisotropic particle velocity distribution. We shall allow for both these factors, although our main attention will be devoted to a plasma whose particles have an almost Maxwellian velocity distribution.

4. Degree to which a plasma is collisional. Depending on the plasma density and temperature and its degree of inhomogeneity, the frequency of binary collisions between the particles may be either smaller or greater than the frequency of the oscillations that are excited. We shall allow for both possibilities in this book.

5. Curvature of the magnetic lines of force. This factor can be characterized by the ratio of the length scale a of the plasma gradient to the radius of curvature a/R. The nature of the instabilities depends strongly on the sign of the curvature and on the relationship between the parameters a/R and $(\rho/a)^2$. We shall consider all the physically different situations that can arise.

6. Effect of a static or quasistatic electric field. This field may have a component along or at right angles to the magnetic field. We shall consider both possibilities.

7. Shear of the magnetic field. In the simplest cases this can be characterized by the angle between magnetic lines of force separated from each other by a distance of order a. This factor will also be taken into account.

We should point out that the assumption of a small β appreciably simplifies the exposition of instability theory. It does, however, mean that one excludes the instabilities inherent in plasmas with high density and temperature. This is unfortunate, since it has now been recognized that it is only in plasmas with a value of β that is not small that the problem of controlled thermonuclear fusion can be successfully attacked (if β is very small, a positive energy balance cannot be achieved). In the theoretical investigation of plasmas with intermediate values of β, one cannot assume that the perturbations of the electric field are electrostatic in nature; a correct treatment is then only possible with a much more complicated mathematical apparatus. A systematic exposition of plasma instability theory with fairly large β would greatly increase the size of this book. We have therefore restricted ourselves to some of the simplest results of this branch of instability theory.

One could construct the theory of gradient instabilities directly on the basis of the equations that take into account simultaneously all the various possibilities. However, we have preferred to adopt a different approach: we first describe the simplest instabilities and then progressively ramify our original description by taking into account additional factors. This is exactly the same as the approach we have adopted in our exposition of the theory of instabilities of a homogeneous plasma (in Volume 1).

The instability mechanisms inherent in an inhomogeneous plasma can be demonstrated in various special cases – when the inhomogeneity of the magnetic field is ignored and also collisions and the electric field. To avoid the problem of having to deal with two species of charges – electrons and ions – one can adopt the approximation of a strongly inhomogeneous plasma and assume that one has purely electron perturbations. This approach is adopted in Chapter 1, which has two purposes – to introduce the reader to the theory of gradient instabilities in general and to summarize the main results on the electron gradient instabilities that can develop in a strongly inhomogeneous plasma.

Chapters 2 and 3 are concerned with the problem of how the results obtained in Chapter 1 are modified when the plasma is not so strongly inhomogeneous. In Chapter 2 it is shown that the purely electron instabilities are replaced by ion-cyclotron and high-frequency electron–ion instabilities if the ratio ρ/a is not too small. If ρ/a is very small, only perturbations whose frequency is low compared with the ion-cyclotron frequency can be excited (Chapter 3).

The instabilities considered in Chapters 1-3 constitute the class of so-called drift, or universal, or current-convective, or beam-drift instabilities of a collisionless plasma. At one time the word "drift" was used in various papers to denote waves with a phase velocity of the order of the Larmor drift velocity (see, for example, the papers of Kadomtsev, Timofeev, and the author quoted at the end of Chapter 3). However, the term "Larmor drift" is rather unfortunate, since no drift of particles actually takes place at all. What is more, the mechanism of the majority of gradient instabilities and, in particular, the Kadomtsev–Timofeev drift instability, is not in any way related to Larmor drift. (The assertion to the contrary in the two papers mentioned above is incorrect.) Therefore, in Chapters 1-3 and subsequently, we have refrained from using terms like "drift instabilities" and "beam-drift instabilities" in view of the lack of physical justification for such terms and the possible misunderstandings they could introduce in the interpretation of the instability mechanisms.

Begining with Chapter 4 we take into account collisions between charged particles. Chapter 4 is devoted to an exposition of the procedure for allowing for collisions, and the gradient instabilities of a collisional plasma are discussed in Chapter 5.

The number of gradient instabilities is further increased if one allows for curvature of the magnetic lines of force. It is not surprising that the analysis of the instabilities is much more complicated in this case, if for no other reason than the great variety of physically different configurations of the magnetic field. To simplify this analysis, the effect of the curvature can be simulated by introducing a fictitious gravitational force into the problem. This approach is adopted in Chapter 6. We discuss the flute and various other forms of gravitational instability and a number of stabilizing effects due to a favorable direction of the curvature of the lines of force.

If there is an electric field present in the plasma at right angles to the magnetic field, two new forms of gradient instability are possible: a centrifugal instability due to the plasma inhomogeneity and the difference between the electric drifts of the electrons and ions and an instability due to the slipping of neighboring layers of the plasma relative to each other. These instabilities are discussed in Chapter 7.

Beginning with Chapter 8, we allow for a shear of the magnetic field. In Chapter 8 we consider the influence of a shear when there is no gravitational force; in Chapter 9, when there is. In both cases the shear gives rise to stabilization effects.

Taken together, the results of Chapters 1-9 give a fairly clear picture of all the main types of instabilities of both collisionless and collisional plasmas. However, we have yet to consider a weakly ionized plasma, for the collisions of charged particles with neutrals are ignored in the first nine chapters. The gradient instabilities of a weakly ionized plasma are discussed in Chapter 10.

Throughout the first ten chapters it is assumed that the approximation of straight magnetic lines of force is valid, hence the title of the first part of the book. The analysis of instabilities which allows for the real curvature of the magnetic lines of force is contained in Part 2, which comprises Chapters 11-18. This analysis is of particular importance for the problem of plasma containment in different types of magnetic trap.

The most rapidly developing instabilities of a plasma in a curved magnetic field can be studied by two essentially different approaches: the electrodynamic, based on the notion of perturbed fields, and the magnetohydrodynamic, which employs the notion of the displacement of the plasma as a whole. In Volume 1 and Part 1 of this volume we have adopted the electrodynamic approach exclusively. This approach is also adopted (Chapter 11) for the investigation of the most rapidly growing gradient perturbations of a plasma in a curved magnetic field. (The principal representative of such instabilities is the flute instability.) The magnetohydrodynamic approach is formulated in Chapter 12. In a number of the following chapters we use this approach to study the fast gradient (magnetohydrodynamic) instabilities of a plasma confined in different types of magnetic trap.

Besides the magnetohydrodynamic instabilities, a magnetically confined plasma can sustain various forms of nonmagnetohydrodynamic (slow gradient) instabilities. Such instabilities are particularly important if the plasma is magnetohydrodynamically stable. Therefore, when analyzing the magnetohydrodynamic instabilities in a plasma in specific types of magnetic trap, we shall also discuss the nonmagnetohydrodynamic instabilities that are possible under these conditions.

This analysis comprises Chapters 13-18. In Chapter 13 we discuss the instabilities of a cylindrical plasma column; in Chapter 14, magnetohydrodynamic; and in Chapter 15, nonmagnetohydrodynamic instabilities of a plasma in adiabatic traps. Chapter 16 is devoted to a discussion of the instabilities of a plasma in multipole traps. Finally, in Chapters 17 and 18 we consider closed traps with magnetic surfaces – without a current in Chapter 17 and with a current in Chapter 18.

The analysis of the instabilities of an inhomogeneous plasma contained in this book is not exhaustive. A more detailed study of the various instabilities can be found by referring to the original papers and the reviews listed at the end of each chapter.

Contents

Part I

INSTABILITIES OF AN INHOMOGENEOUS PLASMA IN THE APPROXIMATION OF STRAIGHT LINES OF FORCE

Chapter 1

Gradient Mechanisms of Excitation of Oscillations and Their Role in Beam–Plasma Interaction Processes

§1.1. Convection of Charges across the Magnetic Field Due to Drift in Crossed Fields

We shall now study new types of instability whose dispersion equations contain the spatial gradients of the steady-state parameters of the plasma and cannot therefore be reduced to the instabilities discussed in Volume 1. The present chapter is an introduction to the theory of these instabilities, which we shall call gradient instabilities. In this chapter we shall discuss the physical meaning of the gradient effects, derive the equations that describe the gradient instabilities, and analyze the instabilities that can develop in beam–plasma systems. We shall consider only purely electron instabilities, allowing for the motion of the ions in subsequent chapters. These purely electron gradient instabilities are possible if the distribution function of the beam has sufficiently large transverse gradients.

1. Particle Drift in Crossed E and B_0 Fields.

If an electric field **E** acts on particles that are in a static magnetic field $\mathbf{B}_0$, the particles are displaced not only along **E** but also in the direction $[\mathbf{E}, \mathbf{B}_0]$. This follows from the equations of motion

$$\frac{d\mathbf{V}}{dt}=\frac{e}{m}\mathbf{E}+[\mathbf{V},\ \boldsymbol{\omega}_B], \qquad \omega_B=\frac{e\mathbf{B}_0}{mc}. \tag{1.1}$$

To see this, take $\mathbf{B}_0 \parallel \mathbf{z}$, express the coordinate-time dependence of **E** (**r**, t) in the form $\mathbf{E}(x, y)\exp\{-i\omega t + ik_z z\}$, and assume that the unperturbed velocity of a particle is parallel to the magnetic field, $\mathbf{V}_0 \parallel \mathbf{B}_0$; in the linear approximation in **E**, Eq. (1.1) then yields the particle velocity:

$$\mathbf{V} = \mathbf{V}_0 + \mathbf{V}'_\perp + V'_z \mathbf{e}_z, \tag{1.2}$$

where **V'** satisfies the equations

$$\left.\begin{aligned} -i(\omega - k_z V_0)\,\mathbf{V}'_\perp &= \frac{e}{m}\mathbf{E}_\perp + [\mathbf{V}'_\perp, \boldsymbol{\omega}_B],\\ -i(\omega - k_z V_0)\,V'_z &= \frac{e}{m}E_z. \end{aligned}\right\} \tag{1.3}$$

Hence

$$\left.\begin{aligned} \mathbf{V}'_\perp &= \frac{ie}{mD}\{\omega'\mathbf{E}_\perp + i[\mathbf{E}_\perp, \boldsymbol{\omega}_B]\},\\ V'_z &= \frac{ie}{m}\cdot\frac{E_z}{\omega - k_z V_0},\\ D &\equiv (\omega - k_z V_0)^2 - \omega_B^2,\\ \omega' &= \omega - k_z V_0. \end{aligned}\right\} \tag{1.4}$$

The component of **V'** in the direction at right angles to **E** and $\mathbf{B}_0$, which is known as the drift in the crossed fields, is especially large in the case of a low-frequency field **E**, when $|\omega - k_z V_0| \ll \omega_B$. At the same time

$$\delta\mathbf{V}' = \frac{e}{m\omega_B}[\mathbf{E}, \mathbf{e}_z] \equiv c\,\frac{[\mathbf{E}, \mathbf{B}_0]}{B_0^2}. \tag{1.5}$$

In this approximation, the drift velocity in the crossed fields is independent of both the mass and the charge of the particles.

2. Convective Contribution to the Charge Density Due to Drift in Crossed Fields. Now suppose that instead of a single particle a beam with density n_0 that depends on the transverse coordinates, $n_0 = n_0(x, y)$, moves along the magnetic field $\mathbf{B}_0$ with velocity $\mathbf{V}_0$. Under the influence of the field **E** of perturbations, the beam density is changed by an amount n', which can be calculated from the continuity equation

$$-i(\omega - k_z V_0)\,n' + ik_z V'_z n_0 + \operatorname{div}(n_0 \mathbf{V}'_\perp) = 0 \tag{1.6}$$

and the expressions (1.4) for the components of the perturbed velocity V'. In the case of a low-frequency ($|\omega - k_z V_0| \ll \omega_B$) electric field $E = -\nabla\psi$, the perturbed density is

$$n' = \frac{e}{m}\left\{\frac{n_0 k_z^2 \psi}{(\omega - k_z V_0)^2} + \frac{1}{\omega_B^2}\nabla_\perp (n_0 \nabla_\perp \psi) - i\frac{[\nabla\psi,\ \nabla n_0]_z}{(\omega - k_z V_0)\,\omega_B}\right\}. \quad (1.7)$$

The first term in the curly brackets is due to the perturbed motion of the particles along the magnetic field with velocity V_z' (longitudinal inertia of the particles); the second to motion of the particles in the direction of $E_\perp$ (transverse inertia). Both these terms are present in the expression for n' even when $\nabla n_0 = 0$, i.e., they are not peculiar to an inhomogeneous plasma. The third term in the curly brackets, due to particle drift in the crossed fields, depends essentially on the plasma inhomogeneity – it is nonvanishing only if

$$\nabla n_0 \neq 0. \quad (1.8)$$

It also follows from (1.7) that the drift in crossed fields does not make a contribution to the perturbed density for all types of perturbation but only when the perturbed electric field has a component at right angles to the density gradient and the static magnetic field:

$$[\nabla\psi,\ \nabla n_0]_z \neq 0. \quad (1.9)$$

Examples of perturbations that satisfy the conditions (1.9) for the cases of planar and cylindrical symmetry are as follows. If $\nabla n_0 \parallel x$, the field of such perturbations must necessarily depend on y, $\partial\psi/\partial y \neq 0$ [$k_y \neq 0$ for $\psi \sim \exp(ik_y y)$]; if $\nabla n_0 \parallel r_\perp$, the field must depend on the azimuthal coordinate, $\partial\psi/\partial\varphi \neq 0$ [$l \neq 0$ for $\psi \sim \exp(il\varphi)$].

The change of the density under the influence of drift in crossed fields is due to the removal of plasma with one density from a given point and the arrival at the same point of plasma with a different density. A process of this kind may be called *convection*.

3. Relationship between the Density and Currents in Small-Scale Perturbations when Convection Effects Are Present. Suppose the transverse wave-

length of the perturbations is fairly small compared with the inhomogeneity scale a of the plasma; then the unperturbed potential can be represented in the form

$$\psi = \psi^{(0)}(\mathbf{r}_\perp)\exp\left(i\int^{\mathbf{r}} \mathbf{k}_\perp(\mathbf{r}_\perp)\,d\mathbf{r}_\perp\right), \tag{1.10}$$

where $k_\perp a \gg 1$, $\psi^{(0)}$ and $k_\perp$ are functions that vary little over the distance $1/k_\perp$. In this approximation the perturbed density (1.7) is

$$n' = \frac{e\psi}{m}\left\{\frac{n_0 k_z^2}{(\omega - k_z V_0)^2} - \frac{n_0 k_\perp^2}{\omega_B^2} + \frac{[\mathbf{k}_\perp, \nabla n_0]_z}{\omega_B(\omega - k_z V_0)}\right\}. \tag{1.11}$$

If convection is neglected – the last term in the curly brackets is unimportant – this expression is related to the perturbed currents by a simple formula that follows from the continuity equation:

$$en' \equiv \rho' = \frac{1}{\omega}\mathbf{k}\mathbf{j}', \tag{1.12}$$

where

$$\left.\begin{aligned} \mathbf{j}_\perp &= -\frac{e^2 n_0 \psi}{m\omega_B^2}\{\mathbf{k}_\perp(\omega - k_z V_0) + i[\mathbf{k}_\perp, \boldsymbol{\omega}_B]\}, \\ j_z' &= \frac{e^2 n_0 \psi}{m}\left\{\frac{k_z\omega}{(\omega - k_z V_0)^2} - \frac{k_\perp^2 V_0}{\omega_B^2}\right\}. \end{aligned}\right\} \tag{1.13}$$

In this case the currents and the density depend on the local values of the unperturbed density but not on the derivatives of the density with respect to the coordinates.

If convection is taken into account, both the relationship (1.12) between the density and the currents and the expressions for the currents are changed. Equation (1.12) for the amplitudes of n' and j' [the coefficients of an exponential function of the type (1.10)] is now replaced (the primes being omitted) by

$$\rho(\mathbf{k}, \mathbf{r}) = \frac{1}{\omega}(\mathbf{k}\mathbf{j}(\mathbf{k}, \mathbf{r}) - i\nabla_\perp \mathbf{j}_\perp(\mathbf{k}, \mathbf{r})); \tag{1.14}$$

we see that a term containing the derivative of the amplitude of the transverse current is added to the expression for the density. Although the contribution of this derivative may be small as $1/ka$ compared with $\mathbf{k}_\perp \mathbf{j}_\perp$, it is in reality much greater since the principal term in $\mathbf{j}_\perp$, which is proportional to $[\mathbf{k}, \boldsymbol{\omega}_B]$, gives zero when multiplied by $k_\perp$. On the other hand, in deriving (1.14) we have omitted the small terms of order $1/ka$ that result from the differentiation of the part of $j_\perp$ parallel to $k_\perp$. When convection is taken into account, the expression for the longitudinal current j_z is also changed, the term

$$j_{z\,\text{conv}}(\mathbf{k},\ \mathbf{r}) = -iV_0\nabla_\perp \mathbf{j}_\perp(\mathbf{k},\ \mathbf{r})/(\omega - k_z V_0) \tag{1.15}$$

being added to the right-hand side of the second equation in (1.13). This additional term describes the effect of *convection of the longitudinal current*.

Taking into account (1.14) and (1.15), we find that the convective contribution to the density is determined by the convective contribution to the longitudinal current and the derivative of the transverse current:

$$\rho_{\text{conv}}(\mathbf{k},\ \mathbf{r}) = \frac{1}{\omega}\left(k_z j_{z\,\text{conv}}(\mathbf{k},\ \mathbf{r}) - i\nabla_\perp \mathbf{j}_\perp(\mathbf{k},\ \mathbf{r})\right). \tag{1.16}$$

4. *Relationship between the Scalar Permittivity and the Permittivity Tensor.* In the case of the oscillations of a homogeneous plasma, Poisson's equation and the equations that describe the plasma yield a dispersion relation of the form

$$\varepsilon_0 \equiv 1 + \sum_\alpha \varepsilon_0^{(\alpha)} = 0, \tag{1.17}$$

where $\varepsilon_0^{(\alpha)}$ is the contribution of the α-th component of the plasma to the scalar permittivity. The quantity $\varepsilon_0^{(\alpha)}$ is related to the corresponding contributions to the components of the permittivity tensor $\varepsilon_{ik}^{(\alpha)}$ by the equation

$$\varepsilon_0^{(\alpha)} = \frac{1}{k^2} k_i \varepsilon_{ik} k_k. \tag{1.18}$$

This result can be regarded as a consequence of the continuity equation (1.12); for writing Poisson's equation in the form

$$1 - \frac{4\pi}{k^2}\sum_\alpha \chi^{(\alpha)} = 0, \tag{1.19}$$

where χ is the coefficient of proportionality between the charge density ρ and the potential ψ, $\rho = \chi\psi$, we find that

$$\varepsilon_0^{(\alpha)} = -\frac{4\pi}{k^2}\chi^{(\alpha)}. \tag{1.20}$$

But, in accordance with (1.12) and the definition $j_i^{(\alpha)} = \frac{\omega}{4\pi i}\varepsilon_{ik}^{(\alpha)}E_k$,

$$\chi^{(\alpha)} = -\frac{1}{4\pi}k_i\varepsilon_{ik}^{(\alpha)}k_k \tag{1.21}$$

for $\mathbf{E} = -i\mathbf{k}\psi$. Equations (1.20) and (1.21) together yield (1.18). We have already drawn attention to this in the appendix to Chapter 1 in Volume 1. This simple relationship between the permittivity scalar and tensor ceases to hold in an inhomogeneous plasma in which convection effects play a role. Recalling that in this case the density satisfies Eq. (1.14), we find that the relationship (1.18) between the scalar ε_0 defined by (1.20) and the tensor ε_{ik} is replaced by

$$\varepsilon_0^{(\alpha)} = \frac{1}{k^2}\left(k_i\varepsilon_{ij} - i\frac{\partial\varepsilon_{ij}}{\partial x_i}\right)k_j. \tag{1.22}$$

In accordance with §1.1.3, the convective contribution to $\varepsilon_0^{(\alpha)}$ is

$$\varepsilon_{0\,\text{conv}}^{(\alpha)} = \frac{1}{k^2}\left(k_z\varepsilon_{zj\,\text{conv}} - i\frac{\partial\varepsilon_{ij}}{\partial x_i}\right)k_j. \tag{1.23}$$

The complete expression for $\varepsilon_0^{(\alpha)}$ with allowance for (1.11) is

$$\varepsilon_0^{(\alpha)} = -\frac{\omega_p^2}{(\omega - k_zV_0)^2}\left(\frac{k_z}{k}\right)^2 + \left(\frac{\omega_p}{\omega_B}\right)^2\left(\frac{k_\perp}{k}\right)^2 - \frac{[\mathbf{k}_\perp, \nabla\omega_p^2]_z}{k^2\omega_B(\omega - k_zV_0)}, \tag{1.24}$$

$$\omega_p^2 = 4\pi e^2 n_0/m.$$

Convection is not the only new effect we encounter in an inhomogeneous plasma in a magnetic field. Some more new features will be discussed in §1.6.

§1.2. Convective Excitation of Oscillations of a Cold Plasma by a Cold Beam with an Inhomogeneous Density

We shall now consider the simplest example of instability of an inhomogeneous plasma due to convection effects. We shall consider a plasma–beam system in a magnetic field. We shall assume that the thermal velocities of the particles of the beam and the plasma are fairly low compared with the mean velocity of the beam. In general, the two-stream instability can arise in such a system. However, if the densities of the beam and the plasma are sufficiently low, so that $\omega_p \ll \omega_B$ and, in addition, the transverse dimension of the plasma is sufficiently small, the two-stream instability must be absent. We discussed this situation in §9.7 in Volume 1, in which it was explained as follows. The two-stream instability is associated with the excitation of perturbations with a fairly long wavelength:

$$k_z \lessapprox \omega/V_0. \tag{1.25}$$

In (1.25), ω is to be understood as the frequency of the characteristic plasma oscillations; for $\omega_p \ll \omega_B$, this is

$$\omega \simeq \omega_p k_z/k. \tag{1.26}$$

Here, $k=\sqrt{k_z^2+k_\perp^2}$ is the total wave number. It is certainly not less than $k_\perp$, and the latter is bounded below by the reciprocal of the transverse scale a of the plasma, i.e.,

$$k \gtrapprox k_{\perp\min} \simeq 1/a. \tag{1.27}$$

The conditions (1.25)-(1.27) show that the two-stream instability is impossible if the density of the plasma and its transverse dimension are sufficiently small:

$$\omega_p < V_0/a. \tag{1.28}$$

In what follows we shall assume that this condition is satisfied. We shall show that the beam–plasma system may nevertheless be unstable. The cause of this instability is the convective transport of beam particles at right angles to the magnetic field.

An equation for the oscillations of the plasma–beam system may be obtained by substituting the value of the perturbed beam and plasma densities from (1.7) into Poisson's equation:

$$-\Delta\psi = 4\pi \sum_{\alpha=1,\,2} en'(\psi). \tag{1.29}$$

In this equation the sum is over two groups of particles: 1, the plasma; 2, the beam. Taking account of the explicit form of (1.7), we find that in the approximation of a strong magnetic field, $\omega_B \gg \omega_p$, the contribution to (1.29) from the transverse inertial motion of the particles is small, as $(\omega_p/\omega_B)^2$. We shall neglect these terms. We shall also neglect the contribution to the beam density from the longitudinal inertial motion of its particles. It is precisely this term that would give rise to the two-stream instability if it were not forbidden by the inequality (1.28). Thus, the contribution of the beam to (1.29) under the assumptions we have made is

$$n_2' = -\frac{ie}{m}\frac{[\nabla\psi, \nabla n_2]_z}{(\omega - k_z V_0)\,\omega_B}. \tag{1.30}$$

We shall assume that our system is confined in the transverse direction by conducting walls and that the plasma density in the region between the walls is constant, $\nabla n_1 = 0$. There is then no convective contribution to the perturbed density and, since we also neglect the transverse inertia, Eq. (1.7) in the adopted approximation yields

$$n_1' = \frac{e}{m}\cdot\frac{n_1 k_z^2}{\omega^2}\psi. \tag{1.31}$$

Combining (1.29)-(1.31), we obtain the desired equation for ψ:

$$\Delta\psi + \left(\frac{\omega_{p1}}{\omega}k_z\right)^2\psi - i\frac{4\pi e^2}{m}\frac{[\nabla\psi, \nabla n_2]_z}{\omega_B(\omega - k_z V_0)} = 0. \tag{1.32}$$

To demonstrate in the simplest manner possible that convective excitation of oscillations is possible, let us consider the case $\nabla n_2 \parallel x$ and assume that the density profile of the beam is linear: $\bar{n}_2 = n_2(1 + \varkappa x)$.

Under this assumption plane waves with $k_x = \pi n \varkappa$, $n = 0, 1, 2, 3, \ldots$ are eigenfunctions. From (1.32) we then obtain the dispersion equation

$$1 - \left(\frac{\omega_{p1}}{\omega}\right)^2\left(\frac{k_z}{k}\right)^2 + \frac{\varkappa k_y \omega_{p2}^2}{k^2\omega_B(\omega - k_z V)} = 0, \tag{1.33}$$

where $V \equiv V_0$, $k^2 = k_x^2 + k_y^2 + k_z^2$. In the special case $\omega \ll k_z V$, Eq. (1.33) yields

$$\omega = \pm\frac{\omega_{p1}\left|\frac{k_z}{k}\right|}{\sqrt{1 - \frac{\varkappa k_y \omega_{p2}^2}{k^2 k_z V \omega_B}}}. \tag{1.34}$$

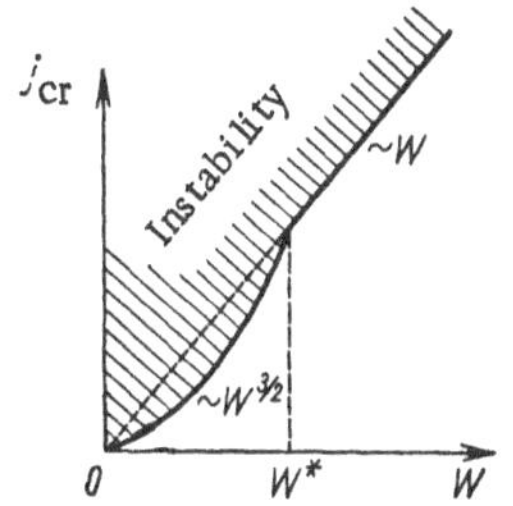

Fig. 1.1. Dependence of the critical current on the particle energy W. The shaded region in which $j > j_{cr}$, corresponds to instability.

We see that perturbations grow if

$$\frac{k_y \varkappa \omega_{p2}^2}{k^2 k_z V \omega_B} > 1. \tag{1.35}$$

The maximal growth rate is approximately

$$\gamma \simeq \frac{\omega_{p1}}{\omega_B} \cdot \frac{\varkappa \omega_{p2}^2}{k^2 V}. \tag{1.36}$$

In strong magnetic fields, the instability boundary of a beam in a plasma must satisfy a relation of the type (1.28); in weak magnetic fields, a relation of the type (1.35). If we introduce quantities like the current density of the beam, $j \sim n_2 V$, and the energy of its particles, $W \sim V^2/2$, then, using (1.28) and (1.35), we conclude that the critical current must depend on the particle energy in the manner shown in Fig. 1.1:

for small w,

$$j_{cr} \sim W^{3/2} \quad \text{(two-stream excitation)} \tag{1.37}$$

for large w,

$$j_{cr} \sim W \quad \text{(convective excitation).} \tag{1.38}$$

The region of large currents, $j > j_{cr}$, corresponds to instability. The dependence (1.37) is replaced by (1.38) when $W = W^*$, where

$$W^* \sim \sqrt{B_0}. \tag{1.39}$$

§1.3. Instability of a Stream with an Inhomogeneous Velocity Profile

In §1.2 we discussed the possibility of convective excitation of oscillations under conditions when there are two groups of particles – fast (in the beam) and slow (in the plasma) – at one and the same point of space. The decisive factor that makes such an instability possible is the presence of particles with different velocities and not the fact that, locally, the plasma has a two-component nature. To see this, let us consider the instability of a single-component system, that is, a plasma consisting of a stream of particles (of one species) with a velocity $\mathbf{V}_0 \parallel \mathbf{B}_0$ that depends on the transverse coordinate,

$$\nabla_{\perp} V_0 \neq 0. \tag{1.40}$$

This condition means that the system contains particles with different velocities.

A stream with a inhomogeneous velocity profile may be unstable even if the convection effect can be ignored. This was demonstrated in Volume 1 for the case when there is no static magnetic field. An instability of this kind is similar to the two-stream instability (the instability of spatially separated beams). However, if the magnetic field is sufficiently strong, $\omega_B > \omega_p$, this instability disappears (see the problem). Below, the magnetic field is assumed strong ($\omega_B > \omega_p$), so that the only possible cause of instability is convection.

The equation for the oscillations of a stream with an inhomogeneous velocity profile can be obtained in the same way as in §1.1, i.e., by using Poisson's equation, the continuity equation, and the equations of motion. In the equations of motion of type (1.1) we must now remember that V_0 depends on the transverse coordinate. The first equation of the system (1.3) remains unaltered, but the second takes the form

$$-\mathrm{i}(\omega - k_z V_0) V'_z + (\mathbf{V}'_{\perp}\nabla) V_0 = \frac{e}{m} E_z. \tag{1.41}$$

As a result, we obtain an equation for the potential ψ which for $\nabla V_0 \parallel \mathbf{x}$ and $\nabla n_0 = 0$ is

$$\frac{d^2\psi}{dx^2} - \psi\left[k_y^2 + k_z^2 - \frac{\omega_p^2 k_z^2}{(\omega - k_z V_0)^2} + \frac{k_y k_z \omega_p^2}{\omega_{B_e}(\omega - k_z V_0)^2} \cdot \frac{dV_0}{dx}\right] = 0. \tag{1.42}$$

We shall investigate this equation in two limiting cases: a step-like and a smooth (compared with the wavelength along x) velocity profile.

1. Approximation of a Step-like Velocity Profile. Assuming $\frac{\omega}{k_z} > (V_0, \omega_{p_e}/k_y)$ in Eq. (1.42) and solving the latter by the method of surface waves described in §1.7 of Volume 1, we obtain the dispersion equation

$$\omega^2 = \frac{1}{2}\frac{k_y}{|k_y|}\cdot\frac{\omega_{p_e}^2}{\omega_{B_e}}k_z(V_1 - V_2). \tag{1.43}$$

If the sign of k_z/k_y is chosen approximately, the right-hand side of (1.43) is negative. Perturbations of this kind grow and have the growth rate

$$\gamma = \left[\frac{1}{2}\omega_{p_e}^2\left|\frac{k_z(V_2 - V_1)}{\omega_{B_e}}\right|\right]^{1/2}. \tag{1.44}$$

2. Small-Scale Perturbations in a Stream with a Smooth Velocity Profile. If the wavelength λ_x of the perturbations is small compared with the characteristic distance over which the velocity changes, the solution of Eq. (1.42) can be found by using the quasiclassical approximation of §5.4 in Volume 1. The local dispersion equation that follows from (1.42) is

$$k_\perp^2 + k_z^2 - \frac{\omega_{p_e}^2}{(\omega - k_z V)^2}\left(k_z^2 - \frac{k_y k_z}{\omega_{B_e}}\cdot\frac{dV}{dx}\right) = 0; \qquad V \equiv V_0. \tag{1.45}$$

The perturbations grow if

$$\frac{k_z}{k_y} < \frac{1}{\omega_{B_e}}\cdot\frac{dV}{dx}. \tag{1.46}$$

The maximal growth rate of the perturbations is approximately

$$\gamma \simeq \frac{\omega_{p_e}}{\omega_{B_e}}\cdot\frac{dV}{dx}. \tag{1.47}$$

3. Stabilizing Role of a Density Gradient. A density gradient can stabilize the stream. To see this, we include in (1.45) a term with ∇n_0 [it has the same form as in Eq. (1.24)]:

$$k^2 - \frac{\omega_{p_e}^2}{(\omega - k_z V)^2}\left(k_z^2 - \frac{k_y k_z}{\omega_{B_e}}\cdot\frac{dV}{dx}\right) + \frac{\varkappa_n k_y \omega_{p_e}^2}{\omega_{B_e}(\omega - k_z V)} = 0. \tag{1.48}$$

From (1.48) we obtain the stabilization condition

$$\frac{d \ln n_0}{d \ln V} > \frac{k_y V}{\omega_{p_e}}. \tag{1.49}$$

At the limits of applicability of the approximations adopted here, that is, when $\omega_{p_e} \simeq \omega_{B_e}$, $V \simeq v_{T_e}$, and $k_y \simeq 1/\rho_e$, the condition (1.49) means effectively $d \ln n_0/d \ln V > 1$.

Problem. Suppose that an electron stream with a step-like velocity profile ($\nabla V \parallel \mathbf{x}$) is in a magnetic field, $\mathbf{B}_0 \parallel \mathbf{V} \parallel \mathbf{z}$. Find the critical magnetic field strength at which the stream is stabilized against perturbations with $k_y = 0$.

Solution. For $k_y = 0$ and arbitrary ω/ω_B, Poisson's equation reduces to the following differential equation for ψ:

$$\frac{\partial}{\partial x}\left(\varepsilon_\perp \frac{\partial \psi}{\partial x}\right) - k_z^2 \varepsilon_\parallel \psi = 0, \tag{1.50}$$

where

$$\left.\begin{aligned} \varepsilon_\perp &= 1 - \frac{\omega_p^2}{(\omega - k_z V)^2 - \omega_B^2}, \\ \varepsilon_\parallel &= 1 - \frac{\omega_p^2}{(\omega - k_z V)^2}. \end{aligned}\right\} \tag{1.51}$$

We have omitted the subscript e from ω_{p_e} and ω_{B_e}.

As in §1.7 in Volume 1, we shall assume that the stream velocity and density change abruptly at some $x = x_0$ from the values V_1 and n_1 to V_2 and n_2. Let us consider the stability of such a stream against perturbations of the type of surface waves (localized near the interface). As in §1.7 of Volume 1, we obtain the dispersion equation

$$\sqrt{\varepsilon_\perp^{(1)} \varepsilon_\parallel^{(1)}} + \sqrt{\varepsilon_\perp^{(2)} \varepsilon_\parallel^{(2)}} = 0. \tag{1.52}$$

We shall assume that the density on both sides of the interface is the same, $n_2 = n_1$, and take a frame of reference in which $V_2 = -V_1$. Equation (1.52) then yields a biquadratic equation for ω:

$$\omega^4 - \omega^2 [2 (k_z V)^2 + \omega_p^2 + \omega_B^2] + (k_z V)^4 - (k_z V)^2 (\omega_p^2 + \omega_B^2) + \frac{1}{2} \omega_B^2 (\omega_p^2 + \omega_B^2) = 0. \tag{1.53}$$

It follows from (1.53) that the instability considered in §1.7 of Volume 1 occurs only if

$$\omega_p > \omega_B. \tag{1.54}$$

If this is not the case, Eq. (1.52) has no solutions at all, let alone solutions with $\mathrm{Im}\, \omega > 0$. This means that electron surface waves with $k_y = 0$ exist only in a sufficiently weak magnetic field.

§1.4. Role of Convection Effects in Resonant Interaction Processes between Particles and Oscillations

1. Permittivity with Allowance for the Convection Effect for a Beam with a Longitudinal Spread of the Particle Velocities. A plasma with a longitudinal spread of the particle velocities can be regarded as a collection of an infinite number of cold beams each of which has its own density $n_0^{(\alpha)}$ and velocity $V_0^{(\alpha)}$. On this basis, the perturbed density of such a plasma can be represented as a sum of expressions of the type (1.7) over all beams:

$$n'(\psi) = \sum_{\alpha} n'(V_0^{(\alpha)},\ n_0^{(\alpha)},\ \psi). \tag{1.55}$$

Assuming that to each V_0 there corresponds an $n_0(V_0)$ and replacing the summation over α in (1.55) by summation over the velocities V_0, we obtain

$$n'(\psi) = \sum_{V_0} n'(n_0(V_0),\ V_0,\ \psi). \tag{1.56}$$

We then go over from summation over V_0 to integration over the velocity intervals dv_z. Let $f_0(v_z)\,dv_z$ be the number of particles in the interval dv_z. Then (1.56) is transformed into (a similar procedure was used in the problem to §2.1 in Volume 1)

$$n'(\psi) = \int dv_z n'(f_0(v_z),\ \psi). \tag{1.57}$$

It remains to substitute into this equation the explicit form of the function $n'(f_0(v_z))$ from (1.7). The notation $n'(f_0)$ means that in (1.7) n_0 is replaced by f_0 and V_0 by v_z. Neglecting transverse inertia, we obtain

$$n' = \frac{e}{m} \int dv_z \left\{ \frac{k_z^2 f_0}{(\omega - k_z v_z)^2}\psi - i\frac{[\nabla\psi,\ \nabla f_0]_z}{\omega_B(\omega - k_z v_z)} \right\}. \tag{1.58}$$

In the approximation of small-scale perturbations, this n' corresponds in accordance with (1.20) to the permittivity

$$\varepsilon_0^{(\alpha)} = -\frac{4\pi e^2}{mk^2} \int \left\{ \frac{k_z^2 f_0}{(\omega - k_z v_z)^2} - \frac{\frac{\partial f_0}{\partial x} k_y}{\omega_B(\omega - k_z v_z)} \right\} dv_z \equiv \frac{4\pi e^2}{mk^2} \int \frac{dv_z}{\omega - k_z v_z} \left(k_z \frac{\partial f_0}{\partial v_z} + \frac{k_y}{\omega_B} \cdot \frac{\partial f_0}{\partial x} \right). \tag{1.59}$$

2. Convective Excitation of Oscillations of a Plasma by a "Hot" Beam. We now proceed to the direct analysis of the interaction of a plasma with an inhomogeneous low-density beam with a large spread of the longitudinal particle velocities. As in §1.3.2 and §1.2, we shall consider small-scale perturbations, $k_\perp a \gg 1$, localized in a region that is small compared with the dimensions of the beam. We shall assume that the beam has a low density; we shall see that the growth rate of the perturbations is then small compared with the frequency, $\gamma \ll \mathrm{Re}\,\omega$. Under these conditions, the local approximation can still be used if γ is not too small, i.e., $\gamma \gg v_{cr}/a$, where $v_{cr} = \partial\omega/\partial k_x$ (see the condition in §5.3 of Volume 1). If this condition is not satisfied, one must solve the problem of the excitation of the characteristic plasma oscillations.

The local dispersion equation that describes the oscillations of a system consisting of a homogeneous plasma and an inhomogeneous beam has the form

$$\varepsilon_0^{(0)} + \varepsilon_0^{(1)} = 0, \tag{1.60}$$

where

$$\varepsilon_0^{(0)} = 1 - \frac{\omega_{p_0}^2}{\omega^2}\cos^2\theta, \tag{1.61}$$

and $\varepsilon_0^{(1)}$ is determined by Eq. (1.59). The subscript 0 refers to the plasma; the subscript 1, to the beam; $\theta \equiv \tan^{-1}(k_\perp/k_z)$ is the angle between the wave vector and the magnetic field.

Assuming that $\varepsilon_0^{(1)}$ is small, we find the frequency and growth rate of the perturbations:

$$\mathrm{Re}\,\omega = \pm\,\omega_{p_e}\cos\theta, \tag{1.62}$$

$$\gamma = -\frac{\mathrm{Im}\,\varepsilon_0^{(1)}}{\partial\varepsilon_0^{(0)}/\partial\,\mathrm{Re}\,\omega}. \tag{1.63}$$

With allowance for (1.59),

$$\gamma = \frac{\pi}{2}\cdot\frac{\omega}{|k_z|}\cdot\frac{4\pi e^2}{mk^2}\left(k_z\frac{\partial f_0^{(1)}}{\partial v_z} + \frac{k_y}{\omega_B}\cdot\frac{\partial f_0^{(1)}}{\partial x}\right)_{v_z=\frac{\omega}{k_z}}. \tag{1.64}$$

Fig. 1.2. Distribution function of the longitudinal velocities of a system consisting of a plasma and a hot beam: $f_0^{(0)} + f_0^{(1)}$. In the region of resonant velocities $v_z = \omega/k_z$, we have $\partial f_0^{(1)}/\partial v_z = 0$ in cases a and b and $\partial f_0^{(1)}/\partial v_z < 0$ in case c.

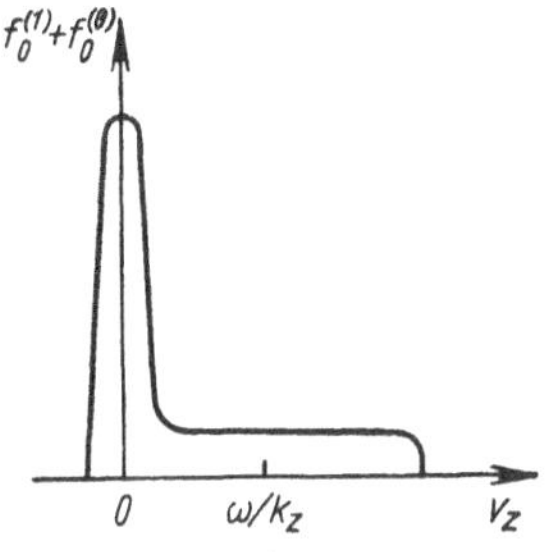

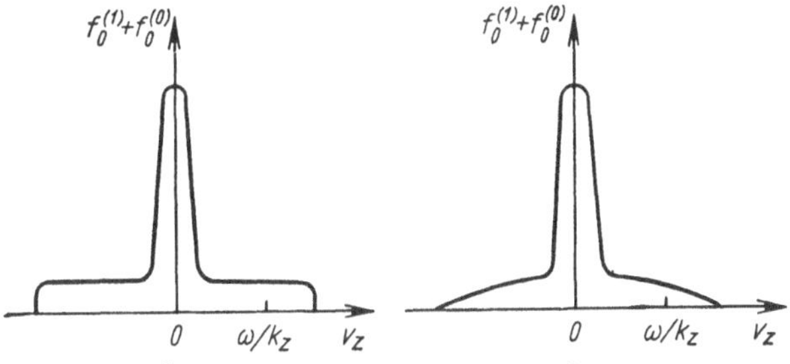

If convection is ignored (the term with $\partial f_0/\partial x$), the growth rate is positive (there is instability), provided the distribution function of the beam has a positive derivative, $\partial f_0^{(1)}/\partial v_z > 0$ (it is assumed that $\omega/k_2 > 0$) in the region of resonant velocities. If convection is taken into account, excitation can occur even if $\partial f_0^{(1)}/\partial v_z \leq 0$. For example, if $\partial f_0^{(1)}/\partial v_z = 0$ (see Fig. 1.2a) it follows from (1.64) that

$$\gamma = \frac{1}{2} \cdot \frac{4\pi e^2 \omega k_y}{mk^2 |k_z| \omega_B} \left(\frac{\partial f_0^{(1)}}{\partial x}\right)_{v_z = \frac{\omega}{k_z}}, \tag{1.65}$$

so that $\gamma > 0$ if k_z/k_y has the appropriate sign.

Equation (1.65) shows that the excitation of oscillations by the convective mechanism is not due to the presence of a mean directed velocity of the beam but simply to the presence of particles with a resonant value of the velocity. In particular, the result (1.65) is not affected if the v_z distribution has the form shown in Fig. 1.2b. In this case it is inappropriate to speak of a directed beam; rather, one must say there is a group of sufficiently heated particles.

If $\partial f_0^{(1)}/\partial v_z < 0$, instability occurs if the convection is stronger than the usual Landau damping, i.e., if

$$\left|\frac{k_y}{\omega_B}\cdot\frac{\partial f_0^{(1)}}{\partial x}\right| > \left|k_z\frac{\partial f_0^{(1)}}{\partial v_z}\right|. \tag{1.66}$$

In particular, if the fast particles have a Maxwellian distribution, $f_0^{(1)} \sim n_1(x)\,e^{-mv_z^2/2T_\parallel}$ (see Fig. 1.2c), the condition (1.66) entails

$$\left|\frac{k_y\varkappa}{\omega_B}\right| > \frac{m\omega}{T_\parallel}, \qquad \varkappa = \frac{\partial \ln n_1}{\partial x}, \tag{1.67}$$

and in this case (1.64) yields

$$\gamma = -\left(1-\frac{k_y\varkappa T_\parallel}{m\omega_B\omega}\right)\frac{\sqrt{\pi}}{|k_z|\,v_{T_\parallel}}\left(\frac{\omega}{kd_1}\right)^2, \qquad d_1^2 = \frac{T_\parallel}{4\pi n_1 e^2}. \tag{1.68}$$

It can be seen that oscillations whose frequency is sufficiently low,

$$\omega < \left|\frac{\varkappa T_\parallel k_y}{m\omega_B}\right|, \tag{1.69}$$

are excited. In the example considered, the frequency is determined by the expression (1.62), so that the condition (1.67) gives

$$\cos\theta < \frac{\varkappa k_y T_\parallel}{m\omega_B\omega_p}. \tag{1.70}$$

However, $\cos\theta$ must not be too small, $\cos\theta > (m_e/m_i)^{1/2}$, since it would otherwise be necessary to take into account the ions. Therefore $\varkappa$ must also be not too small, i.e., the (electron) instability we are considering is possible only if there is a sufficiently large density gradient.

Note finally that if $T_\parallel = T_\perp$ the instability condition (1.69) can be represented in the form

$$\frac{\omega}{k_y V_\perp} < 1, \tag{1.71}$$

where $V_\perp \equiv \varkappa T/m\omega_B$ can be interpreted as the macroscopic velocity of the given component of the plasma at right angles to the magnetic field; this is also called the mean velocity of the Larmor currents (we shall consider $V_\perp$ in more detail in §1.6). Expressed in this form, the instability condition can be formulated as follows: perturbations grow whose transverse phase velocity ω/k_y is in the direction of $V_\perp$ and does not exceed $V_\perp$. In using this rule, one must remember that the excitation mechanism is not in fact related to the presence of $V_\perp$ and is manifested even if $V_\perp = 0$ ($T_\perp = 0$) if $T_\parallel \neq 0$. In addition, the instability condition does not reduce to one of the type (1.71) if the longitudinal velocity distribution of the particles is not Maxwellian.

§1.5. Heat Convection and Instability Due to a Temperature Gradient

Suppose there is a single-component plasma with $\nabla n_0 = 0$ and $\nabla T \neq 0$. The presence of a temperature gradient may be responsible for an instability of the plasma. We can see this by taking the example of a plasma in a strong magnetic field, $\omega_B \gg \omega_p$.

Let us consider perturbations with $k_z v_T \ll \omega \ll \omega_B$. In the expression (1.59) for $\varepsilon_0^{(\alpha)}$ we assume that $f_0(v_z)$ is Maxwellian; we make an expansion in a series in $k_z v_z/\omega$ and substitute the result into the local dispersion equation $1 + \varepsilon_0^{(\alpha)} = 0$, $\alpha = 0$. Then

$$1-\left(\frac{\omega_p}{\omega}\cos\theta\right)^2\left(1-\frac{\omega_T}{\omega}\right)=0, \tag{1.72}$$

where $\omega_T \equiv \frac{\partial T}{\partial x}\cdot\frac{k_y}{m\omega_B}$. If $\omega \ll \omega_T$, Eq. (1.72) yields a cubic equation for ω:

$$\omega^3 = -\omega_p^2\omega_T\cos^2\theta, \tag{1.73}$$

from which we find that one of the roots corresponds to instability with the growth rate

$$\gamma = \frac{\sqrt{3}}{2}\left|\omega_p^2\omega_T\cos^2\theta\right|^{1/3}. \tag{1.74}$$

This result satisfies the condition $\gamma < \omega_T$ for $\cos\theta < \omega_T/\omega_p$. At the limit of applicability, i.e., at $\cos\theta \simeq \omega_T/\omega_p$, we have approximately

$$\gamma \simeq \omega_T. \tag{1.75}$$

Equation (1.72) can also be obtained by means of the following system of hydrodynamic equations:

$$\left.\begin{aligned} -i\omega n' + ik_z V_z' &= 0, \\ -i\omega V_z' &= -\frac{ie}{m} k_z \psi - ik_z \frac{T'}{m}, \\ -i\omega T' + (\mathbf{V}_E \nabla) T_0 &= 0. \end{aligned}\right\} \tag{1.76}$$

This system is none other than the linearized equations of the moments of the distribution functions for n, $V_{\parallel}$, and $T_{\parallel}$, in which we have substituted $\mathbf{V}_{\perp} = \mathbf{V}_E \equiv c\,\frac{[\mathbf{E}, \mathbf{e}_z]}{B_0}$, and $\nabla n_0 = 0$ and omitted the terms of order $k_z v_T/\omega$. Using (1.76) to express the density n' in terms of the potential ψ of the perturbation and substituting the density into Poisson's equation, we arrive at (1.72).

This approach makes it clear that the instability is due to the *convection of heat* – the transport of the longitudinal energy of the particles at right angles to the magnetic field due to their drift in the crossed fields.

§1.6. Role of Larmor Currents (Angular Asymmetry of the Steady-State Distribution Function)

Hitherto we have discussed instabilities whose development does not require the particles to have unperturbed transverse velocities. We have only been able to neglect the transverse motion because we have throughout considered only perturbations with wavelength appreciably greater than the Larmor radius of the particles, $k_{\perp}v_{\perp}/\omega_B \ll 1$. We now turn to the investigation of another class of instabilities whose mechanism is due to just these transverse velocities of the particles. We shall see that the growth of such instabilities requires the presence of particles whose velocities $v_{\perp}$ satisfy the condition $k_{\perp}v_{\perp}/\omega_B > 1$.

It is clear that the mere fact that the particles have transverse velocities is not by itself sufficient to cause instability. As in the examples discussed above, in which the energy of the oscillations is drawn from the longitudinal energy of the particles, the true cause of the instabilities which we now consider is the inhomogeneity of the plasma. On the other hand, an inhomogeneity of the plasma does not necessarily lead to instabilities. One must also have a mechanism for transforming the energy of the steady-state motion of the particles into the energy of the perturbations. One such mechanism leading to instability is the convection effect discussed §1.1. It is this effect that is responsible for the instabilities considered in §§1.2-1.5.

Turning to the study of instabilities whose energy is derived from the transverse energy of the particles, we must first of all elucidate the mechanism of these instabilities.

Consider a group of particles with fairly large transverse velocities, assuming that their Larmor radius is large compared with the transverse wavelength:

$$\frac{k_\perp v_\perp}{\omega_B} \gg 1. \tag{1.77}$$

Such particles traverse many wavelengths before their trajectories begin to be curved. The curvature of the particle trajectories can be entirely ignored if during the time of a gyration in the Larmor orbit the perturbed field makes sufficiently many oscillations and, as a result of the instability, appreciably changes its amplitude, i.e., if

$$\mathrm{Re}\,\omega \gg \omega_B,\ \gamma \gg \omega_B. \tag{1.78}$$

In this approximation the static magnetic field does not affect the perturbed motion of the particles and is manifested only in the nature of the steady-state particle velocity distribution. The contribution of our group of particles to the local permittivity then has the form well known in the theory of plasma oscillations without a magnetic field:

$$\varepsilon_0 = \frac{4\pi e^2}{mk^2}\int \frac{\mathbf{k}\partial f_0/\partial \mathbf{v}\ d\mathbf{v}}{\omega - \mathbf{k}\mathbf{v}}, \tag{1.79}$$

or, equivalently,

$$\varepsilon_0 = \frac{4\pi e^2}{mk}\int \frac{\frac{\partial f_k}{\partial v_k}}{\omega - kv_k}\, dv_k, \tag{1.80}$$

where v_k is the projection of the velocity onto the direction of the wave vector, and f_k is the distribution function of the particle velocities in the direction of **k**.

Fig. 1.3. Model to demonstrate the beam-type nature of the particle velocity distribution resulting from spatial inhomogeneity.

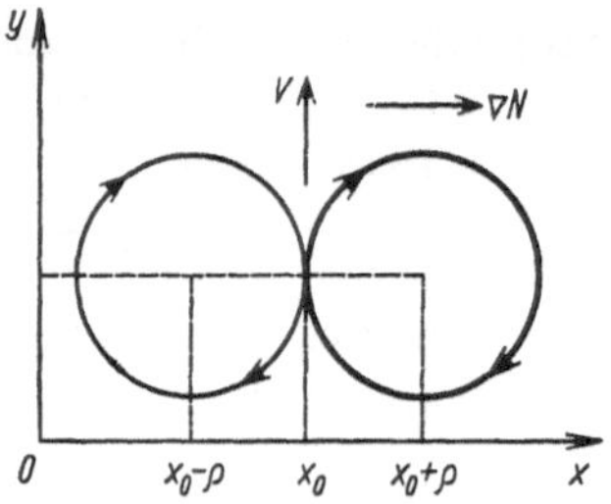

It is the form of the function f_k that determines whether our group of particles can lead to instability. Excitation is possible only if the maximum of f_k does not occur at $v_k = 0$, i.e., in the case of a beam-type distribution. We shall now show that because of a spatial inhomogeneity the particle velocity distribution may have a beam-type nature.

Let us first consider a simple model (Fig. 1.3). Suppose the guiding centers of particles with Larmor radii ρ are situated at the points $x_0 + \rho$ and $x_0 - \rho$. If the density of the Larmor circles N(x) is inhomogeneous, the mean velocity of the particles at the point x_0 is

$$\overline{V}_y = \frac{v_\perp [N(x_0+\rho) - N(x_0-\rho)]}{N(x_0+\rho)+N(x_0-\rho)} \simeq \frac{v_\perp \rho}{N(x_0)} \cdot \frac{\partial N(x_0)}{\partial x_0}. \tag{1.81}$$

This velocity is at right angles to the magnetic field and the density gradient, i.e., we have a transverse beam whose velocity is small compared with the mean particle velocities as the ratio of the Larmor radii to the inverse density gradient:

$$\frac{V_y}{v_\perp} \sim \rho \frac{\partial \ln N}{\partial x}. \tag{1.82}$$

Let us now consider the transverse velocity distribution of the particles more rigorously. In the case of an inhomogeneous plasma, the distribution function satisfies the equilibrium Boltzmann equation

$$\mathbf{v}\nabla f_0 + [\mathbf{v}, \boldsymbol{\omega}_B] \frac{\partial f_0}{\partial \mathbf{v}} = 0 \tag{1.83}$$

or, equivalently, it is a function of the integrals of the equations of motion of the particles,

$$\frac{d\mathbf{v}}{dt} = [\mathbf{v}, \boldsymbol{\omega}_B], \qquad \frac{d\mathbf{r}}{dt} = \mathbf{v}. \tag{1.84}$$

The integrals of the equations (1.84) are, for example,

$$\left.\begin{aligned} \varepsilon_{\perp}=\frac{v_{\perp}^{2}}{2},\ V_z=v_z,\\ X=x+\frac{v_y}{\omega_B},\ Y=y-\frac{v_x}{\omega_B}. \end{aligned}\right\} \tag{1.85}$$

The first two integrals of the motion correspond to the conservation of the transverse energy and the longitudinal velocity of the particles. The integrals $\mathbf{R}$ = (X, Y) indicate that the coordinate of the guiding center of a particle gyrating in the homogeneous magnetic field does not change.

Thus, the steady-state distribution function is

$$f_0(\mathbf{r},\ \mathbf{v})=F(\varepsilon_{\perp},\ V_z,\ X,\ Y), \tag{1.86}$$

where F is some arbitrary function. Assuming that the inhomogeneity is weak and expanding the right-hand side of (1.86) in a series in X–x and Y–y, we obtain approximately

$$f_0(\mathbf{r},\ \mathbf{v})=F(\varepsilon_{\perp},\ V_z,\ x,\ y)+\frac{[\mathbf{v},\ \mathbf{e}_z]}{\omega_B}\nabla F(r). \tag{1.87}$$

The beam-type nature of f_0 (in the direction at right angles to B_0) is due to the presence of the small correction on the right-hand side of (1.87). If F is a Maxwellian function with $\nabla n_0 \parallel \mathbf{x}$ and $\nabla T=0$ and $\mathbf{k}\parallel\mathbf{y}$, the function f_k in (1.80) has the form

$$f_k=\left(\frac{m}{2\pi T}\right)^{1/2} n_0\, e^{-\frac{mv_y^2}{2T}}\left(1+\frac{v_y}{\omega_B}\cdot\frac{\partial \ln n_0}{\partial x}\right). \tag{1.88}$$

If $0 < v_y < \varkappa T/m\omega_B$, the derivative $\partial f_k/\partial v_k$ is positive, and this is equivalent to the presence of a beam. This must lead to the excitation of oscillations whose transverse phase velocity ω/k_y satisfies

$$0<\frac{\omega}{k_y}\cdot\frac{m\omega_B}{\varkappa T}<1. \tag{1.89}$$

Let us now consider an example of an instability which is caused by this asymmetry of the angular distribution of the particles in the space of the transverse velocities. Suppose that in a cold homogeneous plasma of fairly high density ($\omega_p \gg \omega_B$) there is a group of fast particles with transverse temperature T and density n_1 and $\nabla T = 0$, $\nabla n_1 \neq 0$. The perturbations of such a system with frequencies $\omega \gg \omega_B$ and $k_z = 0$ are described by the local dispersion equation

$$1-\left(\frac{\omega_p}{\omega}\right)^2+\frac{1}{(kd_1)^2}\left[1+i\sqrt{\pi}\frac{\omega-k_yV_y}{|k|v_{T1}}\right]=0. \qquad (1.90)$$

Here, $d_1^2 = T/4\pi e^2 n_1$, $V_y = \varkappa T/m\omega_B$. This result can be obtained by using the relations (1.80) and (1.88).

It follows from (1.90) that oscillations are excited if the condition (1.89) is satisfied; if $\mathrm{Re}\,\omega = \omega_p$, this reduces to

$$k_y\frac{\varkappa T}{m\omega_B}>\omega_p. \qquad (1.91)$$

The growth rate is approximately

$$\gamma \simeq \alpha\omega_p^4/(kv_T)^3, \qquad \alpha = n_1/n_0. \qquad (1.92)$$

This quantity must satisfy the condition $\gamma > \omega_B$. Hence, from (1.91) we find that the high-frequency instability we are considering is possible only if

$$\frac{\rho}{a}>\left(\frac{\omega_B}{\omega_p}\right)^{1/3}. \qquad (1.93)$$

Here we have assumed $\alpha \approx 1$.

§1.7. General Expressions for ε_0. Relative Importance of Convection and Larmor Currents

1. Permittivity. In the above we have considered some limiting cases of perturbations: $k_\perp\rho \to 0$, $\omega \leq \omega_B$ (for which convection plays an important role) and $k_\perp\rho \gg 1$, $\omega \gg \omega_B$ (for which the Larmor currents are important). However, to obtain a complete picture of the gradient instabilities this is not sufficient – we must also investigate perturbations with arbitrary $k_\perp\rho$ and ω. In calculating the general expression for the perturbed charge density we cannot adopt the simplified procedures used in §§1.1, 1.4, and 1.6. A more general approach must be developed.

One of the regular methods of finding the perturbed distribution function is the method of trajectory integrals. It has already been discussed in detail in Volume 1 (see §§5.1 and 7.2). Let us recall the basic idea of this method, which we shall then use to calculate the perturbed distribution function of an inhomogeneous plasma in a magnetic field.

First, in the linearized Boltzmann equation we go over from the variables $\mathbf{r}$, $\mathbf{v}$, and t to new variables $\mathbf{r}_0$, $\mathbf{v}_0$, and t, where $\mathbf{r}_0$ and $\mathbf{v}_0$ are the coordinates and the velocity of a particle at $t = t_0$ that is situated at the point $\mathbf{r}$ at the time t and has the velocity $\mathbf{v}$ at this time. We write the transport equation in the form

$$\frac{df}{dt} = \frac{e}{m} \nabla\psi \frac{\partial f_0}{\partial \mathbf{v}} . \tag{1.94}$$

Here it is assumed that the perturbed field is an electrostatic field, $\mathbf{E} = -\nabla\psi$.

Equation (1.94) can be integrated directly. We then make the inverse transition to the variables $\mathbf{r}$, $\mathbf{v}$, and t. With allowance for the time dependence $\psi \sim \exp(-i\omega t)$ and the fact that $f(t = -\infty) = 0$, the result can be represented in the form

$$f(r, \mathbf{v}) = \frac{e}{m} \int\limits_{-\infty}^{t} \exp[i\omega(t - t')] \nabla\psi[\mathbf{r}(t')] (\partial f_0/\partial \mathbf{v}(t'))\, dt'. \tag{1.95}$$

In (1.95) we must substitute the actual form of f_0 and calculate the integral over t'. In the case of an inhomogeneous plasma, f_0 has the form (1.86). Therefore, the derivative in (1.95) is

$$\frac{\partial f_0}{\partial \mathbf{v}} = \mathbf{v}_\perp \frac{\partial F}{\partial \varepsilon_\perp} + \mathbf{e}_z \frac{\partial F}{\partial V_z} + \frac{[\mathbf{e}_z, \mathbf{e}_\alpha]}{\omega_B} \frac{\partial F}{\partial \mathbf{R}_\alpha}, \qquad \mathbf{R}_\alpha = (X, Y, 0). \tag{1.96}$$

Since $\partial F/\partial I$ and $I = (\varepsilon_\perp, V_z, \mathbf{R}_\alpha)$ are constants of the motion, they can be taken in front of the integral over t'. Substituting (1.96) into (1.95), we obtain

$$f = \frac{e}{m} \left\{ \frac{\partial F}{\partial \varepsilon_\perp} \int\limits_{-\infty}^{t} e^{i\omega(t-t')} \mathbf{v}_\perp \nabla\psi\, dt' + \right.$$

$$\left. + \frac{\partial F}{\partial V_z} \int\limits_{-\infty}^{t} e^{i\omega(t-t')} \frac{\partial\psi}{\partial z}\, dt' + \frac{1}{\omega_B} \left[\frac{\partial F}{\partial \mathbf{R}_\perp}, \int e^{i\omega(t-t')} \nabla_\perp \psi\, dt' \right]_z \right\}. \tag{1.97}$$

We represent the coordinate dependence of ψ in the form of a set of plane waves:

$$\psi(\mathbf{r}) = e^{ik_z z} \int d\mathbf{k}_\perp \psi(\mathbf{k}_\perp) e^{i\mathbf{k}_\perp \mathbf{r}_\perp}. \tag{1.98}$$

In addition, we use the transformation

$$\mathbf{v}_\perp \nabla \psi = d\psi/dt + i(\omega - k_z v_z)\psi. \tag{1.99}$$

Then (1.97) is transformed into

$$f = (e/m)\left\{\psi \partial F/\partial \varepsilon_\perp - G \int d\mathbf{k}_\perp \psi(\mathbf{k}_\perp) \exp(i\mathbf{k}_\perp \mathbf{r}_\perp) I(\mathbf{k}, \omega, \mathbf{v})\right\}, \tag{1.100}$$

where

$$G = \omega \frac{\partial F}{\partial \varepsilon_\perp} + k_z \left(\frac{\partial F}{\partial v_z} - v_z \frac{\partial F}{\partial \varepsilon_\perp}\right) + \frac{1}{\omega_B}\left[\frac{\partial F}{\partial \mathbf{R}}, \mathbf{k}\right]_z, \tag{1.101}$$

$$I = -i \int_{-\infty}^{t} \exp\{i\omega(t - t') + i\mathbf{k}[\mathbf{r}(t) - \mathbf{r}(t')]\} dt'. \tag{1.102}$$

The integral I is calculated in §7.2 of Volume 1:

$$I = \sum_{n=-\infty}^{\infty} \frac{J_n(\xi) \exp[i\xi \sin(\alpha - \Psi) - in(\alpha - \Psi)]}{\omega - k_z v_z - n\omega_B}, \tag{1.103}$$

where $\Psi = \tan^{-1} k_y/k_x$, $\xi = k_\perp v_\perp/\omega_B$, $\alpha = \tan^{-1} v_y/v_z$.

Substituting (1.103) into (1.100) and integrating the result over the velocities, we find the perturbed density. The integration over the angle can be performed in the same manner as in the case of a homogeneous plasma (see §7.2 in Volume 1). However, it is first necessary to separate the angular dependence in F by representing F in the form of the series (1.87).

Calculating the perturbed density and using the method outlined in §1.1, we find that the contribution of the group of particles under consideration (with subscript α) to the permittivity is

$$\varepsilon_0^{(\alpha)}(\omega, \mathbf{k}, \mathbf{r}) = -\frac{4\pi e^2}{mk^2}\left\langle \frac{\partial F}{\partial \varepsilon_\perp} - \sum_{n=-\infty}^{\infty} \frac{J_n^2(\xi) G_1}{\omega - n\omega_B - k_z v_z}\right\rangle. \tag{1.104}$$

Here $\langle\ldots\rangle$ stands for $\int(\ldots)v_\perp dv_\perp dz$ and

$$G_1 = G + (n/k^2)\,[\nabla G,\ \mathbf{k}]_z. \qquad (1.105)$$

If F is Maxwellian, (1.104) yields

$$\varepsilon_0^{(\alpha)} = \frac{1}{(kd)^2}\left\{1 + \mathrm{i}\sqrt{\pi}\sum_{n=-\infty}^{\infty}\left[\hat{l} + (\hat{l}-1)\frac{n\omega}{z\omega_B}\right]\times\right.$$

$$\left.\times\frac{\omega}{|k_z|v_T}W\left(\frac{\omega - n\omega_B}{|k_z|v_T}\right)I_n(z)\,\mathrm{e}^{-z}\right\}. \qquad (1.106)$$

Here $z = k_\perp^2 T/m\omega_B^2$; I_n is a Bessel function of imaginary argument; $W(x) = \exp(-x^2)\left[1 + (2\mathrm{i}/\sqrt{\pi})\int_0^x \mathrm{e}^{t^2}dt\right]$; and the operator $\hat{l}$ is given by

$$\hat{l} = 1 - \frac{k_y T}{m\omega\omega_B}\left(\frac{\partial \ln n_0}{\partial x} + \frac{\partial T}{\partial x}\cdot\frac{\partial}{\partial T}\right). \qquad (1.107)$$

We now note some limiting results that follow from our investigation.

A. Low-frequency perturbations, $(\omega,\ k_z v_T') \ll \omega_B$

$$\varepsilon_0^{(\alpha)} = \frac{1}{(kd)^2}\left[1 + \mathrm{i}\sqrt{\pi}\,\hat{l}\,\frac{\omega}{|k_z|v_T}W\left(\frac{\omega}{|k_z|v_T}\right)I_0\mathrm{e}^{-z}\right]. \qquad (1.108)$$

B. Perturbations with $\omega \geq \omega_B$, $k_\perp\rho \geq 1$

$$\varepsilon_0^{(\alpha)} = \frac{1}{(kd)^2}\left[1 + \mathrm{i}\sqrt{\pi}\,\hat{l}\,\frac{\omega}{|k_z|v_T}\sum_{n=-\infty}^{\infty}W\left(\frac{\omega - n\omega_B}{|k_z|v_T}\right)I_n\mathrm{e}^{-z}\right]. \qquad (1.109)$$

C. Perturbations with $\gamma \gg \omega_B$, $k\rho \gg 1$

$$\varepsilon_0^{(\alpha)} = \frac{1}{(kd)^2}\left(1 + \mathrm{i}\sqrt{\pi}\,\hat{l}\,\frac{\omega}{kv_T}\right). \qquad (1.110)$$

For arbitary F and low frequencies of the perturbations, $|\omega - k_z v_z| \ll \omega_B$, Eq. (1.105) yields

$$\varepsilon_0^{(\alpha)} = -\frac{4\pi e^2}{mk^2}\left\langle\frac{\partial F}{\partial \varepsilon_\perp} - \frac{GJ_0^2}{\omega - k_z v_z}\right\rangle. \qquad (1.111)$$

In particular, if $F \propto \exp[-(m/2)(v-V)^2]$ and $V \parallel z$ this means

$$\varepsilon_0^{(\alpha)} = \frac{1}{(kd)^2}\left[1 + i\sqrt{\pi}(\omega - k_z V)\left(\hat{l} - \frac{k_y T}{m(\omega - k_z V)\omega_B}\cdot\frac{\partial V}{\partial x}\cdot\frac{\partial}{\partial V}\right)\times \right.$$

$$\left. \times \frac{I_0(z)\,e^{-z}}{|k_z|v_T} W\left(\frac{\omega - k_z V}{|k_z|v_T}\right)\right]. \qquad (1.112)$$

In the limiting case $T \to 0$, this yields the dispersion equation (1.45).

2. Relative Importance of Convection and Larmor Currents. As we have already mentioned, the distinctive property of an inhomogeneous plasma in a magnetic field is the dependence of its premittivity on the gradients of the steady-state distribution function [see, for example, the expression (1.104)]. The presence of gradient terms in ε_0 is due to two main effects – convection of particles in the perturbed field and asymmetry of the steady-state distribution function with respect to the angles in the space of the transverse velocities. In analyzing the different limiting cases of perturbations, we were guided by a physical principle: for long-wavelength perturbations convection is decisive; for short-wavelength perturbations, asymmetry.

We shall now consider how these two effects can be distinguished for arbitrary frequencies and wavelengths. We shall characterize the contribution of convection to $\varepsilon_0^{(\alpha)}$ by Eq. (1.23). To find the actual expression, we must, proceeding from (1.100), calculate ε_{ij}. The contribution to $\varepsilon_0^{(\alpha)}$ due to the velocity anisotropy of f_0, $\varepsilon_{0\,\text{asym}}^{(\alpha)}$, can be found by calculating the part of the perturbed density due to the second term in (1.87). This is done by means of Eq. (1.95), in which f_0 on the right-hand side must be replaced by the term we have indicated.

Exact equality of the gradient part of $\varepsilon_0^{(\alpha)}$ determined by Eq. (1.104) and the sum $\varepsilon_{0\,\text{conv}}^{(\alpha)} + \varepsilon_{0\,\text{asym}}^{(\alpha)}$ is obtained when $k_z = 0$, i.e., just when $\varepsilon_{0\,\text{conv}}^{(\alpha)}$ does not depend on $\varepsilon_{zj}k_j$ and $\varepsilon_{0\,\text{conv}}^{(\alpha)}$ is itself determined most rigorously. However, if $k_z \neq 0$, the equation $\varepsilon_{0\,\text{grad}}^{(\alpha)} = \varepsilon_{0\,\text{conv}}^{(\alpha)} + \varepsilon_{0\,\text{asym}}^{(\alpha)}$ holds in the following cases: a) $\xi \ll 1$, $\omega \ll \omega_{Bi}$; b) $\xi > 1$. In the first case, we have already pointed out that $\varepsilon_{0\,\text{grad}}^{(\alpha)} = \varepsilon_{0\,\text{conv}}^{(\alpha)}$ and $\varepsilon_{0\,\text{asym}}^{(\alpha)} \to 0$; in the second case, $\varepsilon_{0\,\text{grad}}^{(\alpha)} = \varepsilon_{0\,\text{asym}}^{(\alpha)}$ and $\varepsilon_{0\,\text{conv}}^{(\alpha)} \to 0$. This justifies the approximate manner in which we have taken into account the gradient effects in the foregoing sections.

Bibliography

Papers in which equations of plasma oscillations with allowance for gradient effects are derived:

1. Yu. A. Tserkovnikov, Zh. Eksp. Teor. Fiz., 33:67 (1957). The Boltzmann equation is solved for perturbations with $k_{\parallel} = 0$, $\omega \ll \omega_B$.
2. L. I. Rudakov and R. Z. Sagdeev, Dokl. Akad. Nauk SSSR, 138:581 (1961) [Sov. Phys. – Doklady, 6:415 (1961)]. Perturbations with $k_{\perp}\rho \to 0$, $\omega/\omega_B \to 0$, $k_{\parallel} \neq 0$.
3. N. A. Krall and M. N. Rosenbluth, Phys. Fluids, 5:1435 (1962). Perturbations with $k_{\parallel} = 0$ and arbitary ω/ω_B and $k_{\perp}\rho$.
4. A. B. Mikhailovskii, Nucl. Fusion, 2:162 (1962).
5. B. B. Kadomtsev and A. V. Timofeev, Dokl. Akad. Nauk SSSR, 146:581 (1962) [Sov. Phys. – Doklady, 7:826 (1962)]. In [4, 5] the Boltzmann equation is solved for arbitrary k_z, $k_{\perp}\rho$, ω/ω_B.

Papers in which electron gradient instabilities are investigated:

6. A. B. Mikhailovskii, Zh. Tekh. Fiz., 35:1945 (1965) [Sov. Phys. – Tech. Phys., 10:1498 (1966)]. Excitation of plasma oscillations by a beam with an inhomogeneous density; §§1.2, 1.4.
7. A. B. Mikhailovskii and A. A. Rukhadze, Zh. Tekh. Fiz., 35:2143 (1965) [Sov. Phys. – Tech. Phys., 10:1644 (1966)]. The instability of a stream with an inhomogeneous velocity profile is investigated; §1.3.
8. A. B. Mikhailovskii and É. A. Pashitskii, Dokl. Akad. Nauk SSSR, 165:796 (1965) [Sov. Phys. – Doklady, 10:1157 (1966)]. Instability of the electron component of a plasma resulting from a temperature gradient; §1.5.
9. A. B. Mikhailovskii and K. Jungwirth, Zh. Tekh. Fiz., 36:777 (1966) [Sov. Phys. – Tech. Fiz., 11:581 (1966)]. Excitation of plasma oscillations by a low-density beam; §1.4. A detailed analysis is also made of the excitation of the characteristic oscillations of a plasma cylinder by both the longitudinal and transverse energy of a beam.
10. K. Jungwirth, Czech. J. Phys., B17:498 (1967). Excitation of plasma oscillations by a beam with finite transverse energy; §1.6.
11. K. Jungwirth, Czech. J. Phys., B18:629 (1968). Excitation of plasma oscillations by an azimuthal electron beam.
12. M. V. Nezlin, M. I. Taktakishvili, and A. S. Trubinkov, Zh. Eksp. Teor. Fiz., 55:397 (1968) [Sov. Phys. – JETP, 28:208 (1969)].
13. M. D. Raizer, A. A. Rukhadze, and P. S. Strelkov, Zh. Eksp. Teor. Fiz., 53:1891 (1967) [Sov. Phys. – JETP, 26:1075 (1968)].
14. M. D. Raizer, A. A. Rukhadze, and P. S. Strelkov, Zh. Tekh. Fiz., 38:776 (1968) [Sov. Phys. – Tech. Phys., 13:586 (1968)]. The papers [12-14] contain a discussion of the thresholds of the two-stream instabilities when allowance is made for convection effects; §1.2.

Chapter 2

Ion-Cyclotron and High-Frequency Instabilities of a Plasma with a Finite Larmor Radius of the Ions and Electrons

§2.1. Plasma with Hot Maxwellian Ions

1. Preliminary Comments. Suppose there is an inhomogeneous plasma with a Maxwellian distribution of the ion and electron velocities and $T_i \geq T_e$. Let us first make some approximate estimates of the extent to which the plasma must be inhomogeneous for instabilities to arise with frequencies comparable with the ion-cyclotron frequency ($\omega \gtrsim \omega_{B_i}$).

In accordance with the results of the foregoing chapter, gradient effects in a Maxwellian plasma are important provided the frequency of the perturbations is not too high:

$$\omega \lesssim k_y v_T \rho / a. \tag{2.1}$$

This does not contradict the condition $\omega \gtrsim \omega_{B_i}$ if

$$\rho_i / a > 1 / k_y \rho_i. \tag{2.2}$$

This inequality can be satisfied more readily, the larger is k_y. However, k_y cannot be arbitrarily large: the transverse wavelength of the perturbation must not be much smaller than the Larmor radius of the electrons,

$$k_\perp \rho_e \lesssim 1 \,; \tag{2.3}$$

otherwise, as follows from (1.109), the gradient effects would not be important at all. From (2.2) and (2.3) we find that it is meaningful to consider gradient instabilities with $\omega \gtrsim \omega_{B_i}$ in a plasma with $T_i \simeq T_e$ only if

$$\rho_i/a \gtrsim (m_e/m_i)^{1/2}. \tag{2.4}$$

We shall say that a plasma of this kind is a plasma with a finite ion Larmor radius.

We now note that the gradient effects may be important in perturbations with $\omega \gtrsim \omega_{B_i}$. Since these perturbations have a short wavelength with respect to the ions, $k_\perp \rho_i \gg 1$ [condition (2.2)], the comments made in §1.7 show that in this case ion Larmor currents may be important but that the convection of ion charge is certainly not important. With respect to the electrons, the perturbations have a long wavelength ($k_\perp \rho_e < 1$), and it follows that electron convection may be important but certainly not the electron Larmor currents.

The calculations that follow establish the existence of instabilities due to the above gradient effects. The oscillations are excited by the ion Larmor currents, whereas the existence of a branch of oscillations that can be excited is due to electron convection.

§2. Cyclotron Instability for $k_z = 0$. Let us consider perturbations with $k_z = 0$, $z_i \gg 1$, $z_e < 1$, and $\omega \approx \omega_{B_i}$ in a plasma with $\nabla n_0 \neq 0$, $\nabla T = 0$, $T_i > T_e$. Under these assumptions, (1.108) and (1.109) yield

$$\left.\begin{aligned} \varepsilon_0^{(e)} &= \left(\frac{\omega_{p_e}}{\omega_{B_e}}\right)^2 + \frac{\varkappa k_y \omega_{p_e}^2}{\omega \omega_{B_e} k^2}, \\ \varepsilon_0^{(i)} &= \frac{1}{(k_\perp d_i)^2}\left[1 - \sum_{n=-\infty}^{\infty} \frac{\omega - \omega_{n_i}}{\sqrt{2\pi z_i}\,(\omega - n\omega_{B_i})}\right], \\ \omega_{n_i} &\equiv \frac{k_y \varkappa c T_i}{e_i B_0}, \end{aligned}\right\} \tag{2.5}$$

so that the dispersion equation is

$$1 + \left(\frac{\omega_{p_e}}{\omega_{B_e}}\right)^2 + \frac{1}{(k_\perp d_i)^2}\Big)\left(1 - \frac{\omega_{n_i}}{\omega}\right)\left(1 - \sum_{n=-\infty}^{\infty} \frac{1}{\sqrt{2\pi z_i}} \frac{\omega}{\omega - n\omega_{B_i}}\right) = 0. \tag{2.6}$$

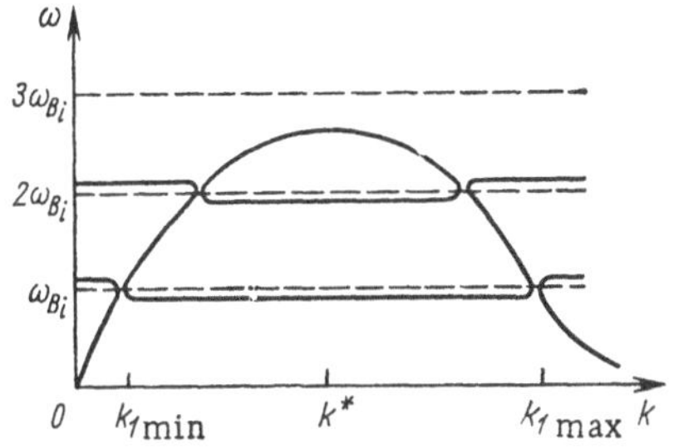

Fig. 2.1. Oscillation branches of an inhomogeneous plasma for $k_z = 0$. Here
$$k^* = \min\left(\frac{1}{d_i}, \frac{m_i}{m_e}\cdot\frac{1}{\rho_i}\right).$$

If the terms with ω_{n_i} are neglected, this equation describes the ion-cyclotron oscillations of a homogeneous plasma considered in §8.4 of Volume 1. It can be seen that the approximation of a homogeneous plasma is applicable only if $n\omega_{B_i} \gg \omega_{n_i}$. If $n\omega_{B_i} \lesssim \omega_{n_i}$, the ion-cyclotron oscillation branches have the dispersion law

$$\omega_n(k) = n\omega_{B_i}\left[1 + \frac{1}{\sqrt{2\pi z_i}}\,\frac{1}{1 - \dfrac{\omega_{n_i}}{n\omega_{B_i}} + (kd_i)^2\left(1 + \dfrac{\omega_{p_e}^2}{\omega_{B_e}}\right)}\right]. \tag{2.7}$$

Apart from the cyclotron branches (2.7), Eq. (2.6) describes a further branch of oscillations whose frequency is far from the cyclotron harmonics:

$$\omega = \zeta(k)\,\omega_{n_i}, \tag{2.8}$$

where

$$\zeta(k) = [1 + (kd_i)^2(1 + \omega_{p_e}^2/\omega_{B_e}^2)]^{-1}.$$

The branches (2.7) and (2.8) intersect if

$$(\rho_i/a)^2 > 4n^2[(\omega_{B_i}/\omega_{p_i})^2 + m_e/m_i], \qquad a \equiv \varkappa^{-1}. \tag{2.9}$$

The solutions of (2.6) for this case are shown in Fig. 2.1. Near $k = k_n$, corresponding to the point of intersection of the branches, the frequency of the oscillations is complex, i.e., the plasma is unstable. We shall now give some relations that characterize the instability in the limiting cases of dense and low-density plasmas.

A. Dense Plasma $\omega_{p_e} > \omega_{B_e}$. Such a plasma is unstable if

$$\rho_i/a > 2(m_e/m_i)^{1/2}. \tag{2.10}$$

If the inequality (2.10) is not strong, a few harmonics are excited. In this case, the characteristic $k_\perp$ is approximately $k_\perp \simeq (m_i/m_e)^{1/2}\rho_i^{-1}$. The growth rate is

$$\gamma \simeq (m_e/m_i)^{1/4}\,\omega_{B_i}. \tag{2.11}$$

If the left-hand side of (2.10) is much greater than the right-hand side, the number of excited harmonics is

$$n \simeq (\rho_i/2a)\,(m_i/m_e)^{1/2}. \tag{2.12}$$

The growth rates of the first and the highest of the excited harmonics are approximately

$$\left.\begin{aligned} \gamma_{n=1} &\simeq (\rho_i/a)^{1/4}\,\omega_{B_i},\\ \gamma_{\max} &\simeq (8\pi)^{-1/4}\,(\rho_i/a)\,(m_i/m_e)^{1/4}\,\omega_{B_i}. \end{aligned}\right\} \tag{2.13}$$

Hence, in particular, we see that the growth rate is of the order of ω_{B_i} if $\rho_i/a \simeq 2\,(m_e/m_i)^{1/4}$. We shall consider the instability of a plasma with $\rho_i/a > 2\,(m_e/m_i)^{1/4}$ in §2.1.3.

B. Plasma with $\omega_{p_e} < \omega_{B_e}$. It follows from (2.9) that the cyclotron instability does not develop if the plasma density is too low:

$$\omega_{p_i}/\omega_{B_i} < a/2\rho_i. \tag{2.14}$$

In particular, the plasma is stable for all ρ_i/a if $\omega_{p_i} < \omega_{B_i}$.

Let us now assume $a/2\rho_i \lesssim \omega_{p_i}/\omega_{B_i} < (m_i/m_e)^{1/2}$. Under these conditions there is an instability with the characteristic growth rate

$$\gamma \simeq (\rho_i/a)^{1/2}\,\omega_{B_i} \simeq (\omega_{B_i}/\omega_{p_i})^{1/2}\,\omega_{B_i} \tag{2.15}$$

and characteristic wave number $k_\perp \simeq 1/d_i$.

3. High-Frequency Instability of a Strongly Inhomogeneous Plasma. Suppose

$$\rho_i/a > (m_e/m_i)^{1/4}\,(1 + \omega_{Be}^2/\omega_{pe}^2)^{-1/2}. \tag{2.16}$$

In Eq. (2.6) we must then include not one but several terms of the sum over n. Assuming $\gamma > \omega_{B_i}$, we go over in (2.6) to the high-frequency approximation, making the substitution

$$\sum_{n=-\infty}^{\infty} (\omega - n\omega_{B_i})^{-1} \to \int (\omega - n\omega_{B_i})^{-1}\, dn. \qquad (2.17)$$

Retaining only the imaginary contribution of the integral (2.17) in (2.6), we obtain

$$1 + \left(\frac{\omega_{pe}}{\omega_{Be}}\right)^2 + \frac{1}{(kd_i)^2}\left(1 - \frac{\omega_{n_i}}{\omega}\right)\left(1 + \frac{i\sqrt{\pi}\,\omega}{kv_{T_i}}\right) = 0. \qquad (2.18)$$

This equation can also be obtained by using Eq. (1.110). One could then show that the imaginary terms in (2.18) are due to the resonant interaction between the wave and the particles whose velocities satisfy the condition $\omega = \mathbf{kv}$.

It can be seen from (2.18) that $\omega \operatorname{Im} \varepsilon_0 < 0$ for $\omega < \omega_n$, i.e., the energy of the resonant particles is transferred to the oscillations. This negative conductivity effect is due to the asymmetry of the transverse velocity distribution of the particles. The real part of the oscillation frequency (2.18) satisfies the condition (2.8), and the growth rate is approximately

$$\gamma_{\max} \simeq \frac{\sqrt{\pi}}{8\sqrt{2}}\left(\frac{\rho_i}{a}\right)^2\left(\frac{m_i}{m_e}\right)^{1/2}\left(1 + \frac{\omega_{Be}^2}{\omega_{p_e}^2}\right)^{-1/2}\omega_{B_i}. \qquad (2.19)$$

If $\rho_i/a \simeq 2\,(m_e/m_i)^{1/4}$ and $\omega_{p_e} < \omega_{B_e}$, this result is identical with (2.18) to within a numerical factor; thus, the approximation of a continuous spectrum joins on smoothly to the approximation of individual harmonics. The qualitative dependence of the growth rate on the parameter ρ_i/a that follows from (2.13) and (2.19) is shown in Fig. 2.2 (for a plasma with $\omega_{p_e} < \omega_{B_e}$).

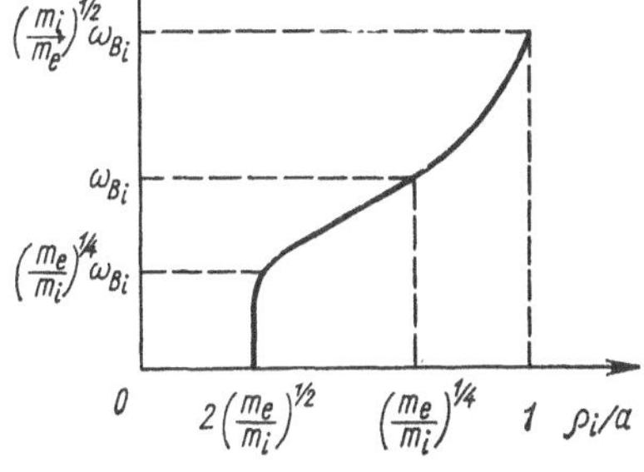

Fig. 2.2. Growth rate of the instability of a plasma with finite ρ_i/a and $\omega_{pe} > \omega_{Be}$.

At the limit of the approximation of a small Larmor radius, $\rho_i \approx a$, we obtain from (2.19), to within a numerical factor,

$$\gamma \simeq \omega_{p_i} (1 + \omega_{pe}^2/\omega_{B_e}^2)^{-1/2}. \tag{2.20}$$

In this case the uncompensated Larmor currents are of the same order as the thermal fluxes, $V_{\perp 0i} \simeq v_{T_i}$ (see §1.6). The case $V_{\perp 0i} \gtrsim v_{T_i}$ corresponds to an azimuthal ion stream. The stability of such a stream is considered in §2.5.

§2.2. Plasma with Hot Maxwellian Electrons

1. Preliminary Comments. We shall consider the excitation of perturbations with $\omega \gtrsim \omega_{B_i}$ in a plasma whose electrons are hotter than the ions, $T_e > T_i$.

In §2.1 we obtained qualitatively the inequality $\rho_i/a > (m_e/m_i)^{1/2}$ [see (2.4)], which shows that gradient excitation of perturbations with $\omega \gtrsim \omega_{B_i}$ is possible in a plasma with $T_i \geq T_e$. For a plasma with $T_e \gg T_i$, this condition must be made more precise as follows. In (2.1) we must substitute the greater of the temperatures, i.e., T_e, and (2.4) is then replaced by

$$\rho_e/a > m_e/m_i. \tag{2.21}$$

This corresponds to a plasma with a finite electron Larmor radius.

If T_e/T_i is sufficiently large, a plasma which is still weakly inhomogeneous as regards the ions, $\rho_i \ll a$, can be such that

$$\rho_e/a \gtrsim \left(\frac{m_e}{m_i}\right)^{1/2}. \tag{2.22}$$

In this case the characteristic gradient frequency $\omega \simeq k_y v_{T_e} \rho_e/a$ of the perturbations with the shortest wavelengths, $k_\perp \rho_e \simeq 1$, greatly exceeds ω_{B_i} and is comparable with the hybrid frequency. Under these conditions high-frequency perturbations, $\omega \gg \omega_{B_i}$, can be excited and, in particular, high-frequency ion-acoustic oscillations.

2. Plasma with $\rho_e/a \geq (m_e/m_i)^{1/2}$. The qualitative considerations of §2.2.1 show that the excitation of perturbations with $\omega \gg \omega_{B_i}$ is to be expected in such a plasma. In accordance with (1.108) and (1.106), the electron and ion permittivities in this case are

$$\left.\begin{aligned} \varepsilon_0^{(e)} &= (kd_e)^{-2}\left[1 + i\sqrt{\pi}\,\hat{l}_e \frac{\omega}{|k_z| v_{T_e}} I_0(z_e)\,e^{-z_e}\right], \\ \varepsilon_0^{(i)} &= -(\omega_{p_i}/\omega)^2. \end{aligned}\right\} \tag{2.23}$$

A. Perturbations with $\omega \ll k_z v_{T_e}$. In this case the dispersion equation $1 + \varepsilon_0^{(e)} + \varepsilon_0^{(i)} = 0$ reduces to

$$\varepsilon_0 = 1 + \frac{1}{(kd_e)^2}\left[1 + \frac{i\sqrt{\pi}\,(\omega - \omega_{ne})}{|k_z| v_{T_e}} I_0(z_e)\,e^{-z_e}\right] - \left(\frac{\omega_{p_i}}{\omega}\right)^2 = 0. \tag{2.24}$$

For simplicity we have here assumed $\nabla T = 0$. Equation (2.24) describes the high-frequency ion-acoustic oscillations with $\nabla n_0 = 0$ discussed in §8.3 of Volume 1.

From (2.24) we find that the growth rate of the oscillations is

$$\gamma = \frac{\sqrt{\pi}\,(\mathrm{Re}\,\omega)^2}{2|k_z| v_{T_e}} I_0(z_e)\,e^{-z_e}\left(\frac{\omega_{n_e}}{\mathrm{Re}\,\omega} - 1\right)\frac{1}{1 + (kd_e)^2}, \tag{2.25}$$

where $\mathrm{Re}\,\omega = k(T_e/m_i)^{1/2}(1 + k^2 d_e^2)^{-1/2}$. If $kd_e \leq 1$, the instability condition ($\gamma > 0$) has the form

$$\rho_e/a > (m_e/m_i)^{1/2}. \tag{2.26}$$

Perturbations with shorter wavelengths are excited for smaller ρ_e,

$$\frac{\rho_e}{a} > \frac{1}{kd_e}\left(\frac{m_e}{m_i}\right)^{1/2}.$$

B. Perturbations with $\omega \gg k_z v_{T_e}$. In accordance with (2.25), the ratio $\gamma/\mathrm{Re}\,\omega$ increases with decreasing $\cos\theta$. Suppose $k < (1/\rho_e, 1/d_e)$. If $\cos\theta \simeq (m_e/m_i)^{1/2}$ in this case, we have

$$\gamma \simeq \mathrm{Re}\,\omega \simeq k(T_e/m_i)^{1/2}. \tag{2.27}$$

By virtue of the condition (2.26) the increment (2.27) exceeds ω_{B_i} even for the minimally possible $k \simeq \varkappa_n$.

For a plasma with $\omega_{pe} < \omega_{Be}$ the expression (2.27) is valid right up to $k_{max} \simeq 1/d_e$, and for such k

$$\gamma_{max} \simeq \omega_{p_i}. \tag{2.28}$$

In a denser plasma, $\omega_{pe} > \omega_{Be}$, one must take k_{max} in (2.27) equal to $1/\rho_e$; then

$$\gamma_{max} \simeq \omega_{hyb} \equiv |\omega_{Be}\omega_{B_i}|^{1/2}. \tag{2.29}$$

If $\cos\theta \leq (m_e/m_i)^{1/2}$, the growth rate decreases. This can be seen by taking $\omega > k_z v_{Te}$ in (2.23).

3. Plasma with $(T_i/T_e)^{1/2} < \rho_0/a < 1$.

A. Excitation of High-Frequency Ion-Acoustic Oscillations. If $\rho_0/a < 1$, the branch of high-frequency ion-acoustic oscillations is excited only if $kd_e \gg 1$. Assuming $kd_e \gg 1$ in (2.24), we find

$$\left.\begin{aligned} &\mathrm{Re}\,\omega = \omega_{p_i}, \\ &\gamma = \frac{\sqrt{\pi}\,\omega_{p_i}^2}{2|k_z|v_{Te}} I_0(z_e)\,e^{-z_e}\left(\frac{\omega_{ne}}{\omega_{p_i}} - 1\right). \end{aligned}\right\} \tag{2.30}$$

The growth rate is positive if

$$k > \frac{1}{d_e}\cdot\frac{a}{\rho_0}, \qquad \rho_0^2 \equiv T_e/m_i\omega_{B_i}^2. \tag{2.31}$$

As a function of k_z, the growth rate is maximal for $k_z \simeq \omega_{p_i}/v_{Te}$; approximately,

$$\gamma_{max} \simeq I_0(z_e)\,e^{-z_e}\omega_{p_i}. \tag{2.32}$$

The condition (2.31) does not contradict the approximation of cold ions, $\omega > kv_{T_i}$, if $k < 1/d_i$, i.e., if

$$\rho_0/a > (T_i/T_e)^{1/2}. \tag{2.33}$$

It follows from (2.32) and (2.31) that in a dense plasma, $\omega_{pe} > \omega_{Be}$, the maximal growth rate is

$$\gamma_{max} \simeq v_{Te}/a, \tag{2.34}$$

and in a less dense plasma,

$$\gamma_{max} \simeq \omega_{p_i}. \tag{2.35}$$

B. Excitation of Ion-Cyclotron Harmonics. As well as the high-frequency perturbations of a plasma with $\rho_0/a > (T_i/T_e)^{1/2}$ considered in §2.2.2A, ion-cyclotron harmonics can also be excited. Neglecting the terms of order ω_{n_i}/ω in (1.109) and assuming $|\omega - n\omega_{B_i}| \gg k_z v_{T_i}$, we find that (2.24) is replaced by the dispersion equation

$$1+\frac{1}{(kd_i)^2}\left(1-\sum_{n=-\infty}^{\infty}\frac{\omega I_n(z_i)\,e^{-z_i}}{\omega-n\omega_{B_i}}\right)+\frac{1}{(kd_e)^2}\left[1+\frac{i\sqrt{\pi}\,(\omega-\omega_{n_e})}{|k_z|\,v_{T_e}}\right]=0. \quad (2.36)$$

The instability condition, $\omega \leqslant \omega_{n_e}$, for $\omega \simeq n\omega_{B_i}$ means

$$\sqrt{z_i} \gtrsim (T_i/T_e)^{1/2}\, a/\rho_0. \quad (2.37)$$

We see that in a plasma with $\rho_0/a > (T_i/T_e)^{1/2}$ perturbations with $z_i < 1$ (and not only with $z_i > 1$) can be excited. In this case, Eq. (2.36) reduces to

$$1-\frac{z_i\omega_{B_i}^2}{\omega^2-\omega_{B_i}^2}-\sum_{|n|\geqslant 2}^{\infty}\frac{\omega I_n(z_i)\,e^{-z_i}}{\omega-n\omega_{B_i}}+\frac{T_i}{T_e}\left[1+\frac{i\sqrt{\pi}\,(\omega-\omega_{n_e})}{|k_z|\,v_{T_e}}\right]=0. \quad (2.38)$$

Here $I_n e^{-z_i}$ is a small parameter; therefore, (2.38) has a solution with $\omega \simeq n\omega_{B_i}$ ($n = 2, 3, \ldots$). If $z_i < 1$, only these harmonics are excited; the perturbations of the first harmonic are damped in this case.

However, if $z \gg 1$, harmonics with all n including $n = 1$ are excited. In this case

$$\gamma \simeq \frac{v_{T_i}}{a}. \quad (2.39)$$

Here we have assumed

$$n \simeq 1,\ k_e v_{T_e} \simeq \omega_{B_i},\ z_i \simeq (a/\rho_i)^2.$$

4. Plasma with $(m_e/m_i)^{1/2} < \rho_0/a < (T_i/T_e)^{1/2}$. In this case the gradient terms of the dispersion equation are important if $\omega \simeq n\omega_{B_i}$, $z_i \gg 1$. In (1.109) and (1.108) we take $|\omega - n\omega_{B_i}| \gg k_z v_{T_i}$, $z_i \gg 1$, $z_e \ll 1$, $\nabla T = 0$. We then obtain an equation of the form (2.36), in which we must assume $z_i \gg 1$. The real part of the oscillation frequency for $k_z v_{T_e} < n\omega_{B_i}$ is near $n\omega_{B_i}$

and is determined by Eq. (8.4) of Volume 1. The growth rate is

$$\gamma=\frac{1}{\sqrt{2z_i}}\cdot\frac{T_i}{T_e}\cdot\frac{(n\omega_{B_i})^2}{|k_z|v_{T_e}}\left(\frac{\omega_{ne}}{n\omega_{B_i}}-1\right)\exp\left[-\left(\frac{n\omega_{B_i}}{k_z v_{T_e}}\right)^2\right]. \qquad (2.40)$$

The plasma is unstable if the condition (2.37) holds. In accordance with the last condition and the original assumption of this subsection, $\rho_0/a < (T_i/T_e)^{1/2}$, only short-wavelength perturbations, $z_i \gg 1$, are excited. The instability is due to the convection of resonant electrons.

As a function of k , the growth rate attains a maximum at $k_z \simeq n\omega_{B_i}/v_{T_e}$. The order of magnitude of γ_{max} is determined by Eq. (2.39).

§2.3. Plasma with a Non-Maxwellian Distribution of the Transverse Ion Velocities

In a plasma with non-Maxwellian ions ($\partial f_{\perp i}/\partial v_\perp > 0$), instabilities for which gradient effects are not important can develop. The most important one is the high-frequency instability ($\gamma > \omega_{B_i}$). considered in §14.1 of Volume 1. The perturbations that are excited have $k_z \neq 0$. We shall now take into account gradient effects and consider perturbations with $k_z = 0$.

Thus, we set $k_z = 0$, $\omega_{B_i} \ll \omega \ll \omega_{B_e}$, and $\gamma > \omega_{B_i}$. To describe the ions we use the high-frequency approximation. We shall neglect the finite value of the electron Larmor radius ($z_e \to 0$). We shall assume that the phase velocity ω/k_y of the oscillations is large compared with the mean velocities $V_{\perp e}$ and $V_{\perp i}$ of the transverse motion of the components, i.e., $\omega/k_y \gg v_{T_i}\rho_i/a$.

Under these assumptions (1.108) and (1.80) yield the dispersion equation

$$1+\left(\frac{\omega_{pe}}{\omega_{B_e}}\right)^2+\frac{\varkappa_n k_y \omega_{pe}^2}{\omega\omega_{B_e}k^2}-\left(\frac{\omega_{p_i}}{k}\right)^2\int\frac{\mathrm{kv}\,\frac{\partial f_{0i}}{\partial\varepsilon_\perp}}{\omega-\mathrm{kv}}\,d\mathrm{v}=0. \qquad (2.41)$$

Here $f_{0i}(\varepsilon_\perp)$ is normalized to unity, and $\varepsilon_\perp \equiv v_\perp^2/2$.

1. Plasma with a δ-Function Ion Distribution. We shall first consider the very simple distribution $f_{0i} = \delta(\varepsilon_\perp - v_0^2/2)$. Assuming also $\omega \ll k_\perp v_0$, we reduce (2.41) to the form

$$1+\frac{\bar{\omega}}{\omega}-\frac{i\omega}{k_\perp v_0}\left(\frac{k_\perp^{(0)}}{k}\right)^2=0. \qquad (2.42)$$

Here $\bar{\omega} = k_y \varkappa_n \omega_{pe}^2 \Big/ \left[k^2 \omega_{Be} \left(1 + \frac{\omega_{pe}^2}{\omega_{Be}^2}\right)\right]$, and $k_\perp^{(0)}$ is determined by the relation $k_\perp^{(0)2} = \left(\frac{\omega_{p_i}}{v_0}\right)^2 \left(1 + \frac{\omega_{pe}^2}{\omega_{Be}^2}\right)^{-1}$.

Equation (2.42) is similar to Eq. (14.1) in Volume 1. The only difference is that instead of the longitudinal inertial motion of the electrons we are now taking into account their convective motion (see §1.1.1).

The roots of Eq. (2.42) are complex. One of them has a positive imaginary part, $\gamma > 0$, which corresponds to instability. The original assumption $\gamma > \omega_{B_i}$ is justified if

$$(\rho_i/a)^{3/2} > (\omega_{B_i}/\omega_{p_i})^2 (1 + \omega_{pe}^2/\omega_{Be}^2). \tag{2.43}$$

This condition plays the role of an approximate instability condition. [The more precise instability condition found by Post and Rosenbluth differs from (2.43) by the presence of a factor 0.38 on the right-hand side.]

The maximal growth rate of the excited perturbations is

$$\gamma_{\max} \simeq \mathrm{Re}\,\omega \simeq \left(\frac{\rho_i}{a}\right)^{3/4} \frac{\omega_{p_i}}{\left(1 + \frac{\omega_{pe}^2}{\omega_{Be}^2}\right)^{1/2}}. \tag{2.44}$$

It is attained for

$$k^2 \simeq \frac{1}{d_0^2} \frac{(\rho_i/a)^{1/2}}{1 + \omega_{pe}^2/\omega_{Be}^2}, \qquad \left(d_0 = \frac{v_0}{\omega_{p_i}}\right). \tag{2.45}$$

2. Spread-out Ion Distribution with $\partial f_{0i}/\partial v_\perp > 0$. We shall investigate the stability of spread-out distributions with $\partial f_{0i}/\partial v_\perp > 0$ for the case of the example

$$f_{0i} = \frac{m(T+\tau)}{T^2} e^{-\frac{mv_\perp^2}{2T}} \left(1 - e^{\frac{-mv_\perp^2}{2\tau}}\right). \tag{2.46}$$

(This distribution was also considered in §14.1 in Volume 1, in which we assumed $\nabla n_0 = 0$.)

If $T > \tau$, the parameter T can be interpreted as a temperature, and $(\tau/m_i)^{1/2}$ then characterizes the magnitude of the velocity below which $\partial f_{0i}/\partial v_\perp > 0$ (Fig. 2.3). If f_{0i} has the form (2.46), Eq.

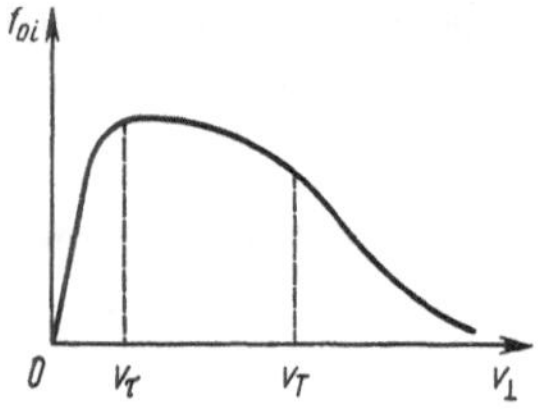

Fig. 2.3. Ion distribution function corresponding to a nonequilibrium distribution of the form (2.46).

(2.42) is replaced by the dispersion equation

$$1+\left(\frac{\omega_{pe}}{\omega_{Be}}\right)^2+\frac{\varkappa_n k_y \omega_{pe}^2}{\omega\omega_{Be}k^2}+\frac{i\sqrt{\pi}}{k^3 d_{\text{eff}}^2}\left[\frac{1}{v_T}\left(\omega-\frac{k_y\varkappa T}{m_i\omega_{B_i}}\right)-\right.$$

$$\left.-\frac{1}{v_1}W\left(\frac{\omega}{kv_1}\right)\left(\omega-\frac{\varkappa_n k_y v_1^2}{2\omega_{B_i}}\right)\right]=0. \tag{2.47}$$

Here $d_{\text{eff}}^2 = T^2/[(T+\tau)m_i\omega_{pi}^2]$ is the square of the effective Debye radius; $v_T = (2T/m_i)^{1/2}$; $v_1 = [2T\tau/m_i(T+\tau)]^{1/2}$. In (2.47) we have assumed $\omega < kv_T$; we have also retained terms of order $k_y V_{0i}/\omega$. If τ and T are of the same order, (2.47) becomes

$$1+\frac{\bar{\omega}}{\omega}-\frac{i\sqrt{\pi}\,\omega}{k^3 v_1}\left[1-\left(\frac{\tau}{T+\tau}\right)^{1/2}\right]\frac{1}{d_{\text{eff}}^2\left(1+\frac{\omega_{pe}^2}{\omega_{Be}^2}\right)}=0. \tag{2.48}$$

This equation differs from (2.42) only by the numerical coefficient (of order unity) in the last term on the left-hand side. Therefore, the results that follow from (2.48) are qualitatively the same as for a δ-function ion distribution (§2.3.1).

The perturbations are insensitive to the detailed form of the distribution function $f_{0i}(v_\perp^2)$ because the dispersion equation does not contain the function f_{0i} itself but only its integral over one of the velocity components (for example, $\int f_{0i}dv_x$, if k is directed along the y axis). For a δ-function distribution and for f_{0i} of the form (2.46), these integrals for small v_y are, respectively,

$$\frac{1}{2\pi}\int f_{0i}\,dv_x=\begin{cases}\dfrac{1}{2\pi v_0}\left(1+\dfrac{v_y^2}{2v_0^2}\right),\\[2ex] \dfrac{1}{\sqrt{\pi}\,v_T}\left(1+\dfrac{T}{\tau}\right)\left[1-\left(\dfrac{\tau}{T+\tau}\right)^{1/2}\right]\times\\[2ex] \times\left[1+\dfrac{m_i v_y^2}{2T}\left(\dfrac{T+\tau}{\tau}\right)^{1/2}\right].\end{cases} \tag{2.49}$$

It can be seen that if $T \simeq \tau$ the two expressions are approximately equivalent, and for any form of f_{0i} the high-frequency instability is described by approximately the same relations as in the case of a δ-function if $\partial \ln f_{0i}/\partial \ln v_\perp \gtrsim 1$ at velocities less than or of the order of the mean velocity.

Let us now consider a plasma with $v_\tau/v_T \ll 1$ (see Fig. 2.3). If $v_\tau/v_T > \rho_i/a$, it follows from (2.47) that the relations (2.43) and (2.45) are replaced by

$$\left.\begin{aligned} &\left(\frac{\rho_i}{a}\right)^{2/3} > \left(\frac{v_T}{v_\tau}\right)^{1/2}\left(\frac{\omega_{Bi}}{\omega_{p_i}}\right)^2\left(1+\frac{\omega_{p_e}^2}{\omega_{B_e}^2}\right),\\ &\gamma \simeq \operatorname{Re}\omega \simeq \left(\frac{\rho_i}{a}\right)^{3/4}\left(\frac{v_\tau}{v_T}\right)^{1/4}\frac{\omega_{p_i}}{\left(1+\frac{\omega_{p_e}^2}{\omega_{B_e}^2}\right)^{1/2}},\\ &k^2 \simeq \frac{1}{d^2}\frac{1}{1+\frac{\omega_{p_e}^2}{\omega_{B_e}^2}}\left(\frac{\rho_i}{a}\frac{v_T}{v_\tau}\right)^{1/2}. \end{aligned}\right\} \tag{2.50}$$

It can be seen from the first relation that the critical value of the parameter ρ_i/a does not depend strongly on v_τ/v_T, $\rho_i/a \sim (v_\tau/v_T)^{3/4}$. The growth rate and the characteristic wave number also depend very weakly on v_τ/v_T. The dependence of the critical value of ρ_i/a on v_τ/v_T is shown schematically in Fig. 2.4.

The results (2.50) were obtained under the assumption that $\omega/kv_\tau < 1$. This is justified if

$$v_\tau/v_T > \rho_i/a. \tag{2.51}$$

At the limit of applicability of (2.50) (for $v_\tau/v_T \simeq \rho_i/a$) the instability condition means

$$\frac{\rho_i}{a} > \left(\frac{m_e}{m_i}\right)^{1/2}\left(1+\frac{\omega_{B_e}^2}{\omega_{p_e}^2}\right)^{1/2}. \tag{2.52}$$

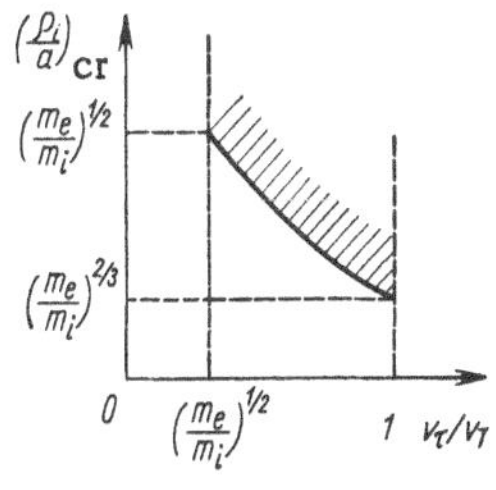

Fig. 2.4. Critical value of ρ_i/a as a function of the extent to which the transverse ion distribution for $\omega_{pe} > \omega_{Be}$ is nonequilibrium. The instability region is shaded.

In order of magnitude this condition is the same as the instability condition (2.9) of a Maxwellian plasma. If $v_\tau/v_T < \rho_i/a$, the fact that the transverse distribution is not an equilibrium distribution ceases to be important.

§2.4. Plasma with a Longitudinal Current

In a plasma with a longitudinal current, instabilities of a two-stream type for which gradient effects are not important can develop. However, we shall here be concerned with situations in which these instabilities do not arise.

1. Cold Plasma ($V > v_{T_e}$) of Small Width $a < V/\omega_{p_e}$ in a Strong Magnetic Field ($\omega_{p_e} \ll \omega_{B_e}$). In §1.2 we have noted that the two-stream instability does not develop in a beam-plasma system if the plasma is of finite width and its density is too low [the condition (1.28)]. The same situation also obtains if the electrons of the plasma move relative to the ions: in a spatially unlimited plasma the Buneman instability should arise, whereas a bounded plasma is stable if the condition (1.28) is satisfied (see §12.2 in Volume 1). To obtain a true picture of the instabilities, we must take into account gradient effects, in the same way as this was done in §1.2. We then obtain the dispersion equation [cf. (1.33)].

$$1-\left(\frac{\omega_{p_i}}{\omega}\right)^2+\frac{\varkappa_n k_y \omega_{p_e}^2}{k^2\omega_{B_e}(\omega-k_z V)}=0. \tag{2.53}$$

The resulting instability condition reduces to (1.35). As a result our analysis in §1.2 of the instability boundaries remains valid.

If (1.35) is not a strong inequality, the growth rate is approximately

$$\gamma \simeq \omega_{p_i}. \tag{2.54}$$

Taking into account (1.35), we note that this result satisfies the condition $\gamma < k_z V$ if $\omega_{p_i} > \omega_{B_i}$, i.e., just when the effect of a magnetic field on the motion of the ions can be ignored (the approximation $|\omega| > \omega_{B_i}$).

2. Plasma with a Current for $\partial f_0^{(e)}/\partial v_z \leq 0$. Suppose the mean velocity of the electron component differs from zero ($V \neq 0$), but the function $f_0^{(e)}$ does not have a maximum for $v_z \neq 0$

$(\partial f_0^{(e)}/\partial \ln v_z \leq 0)$, so that two-stream type instabilities do not develop. A distribution of this type can arise, for example, as a result of the development of the high-frequency ion-acoustic instability when $V \simeq v_{T_e}$ and $\omega_{p_e} < \omega_{B_e}$ followed a process of quasilinear relaxation.

Let us investigate the gradient instabilities of such a plasma. To be specific, we shall consider an electron distribution in the form of a step:

$$f_0^{(e)}(v_z) = \begin{cases} n_0/2u, & V-u<v_z<V+u;\ u>V, \\ 0, & v_z<V-u,\ v_z>V+u. \end{cases} \qquad (2.55)$$

Assuming $\omega/k_z \ll (u, V)$, we arrive at the dispersion equation

$$1-\left(\frac{\omega_{p_i}}{\omega}\right)^2+\frac{\omega_{p_e}^2}{k^2(u^2-V^2)}+\frac{\varkappa_n k_y}{k^2}\,\frac{\omega_{p_e}^2}{2\omega_{B_e}|k_z|u}\left(\ln\left|\frac{u+V}{u-V}\right|+i\pi\right)=0. \qquad (2.56)$$

Thus, we obtain the following.

A. Kinetic Excitation of High-Frequency Ion-Acoustic Oscillations when $u \gg V$. If $u \gg V$, the logarithm in (2.56) can be ignored, and the solution then has the form

$$\left.\begin{aligned} \mathrm{Re}\,\omega &= (m_e/m_i)^{1/2}ku\,[1+(ku/\omega_{p_e})^2]^{-1/2}, \\ \gamma &= (\pi/4)\,\varkappa_n k_y\,(\mathrm{Re}\,\omega)^3/k^2\,|k_z|\,\omega_{B_e}u. \end{aligned}\right\} \qquad (2.57)$$

This instability is similar to the instability considered in §2.2.2A. However, for it to develop, we do not need a threshold condition of the type (2.26) to be satisfied. The latter is obtained from the inequality $|\partial f_0/\partial v_z| < (k_y/k_z\omega_B)\,\partial f_0/\partial x$, and this inequality is satisfied automatically if $f_0^{(e)}$ has the form (2.55).

However, the degree of inhomogeneity does affect the growth rate of the oscillations. Introducing the parameter $\rho_0 \equiv (m_e/m_i)^{1/2}\cdot u/\omega_{B_i}$, which is equivalent to the effective ion Larmor radius (see §2.2), we find from (2.57) that for $\rho_0/a < 1$

$$\gamma_{max} \simeq (\rho_0/a)\,\omega_{p_i}. \qquad (2.58)$$

If $\rho_0/a \ll 1$, this growth rate is appreciably less than (2.25).

B. Hydrodynamic Instability for $u \simeq V$. If u and V are comparable, we must also retain the term with the logarithm in (2.56).

Assuming $V \gg a\omega_{B_i}(m_i/m_e)^{1/2}$, i.e., $\rho_0/a \geq 1$, we find that perturbations with $k_z/k_y < \varkappa_n V/\omega_{B_e}$ are excited with the growth rate

$$\gamma \simeq \mathrm{Re}\,\omega \simeq (m_e/m_i)^{1/2} k_\perp V \lesssim \omega_{p_i}. \tag{2.59}$$

This instability is similar to that considered in §2.2.2B. In this region of parameters the existence of a directed motion of the electrons does not play an important role.

§2.5. Plasma with a Transverse Current

In Chapter 13 in Volume 1 we considered the instabilities of a plasma with a transverse current when gradient effects are ignored. If $\nabla n_0 \neq 0$, we must add to the dispersion equation for current instabilities a term due to electron convection. If the ions and electrons are cold, we then obtain

$$1+\left(\frac{\omega_{p_e}}{\omega_{B_e}}\right)^2+\frac{\varkappa_n k_y \omega_{p_e}^2}{k^2\omega\omega_{B_e}}-\left(\frac{\omega_{p_e}}{\omega}\cos\theta\right)^2-\frac{\omega_{p_i}^2}{(\omega-k_yV)^2}=0. \tag{2.60}$$

We see that convection is important if

$$\cos\theta \lesssim (\varkappa_n V/\omega_{B_e})^{1/2}. \tag{2.61}$$

As a characteristic value of $\cos\theta$, we must here take $\cos\theta \simeq (m_e/m_i)^{1/2}$. We then obtain a condition for convection to be important:

$$V \gg \omega_{B_i} a. \tag{2.62}$$

The opposite inequality gives the limit of applicability of the homogeneous plasma approximation.

Suppose the inequality (2.62) is strong. In (2.60) we can then neglect the contribution of the longitudinal electron motion, $\cos^2\theta \to 0$. It then follows from (2.60) that

$$\gamma_{\max} \simeq (\varkappa_n V/\omega_{B_i})^{1/6}\,\omega_{p_i}\,(1+\omega_{p_e}^2/\omega_{B_e}^2)^{-1/2}. \tag{2.63}$$

This maximal value of the growth rate is attained when k and Re ω are related by

$$\mathrm{Re}\,\omega \simeq k_y V \approx (\varkappa_n V/\omega_{B_i})^{1/2}\,\omega_{p_i}\,(1+\omega_{p_e}^2/\omega_{B_e}^2)^{-1/2}. \tag{2.64}$$

The growth rate (2.63) is somewhat greater than in the case $\nabla n_0 = 0$ and $\cos\theta \simeq (m_e/m_i)^{1/2}$ by a factor of order $(\varkappa_n V/\omega_{B_i})^{1/6}$.

An instability with $\cos\theta = 0$ is not possible for any sign of $\varkappa_n$; for instability it is necessary that

$$\varkappa_n V/\omega_{B_e} < 0. \tag{2.65}$$

In the case of azimuthally gyrating ions, $V = -r\omega_{B_i}$. It therefore follows from (2.65) that a plasma whose density decreases along the radius is unstable:

$$\partial \ln n_0/\partial r < 0. \tag{2.66}$$

Electron convection also affects the stability of a plasma with a finite temperature of the electrons and ions and, in particular, if $T > m_i V^2$. If the condition (2.62) is satisfied, this inequality in the case of hot electrons entails

$$\rho_0/a > (m_e/m_i)^{1/2}. \tag{2.67}$$

At the same time we can, as before, assume $\rho_e/a \ll 1$ and use the expressions for $\varepsilon_0^{(e)}$ obtained in §1.7.

In the case of hot ions, the inequality $T_i \geq m_i V^2$ when the condition (2.62) holds entails

$$\rho_i \gtrsim a. \tag{2.68}$$

This corresponds to a plasma whose ion Larmor radius is of the same order as the inhomogeneity scale. It is by no means always possible to solve the Boltzmann equation for such a plasma. For some special cases of distributions of the particles along the radius a solution has been found by Shima and Fowler and Jungwirth.

Bibliography

1. A. B. Mikhailovskii and A. V. Timofeev, Zh. Eksp. Teor. Fiz., 44:919 (1963) [Sov. Phys. – JETP, 18:1077 (1964)]. The possibility of excitation of ion-cyclotron oscillations in a plasma with an inhomogeneous density (§§2.1 and 2.2) is pointed out.
2. A. V. Timofeev, Dokl. Akad. Nauk SSSR, 152:84 (1963) [Sov. Phys. – Doklady, 8:890 (1964)]. A study is made of the excitation of high-frequency ion-acoustic oscillations in an inhomogeneous plasma with hot electrons (§2.2).

3. A. B. Mikhailovskii, Nucl. Fusion, 5:125 (1965). Studies are made of the ion-cyclotron and high-frequency instabilities of a plasma with $T_i > T_e$ (§2.1)., and it is pointed out that high-frequency perturbations can be excited in a plasma with non-Maxwellian ions (gradient-cone instability), §2.3.
4. Y. Shima and T. K. Fowler, Phys. Fluids, 8:2245 (1965). The instabilities of plasmas with hot ions (§§2.1 and 2.3) are investigated. The influence of electron convection on the stability of an azimuthal ion stream (§2.5) is considered.
5. R. F. Post and M. N. Rosenbluth, Phys. Fluids, 9:730 (1966). The limit of the gradient-cone instability (§2.3) is found. The importance of this instability in the problem of plasma containment in adiabatic traps is discussed.
6. V. V. Arsenin, Dokl. Akad. Nauk SSSR, 156:766 (1964) [Sov. Phys. – Doklady, 9:456 (1964)].
7. A. B. Mikhailovskii, Zh. Tekh. Fiz., 35:1945 (1965) [Sov. Phys. – Tech. Phys., 10:1498 (1966)].
8. V. V. Vladimirov, Dokl. Akad. Nauk SSSR, 162:785 (1965) [Sov. Phys. – Doklady, 10:519 (1965)].
9. E. E. Lovetskii and A. A. Rukhadze, Nucl. Fusion, 6:9 (1966). In [6-9] studies are made of the high-frequency gradient instabilities of a plasma with a longitudinal current (§2.4).
10. A. B. Mikhailovskii and V. S. Tsypin, ZhETF Pis. Red., 3:247 (1966) [JETP Letters, 3:158 (1966)]. The high-frequency gradient instability in a plasma with a transverse current (§2.5) is discussed.
11. A. B. Mikhailovskii, Dokl. Akad. Nauk SSSR, 169:554 (1966) [Sov. Phys. – Doklady, 11:603 (1967)].
12. A. A. Galeev, Zh. Prikl. Mekh. Tekh. Fiz., 7(2):7 (1966). [J. Appl. Mech. Tech. Phys., 7(2):4 (1966)]. In [11, 12] the gradient-cone instability (§2.3) is discussed.
13. M. N. Rosenbluth, Plasma Physics, IAEA (1966), p. 485. Includes a discussion of the gradient-cyclotron instability (§2.1).
14. J. Vaclavik, Czech. J. Phys., B15:832 (1965). The high-frequency instability of a plasma with hot electrons (§2.2) is discussed.
15. K. Jungwirth, Plasma Phys., 10:595 (1968). The instabilities of a plasma with an azimuthal ion stream (§2.5) are investigated.
16. I. S. Baikov and A. A. Rukhadze, Zh. Tekh. Fiz., 35:1913 (1965) [Sov. Phys. – Tech. Phys., 10:1477 (1966)]. A study is made of the excitation of high-frequency oscillations in counter streaming beams of an inhomogeneous plasma.

Chapter 3

Weakly Inhomogeneous Collisionless Plasma

§3.1. Plasma with Inhomogeneous Density

In this chapter we shall investigate low-frequency instabilities, $\omega \ll \omega_{B_i}$. We have noted in §2.1 that such instabilities are the only ones possible if the plasma is weakly inhomogeneous, $\rho_i/a \ll (m_e/m_i)^{1/2}$.

Low-frequency perturbations can also grow in a plasma with finite ρ_i/a, i.e., under conditions when ion-cyclotron and high-frequency perturbations can be excited. Although their growth rate is low, the low-frequency instabilities can lead to more appreciable turbulent losses of the plasma than instabilities with $\gamma \geq \omega_B$, since the rate of these losses depends not only on the growth rate but also on the wavelength and, generally speaking, the low-frequency instabilities have long wavelengths. It follows that the low-frequency instabilities discussed below must be taken into account in an analysis of the behavior of a plasma with not only a small but also a finite ρ_i/a.

As in Chapters 1 and 2, we shall use the collisionless approximation, deferring the study of the effect of binary collisions until Chapter 4. Collisions are not important if the density is not too high and the temperature not too low (estimates are given in Chapter 4).

We shall assume a homogeneous magnetic field. This means that besides the effects of a finite β we also ignore the curvature and shear of the field. (In Chapter 6 we simulate the effect of curvature of the lines of force and in Chapter 11 and the following chapters we take into account the curvature properly. The concept of a shear and the effects related to a shear will be the subject of Chapter 8.)

In this section we shall consider the instabilities of a plasma with an inhomogeneous density, $\nabla n_0 \neq 0,\ \nabla T = 0$, assuming $\beta \ll m_e/m_i,\ \Pi_e \equiv (\omega_{p_e} a/c)^2 \ll 1$. Under these assumptions the problem reduces to an investigation of the dispersion equation that follows from (1.108):

$$\varepsilon_0 = 1 + \sum_{\alpha=i,\ e} \left(1 - \frac{\omega_{n\alpha}}{\omega}\right) \frac{1}{(kd_\alpha)^2} \left[1 + i\sqrt{\pi} \frac{\omega}{|k_z| v_{T\alpha}} \times \right.$$
$$\left. \times W\left(\frac{\omega}{|k_z| v_{T_\alpha}}\right) I_0(z_\alpha) e^{-z_\alpha} \right] = 0. \tag{3.1}$$

Here

$$\omega_{n\alpha} = k_y \varkappa_n T_\alpha / m_\alpha \omega_{B_\alpha},$$
$$\varkappa_n = \partial \ln n_0 / \partial x.$$

1. Hydrodynamic Instability. We shall begin by considering perturbations with $\omega \gg (k_z v_{T_e},\ k_z v_{T_i}),\ (z_e,\ z_i) \ll 1$. We shall assume that the ratio $\omega_{p_e}/\omega_{B_i}$ is large. Under these assumptions, Eq. (3.1) yields

$$\left(\frac{\omega_{p_i}}{\omega_{B_i}}\right)^2 \left(1 - \frac{\omega_{n_i}}{\omega}\right) - \left(\frac{\omega_{p_e}}{\omega}\right)^2 \left(1 - \frac{\omega_{n_e}}{\omega}\right) \cos^2\theta = 0. \tag{3.2}$$

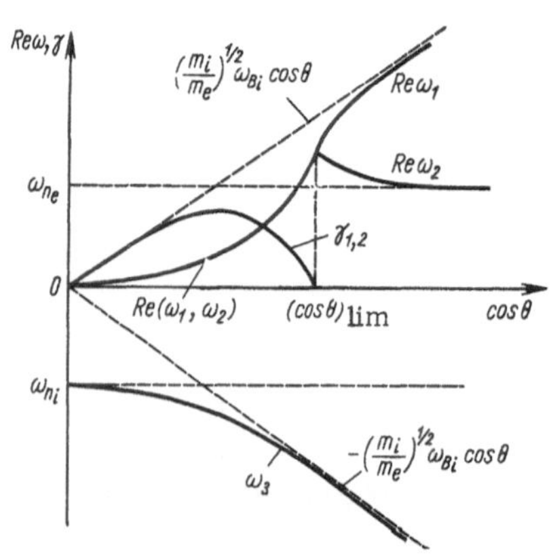

Fig. 3.1. Branches of low-frequency oscillations of a plasma with inhomogeneous density for $|\omega| \gg k_z v_{T_e},\ z_i \ll 1$ (the approximations $\beta \ll m_e/m_i,\ k \gg \omega_{p_e}/c$).

The branches of oscillations described by this equation are shown schematically in Fig. 3.1. For $T_e = T_i$, two of the three roots are complex if

$$\cos\theta < (\cos\theta)_{\lim} \equiv 2.3\left(\frac{m_e}{m_i}\right)^{1/2} \varkappa_n k_y \rho_i^2. \tag{3.3}$$

The maximal growth rate is approximately

$$\gamma_{\max} \simeq \omega_{n_e}. \tag{3.4}$$

2. Kinetic Instability for $z_i \ll 1$. Equation (3.1) also has solutions with $\gamma > 0$ if $\omega \ll k_z v_{T_e}$. In this case the instability is due to the interaction of resonant electrons with the oscillations.

A. Plasma with $T_e \gg T_i$. We shall consider this instability first in the simplest case of cold ions, $T_e \gg T_i$. Letting $T_i \to 0$ in (3.1) and taking $\omega \ll k_z v_{T_e}$, we arrive at the dispersion equation

$$1 + \frac{k_\perp^2 T_e}{m_i \omega_{B_i}^2} - \frac{\omega_{n_e}}{\omega} - \frac{k_z^2 T_e}{m_i \omega^2} + i\sqrt{\pi}\,\frac{\omega}{|k_z|\, v_{T_e}}\left(1 - \frac{\omega_{n_e}}{\omega}\right) = 0. \tag{3.5}$$

The various terms on the left-hand side of this equation arise as follows. The unity is due to the Boltzmann response of the electrons ($n_e' = -(e_e/T_e)\, n_0\psi$); the term $k_\perp^2 T_e/m_i\omega_{B_i}^2$, to the transverse inertia of the ions; $-\omega_{n_e}/\omega$, to the convective drift of the ions; $k_z^2 T_e/m_i\omega^2$, to the inertial motion of the ions along the magnetic field. The imaginary term takes into account the resonant interaction of the electrons with the oscillations; it contains two terms corresponding to the velocity and spatial derivatives of the electron distribution function (see §1.4).

If the term with k_z^2 is ignored and also the small imaginary terms, Eq. (3.5) yields

$$\mathrm{Re}\,\omega = \frac{\omega_{n_e}}{1 + \dfrac{k_\perp^2 T_e}{m_i \omega_{B_i}^2}}. \tag{3.6}$$

It can be seen that the frequency of this oscillation branch satisfies the condition

$$0 < \omega/\omega_{n_e} < 1. \tag{3.7}$$

Resonant electrons that interact with such a wave must give up their energy to the wave, i.e., cause the oscillations to grow. The growth rate is

$$\gamma = \sqrt{\pi} \frac{(\mathrm{Re}\,\omega)^2}{|k_z| v_{T_e}} \frac{k_\perp^2 T_e}{m_i \omega_{B_i}^2} \frac{1}{1 + \frac{k_\perp^2 T_e}{m_i \omega_{B_i}^2}} . \tag{3.8}$$

The excitation is due to the convection of resonant electrons (cf. §1.4). Note that although Eq. (3.6) contains ω_{n_e}, expressed in terms of the mean velocity $V_{\perp 0e}$ of the Larmor currents, the latter do not play any role at all in the given case.

This branch of oscillations is not present in the approximation of a homogeneous plasma, for which $\mathrm{Re}\,\omega \to 0$ as $\nabla n_0 \to 0$. It is due to ion convection. Retaining the term $\sim k_z^2$ in (3.5), we can establish the correspondence between this branch and the oscillation branches of a homogeneous plasma. In this case the solutions (3.5) for $\mathrm{Re}\,\omega$ have the form

$$\mathrm{Re}\,\omega = \frac{1}{2}(1 + k_\perp^2 \rho_0^2)^{-1} \left[\omega_{n_e} \pm \sqrt{\omega_{n_e}^2 + 4k_z^2 T_e (1 + k_\perp^2 \rho_0^2)/m_i}\right]; \quad \rho_0^2 \equiv T_e / m_i \omega_{B_i}^2 . \tag{3.9}$$

These solutions are shown in Fig. 3.2. It can be seen that with increasing k_z the branch (3.6) goes over into the low-frequency ion-acoustic branch described by Eq. (8.31) of Volume 1. The interval of $\cos\theta$ for which the solutions (3.9) differ appreciably from the ion-acoustic solutions is bounded above when $k_\perp \rho_0 < 1$ by the

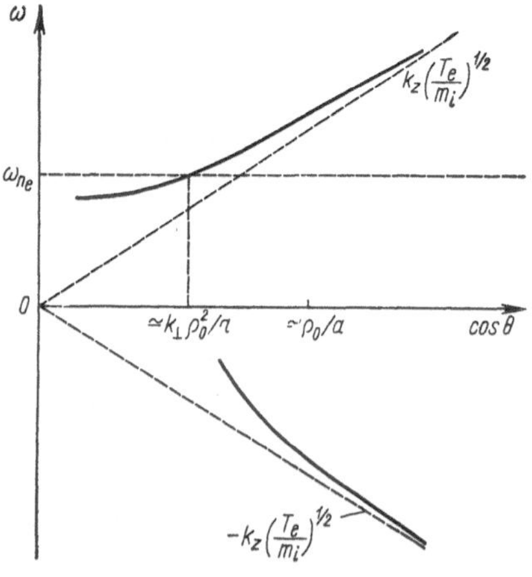

Fig. 3.2. Oscillation branches of an inhomogeneous plasma with $T_e \gg T_i$ for $\omega < k_z v_{T_e}$.

condition (see Fig. 3.2)

$$\cos\theta < \rho_0/a_\perp. \tag{3.10}$$

However, an instability is possible only if $\cos\theta$ is much smaller:

$$\cos\theta < k_\perp \rho_0^2/a_\perp. \tag{3.11}$$

As regards the second root of (3.9) (with $\omega/\omega_{n_e} < 0$), it corresponds to damped oscillations at both large and small values of $\cos\theta$. This is because the sign of $\omega \operatorname{Im} \varepsilon_0$ is positive for perturbations with $\omega/\omega_{n_e} < 0$.

B. Plasma with $T_e = T_i$. For perturbations with $v_{T_i} \ll \omega/k_z \ll v_{T_e}$, $z_i \ll 1$ and $T_e = T_i$, the equation that follows from (3.1) is not (3.5) but

$$1 + z_i\left(1 + \frac{\omega_{n_e}}{\omega}\right) - \frac{\omega_{n_e}}{\omega} - \frac{k_z^2 T}{m_i \omega^2}\left(1 + \frac{\omega_{n_e}}{\omega}\right) + i\sqrt{\pi}\frac{\omega}{k_z v_{T_e}}\left(1 - \frac{\omega_{n_e}}{\omega}\right) = 0. \tag{3.12}$$

If the inequality (3.11) is strong, then

$$\left.\begin{aligned} \operatorname{Re}\omega &= \omega_{n_e}(1 - 2z_i), \\ \gamma &= \frac{2\sqrt{\pi}\,(\omega_{n_e})^2}{|k_z|\,v_{T_e}}\,z_i. \end{aligned}\right\} \tag{3.13}$$

At large k_z, the instability region is bounded by a condition analogous to (3.11):

$$\cos\theta < \varkappa \rho_i z_i^{1/2}. \tag{3.14}$$

As k_z decreases, the kinetic increment (3.13) attains a maximum of order

$$\gamma_{\text{max.kin}} \simeq \omega_{n_e} z_i \tag{3.15}$$

at

$$k_{z\,\text{opt}} \simeq \omega_{n_e}/v_{T_e}. \tag{3.16}$$

With a further decrease of k_z, the kinetic growth rate becomes exponentially small ($\sim\exp[-(\omega_{n_e}/k_z v_{T_e})^2]$). If k_z is so small that the condition (3.3) is satisfied, the hydrodynamic instability con-

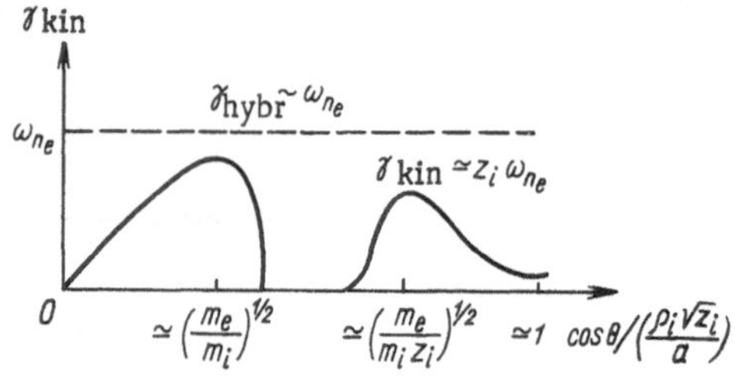

Fig. 3.3. Correspondence between the hydrodynamic and kinetic instabilities for $z_i \ll 1$.

sidered in §3.1.1 develops. The correspondence between the kinetic and the hydrodynamic instabilities is shown in Fig. 3.3.

3. Short-Wavelength Perturbations, $z_i \geq 1$. If z_i increases to values of the order of unity, the gap between the hydrodynamic and kinetic instabilities in Fig. 3.3 closes. At $z_i \simeq 1$, the relations (3.4) and (3.13) yield approximately

$$\gamma \simeq \varkappa v_{T_i}. \tag{3.17}$$

A numerical calculation of the growth rate for some values of z_i has been made by Kadomtsev and Timofeev (Fig. 3.4).

If $z_i \gg 1$, there remains only the kinetic instability. In this case the dispersion equation (3.1) takes the form

$$2 - \frac{\omega_{ne}}{\omega\sqrt{2\pi z_i}} - i\sqrt{\pi}\frac{\omega_{n_e}}{|k_z|v_{T_e}} = 0. \tag{3.18}$$

Here, we have also assumed $v_{T_i} \ll \omega/k_z \ll v_{T_e}$, $\omega_{ne} \gg \omega$. The maximal growth rate is attained for

$$k_z\,\mathrm{opt} = \frac{\sqrt{\pi}}{2}\,\omega_{ne}/v_{T_e}. \tag{3.19}$$

At the same time

$$\gamma_{\max} = \mathrm{Re}\,\omega = \varkappa v_{T_i}/8\sqrt{\pi}. \tag{3.20}$$

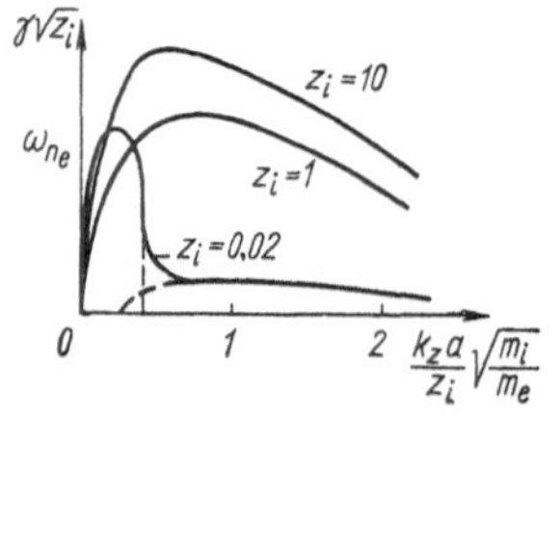

Fig. 3.4. Growth rate γ as a function of the longitudinal wave number k_z for instabilities of a plasma with an inhomogeneous density.

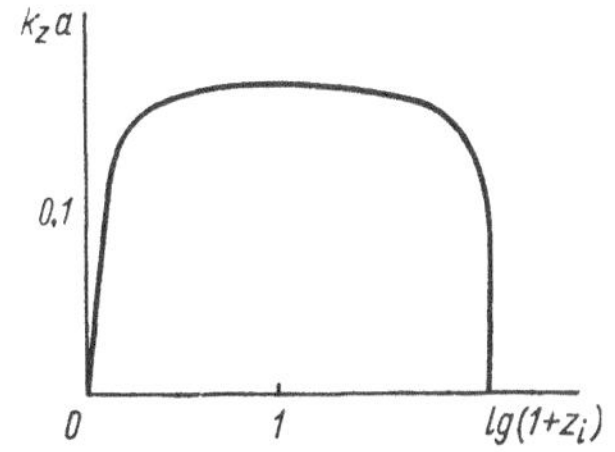

Fig. 3.5. Boundary of the instability region on the (k_z, z_i) plane. Here $a = 1/\varkappa_n$.

It can be seen that in the limit of large z_i the maximal growth rate is independent of the wave number. The dependence $\gamma(k_z)$ for z_i = 10 is shown in Fig. 3.4.

4. Limit of the Instability of a Plasma against Perturbations with $z_i \geq 1$. If z and $\omega/k_z v_{T_i}$ are finite, the dispersion equation (3.1) has the form

$$2+\frac{i\sqrt{\pi}\,(\omega-\omega_{n_e})}{|k_z|\,v_{T_e}}+i\sqrt{\pi}\,\frac{\omega+\omega_{n_e}}{|k_z|\,v_{T_i}}\,I_0(z_i)\,e^{-z_i}W\left(\frac{\omega}{k_z v_{T_i}}\right)=0. \quad (3.21)$$

Assuming here that the frequency is real, we find the instability boundary. The instability and stability regions are shown in Fig. 3.5. The figure shows that the plasma can be unstable only if its longitudinal dimension L is sufficiently large compared with its transverse dimensions a, L > 20a (we assume that $L \simeq \pi/k_z$, $a \simeq 1/\varkappa$).

§3.2. Plasma with Inhomogeneous Temperature

The instabilities considered in §3.1 can arise not only in a plasma with $\eta \equiv \frac{\partial \ln T}{\partial \ln n_0} = 0$, but also if $\eta \neq 0$. Moreover, in a plasma with $\eta \neq 0$ we have some characteristic new types of instability. We now proceed to investigate these instabilities.

1. Excitation of Perturbations with $\omega \simeq k_z v_{T_i}$

A. Long-Wavelength Perturbations, $z_i \ll 1$. In the expressions (1.108) for $\varepsilon_0^{(i)}$ and $\varepsilon_0^{(e)}$ we assume $z_i \ll 1$ and $\omega \ll k_z v_{T_e}$. We neglect the terms of order z_i and $\omega/k_z v_{T_e}$. Then the dispersion equation $\varepsilon_0^{(i)} + \varepsilon_0^{(e)} = 0$ becomes

$$1+\frac{T_i}{T_e}-\frac{\omega\omega_{T_i}}{k_z^2 v_{T_i}^2}+i\sqrt{\pi}\,\frac{\omega}{|k_z|\,v_{T_i}}\,W\left(\frac{\omega}{|k_z|\,v_{T_i}}\right)\left[1-\frac{\omega_{n_i}}{\omega}\left(1-\frac{\eta}{2}\right)-\frac{\omega_{T_i}\omega}{k_z^2 v_{T_i}^2}\right]=0,$$

$$\omega_{T_i}=\frac{k_y c}{e_i B_0}\cdot\frac{\partial T_i}{\partial x}. \quad (3.22)$$

If $|\eta| \ll 1$, we have $k_z v_{T_e} \ll \omega \ll \omega_{T_i}$, and hence

$$\omega^3 + \frac{k_z^2 T}{m_i}\omega_{T_i} = 0, \quad T_e = T_i \equiv T. \tag{3.23}$$

One of the three solutions of this equation has a positive imaginary part, $\gamma > 0$, equal to

$$\gamma = \frac{\sqrt{3}}{2}\left|\frac{k_z^2 T}{m_i}\omega_{T_i}\right|^{1/3}. \tag{3.24}$$

At the limits of applicability of this result (for $k_z v_{T_i} \simeq \omega_{T_i}$) we have approximately

$$\gamma \simeq \omega_{T_i} \simeq k_z v_{T_i}. \tag{3.25}$$

It can be seen that in a plasma with an inhomogeneous temperature, perturbations can grow whose longitudinal phase velocity is of the order of the velocity of ion-acoustic perturbations, $|\omega|/k_z \simeq v_T$. The characteristic value of $\cos\theta$ for such perturbations is

$$\cos\theta \simeq \rho_i/a. \tag{3.26}$$

In a plasma with $\eta = 0$ perturbations with such short wavelengths along B_0 do not grow.

This instability can be regarded as the result of the excitation of the ion-acoustic oscillation branches (§8.3 of Volume 1) by the convective motion of the ions (cf. §1.5). Equation (3.23) is the limiting case of the equation

$$1 - \frac{k_z^2 T_e}{m_i\omega^2}\left(1 - \frac{\omega_{T_i}}{\omega}\right) = 0. \tag{3.27}$$

At relatively large k_z, the frequencies of the ion-acoustic oscillations and the convective branch $\omega = \omega_{T_i}$ are very different and all three roots are real. At small k_z, the oscillation branches "intersect," and the roots become complex.

To find the instability boundary at large k_z, we assume that ω in (3.22) is real and equate the real and imaginary parts of the equations to zero. As a result, we find

$$\left.\begin{aligned} |k_{z.\text{lim}}| &= \frac{1}{2}\frac{|\omega_{T_i}|}{v_{T_i}}\left(1 - \frac{2}{\eta}\right)^{1/2}, \\ \omega &= \frac{\omega_{T_i}}{2}\left(1 - \frac{\eta}{2}\right) = v_{T_i} k_{z.\text{lim}}\left(1 - \frac{2}{\eta}\right)^{1/2}. \end{aligned}\right\} \tag{3.28}$$

These conditions can be satisfied if

$$\eta < 0 \quad \text{or} \quad \eta > 2. \tag{3.29}$$

The relations (3.29) are necessary conditions for instability. The results of the present subsection are valid until $z_i \simeq 1$. At the limit of applicability of the approximation of small z_i

$$\gamma \simeq v_{T_i}/a, \qquad k_z \simeq 1/a. \tag{3.30}$$

B. Short-Wavelength Perturbations, $z_i \geq 1$. For arbitrary z_i, Eq. (3.22) does not follow from (1.108) but

$$1 + \frac{T_i}{T_e} - \frac{\omega\omega_{T_i}}{(k_z v_{T_i})^2} I_0 e^{-z_i} + i\sqrt{\pi} \times$$
$$\times \frac{\omega I_0 e^{-z_i}}{|k_z| v_{T_i}} W\left(\frac{\omega}{k_z v_{T_i}}\right)\left[1 - \frac{\omega_{n_i}}{\omega} + \frac{\omega_{T_i}}{2\omega} \times\right.$$
$$\left.\times\left(1 + 2z_i \frac{I_0 - I_1}{I_0}\right) - \frac{\omega_{T_i}\omega}{(k_z v_{T_i})^2}\right] = 0. \tag{3.31}$$

As with (3.29), we find a necessary instability condition:

$$\eta < 0$$

or

$$\eta > 2\left(1 + 2z_i \frac{I_0 - I_1}{I_0}\right)^{-1}. \tag{3.32}$$

These relations are shown in Fig. 3.6. It is clear from (3.32) and the figure that for $z_i \geq 1$ an instability can develop in a plasma with

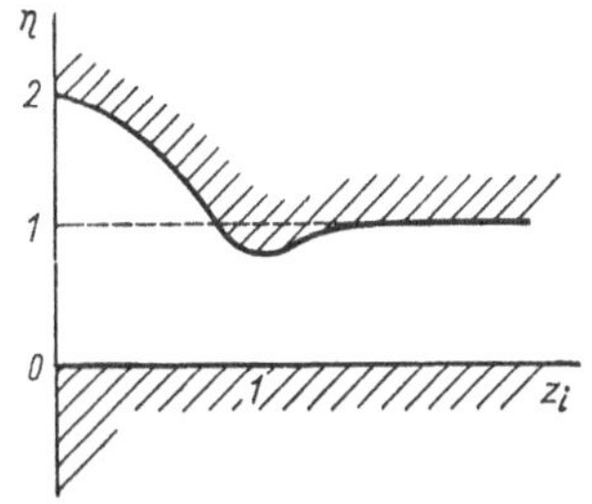

Fig. 3.6. For $\eta \equiv \partial \ln T/\partial \ln n_0$ in the shaded regions an instability with $\gamma = k_z v_{T_i}$ can be excited.

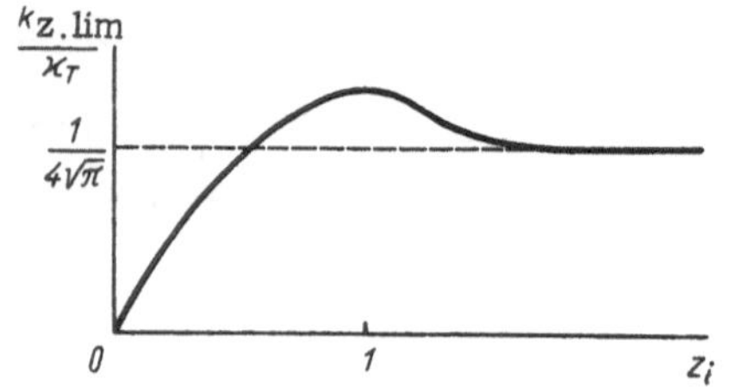

Fig. 3.7. Dependence of $k_{z.\lim}$ on z_0 for $|\eta| \gg 1$. The region under the curve corresponds to instability.

$\eta \geq 1$. The instability boundary that follows from (3.31), $k_{z.\lim} = k_{z.\lim}(z_i)$, is shown in Fig. 3.7 for $|\eta| \gg 1$.

The growth rate of perturbations with $z_i \gg 1$ can be estimated by assuming $\gamma \simeq k_z v_{T_i}$, where k_z is slightly less than $k_{z.\lim}$. Then

$$\gamma \leqslant v_{T_i}/(4\sqrt{\pi}a). \tag{3.33}$$

Let us now consider the limits of applicability of Eq. (3.31). It ceases to hold if $k \geq \min(d_i^{-1}, \rho_e^{-1})$. This means that in the case of a plasma with $\omega_{pe} > \omega_{Be}$ Eq. (3.31) can be used if $z_i < m_i/m_e$, but in a plasma of lower density ($\omega_{pe} < \omega_{Be}$), only if $z_i \leqslant \left(\frac{\omega_{pe}}{\omega_{Be}}\right)^2 \frac{m_i}{m_e}$.

2. Perturbations with $z_i > 1$, $\omega \gtrsim k_z v_{T_e}$. Setting $z_i \gg 1$, $z_e \ll 1$, and $(kd_e)^2 \ll 1$ in (1.108) and neglecting terms of order $z_i^{-1/2}$, we obtain an equation that differs from (3.22) only by the replacement of the ion by the electron subscripts:

$$2 - \frac{\omega_{T_e}\omega}{k_z^2 v_{T_e}^2} + i\sqrt{\pi}\frac{\omega}{|k_z| v_{T_e}} W\left(\frac{\omega}{k_z v_{T_e}}\right)\left[1 - \frac{\omega_{ne}}{\omega}\left(1 - \frac{\eta}{2}\right) - \frac{\omega_{T_e}\omega}{(k_z v_{T_e})^2}\right] = 0. \tag{3.34}$$

Therefore, in analyzing (3.34) we can use the results of the foregoing subsection. In particular, we find that in a plasma with $|\eta| \gg 1$ for which $k_z v_{Te} < \omega_{Te}$, a hydrodynamic instability can develop with the growth rate [cf. (3.24)]

$$\gamma = \frac{\sqrt{3}}{2}\left|\frac{k_z^2 T}{m_e}\omega_{Te}\right|^{1/3}. \tag{3.35}$$

The growth rate attains a maximum at

$$k_{z.\mathrm{opt}} \simeq \frac{\omega_{Te}}{v_{T_e}} \simeq \varkappa_T' k_\perp \rho_e, \tag{3.36}$$

and is approximately

$$\gamma_{\max} \simeq \omega_{T_e}. \tag{3.37}$$

Equation (3.35) can be regarded as a consequence of the equation

$$1 - \frac{k_z^2 T_i}{m_e \omega^2}\left(1 - \frac{\omega_{T_e}}{\omega}\right) = 0. \tag{3.38}$$

Therefore, the instability we are considering can be interpreted as the result of the interaction of the electron-acoustic oscillations (§8.4.2 of Volume 1) with the convective branch of the electrons $\omega = \omega_{T_e}$ (see the similar arguments of the foregoing subsection).

For a plasma with $\omega_{p_e} > \omega_{B_e}$, Eq. (3.34) is valid right up to $z_e \simeq 1$. At the limits of its applicability, $z_e \simeq 1$, we obtain estimates from (3.35) and (3.36) that are similar to (3.30):

$$\left.\begin{aligned} \gamma &\simeq \varkappa_T v_{T_e}, \\ k_{z.\mathrm{opt}} &\simeq \varkappa_T. \end{aligned}\right\} \tag{3.39}$$

Numerically, the growth rate (3.39) is greater than (3.30) by a factor of $(m_i/m_e)^{1/2}$. The longitudinal wave numbers are the same in the two cases, but the transverse wave numbers differ by the factor $(m_i/m_e)^{1/2}$. In the case of a plasma that is unstable against these perturbations, the parameter η must satisfy the conditions (3.29).

In the case of a plasma with $\omega_{p_e} < \omega_{B_e}$, the limits of applicability (3.34) are restricted by the condition $kd_e \leq 1$. If $kd_e \simeq 1$, (3.39) is replaced by

$$\left.\begin{aligned} \gamma &\simeq \frac{\omega_{pe}}{\omega_{Be}} \frac{v_{T_e}}{a}, \\ k_{z.\mathrm{opt}} &\simeq \varkappa_T \frac{\omega_{pe}}{\omega_{Be}}. \end{aligned}\right\} \tag{3.40}$$

3. Dependence of $k_{z.\mathrm{lim}}(z_i)$ on η. Calculating $k_{z.\mathrm{lim}}(z_i)$ for different values of the parameter η, we can find the maximal value of this function of z_i and plot the graph of max $k_{z.\mathrm{lim}}(\eta)$. It has the form shown in Fig. 3.8. In this figure the parameter $k_{z.\mathrm{lim}}$ is divided by the "effective" transverse inhomo-

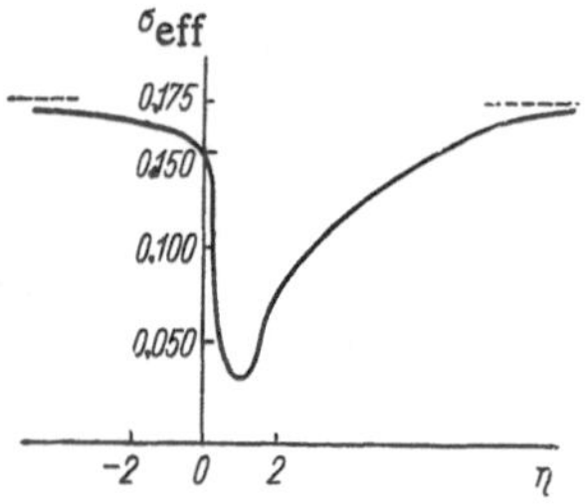

Fig. 3.8. Maximal wave number k_z that bounds the instability region of the plasma as a function of the parameter $\eta \equiv \partial \ln T/\partial \ln n_0$. Here $\sigma_{eff} \equiv \max \frac{|k_{z,\lim}|}{|\varkappa_T|+|\varkappa_n|}$

geneity scale of the plasma. It is clear from the figure that oscillations can be excited only if the longitudinal scale of the plasma is at least an order of magnitude greater than the transverse scale.

§3.3. Plasma with Inhomogeneous Density for $\beta > m_e/m_i$

If $\beta > m_e/m_i$ and $\nabla T = 0$ the approximation of electrostatic perturbations does not give an exhaustive description of the plasma instabilities. If we dispense with the electrostatic assumption, the dispersion equation (3.1) is replaced by

$$\varepsilon_{\|}[1-(\omega/ck_z)^2\varepsilon_{\perp}]+(k_{\perp}/k_z)^2\varepsilon_{\perp}=0. \tag{3.41}$$

Here

$$\begin{gathered}\varepsilon_{\|}=\sum_{i,e}(k_zd)^{-2}(1-\omega_n/\omega)\left[1+i\sqrt{\pi}\,xW(x)\,I_0(z)\,e^{-z}\right],\\ \varepsilon_{\perp}=1+\sum_{i,e}(kd)^{-2}(1-\omega_n/\omega)(1-I_0e^{-z}),\\ x=\omega/|k_z|v_T.\end{gathered} \tag{3.42}$$

We shall now consider the consequences that flow from this equation, assuming $T_e = T_i \equiv T$.

1. Long-Wavelength Perturbations. We shall begin by investigating long-wavelength perturbations, $z_i \ll 1$, with phase velocity ω/k_z less than the thermal velocity of the electrons but greater than that of the ions, $v_{T_i} \ll \omega/k_z \ll v_{T_e}$. Under these assumptions, Eq. (3.41) becomes

$$\left(1-\frac{\omega_{ne}}{\omega}\right)\left(1+\frac{i\sqrt{\pi}\,\omega}{|k_z|v_{T_e}}\right)\left[1-\frac{\omega^2}{c_A^2k_z^2}\left(1-\frac{\omega_{n_i}}{\omega}\right)\right]+z_i\left(1-\frac{\omega_{n_i}}{\omega}\right)=0, \tag{3.43}$$

where $c_A^2=B_0^2/4\pi n_0m_i$.

It can be seen that the last term on the left-hand side of the equation is small as z_i. If we neglect this term, we obtain

$$\omega_{1,2} = -\frac{|\omega_{n_i}|}{2} \pm \sqrt{\left(\frac{\omega_{n_i}}{2}\right)^2 + (k_z c_A)^2}, \tag{3.44}$$

$$\omega_3 = \omega_{n_e}. \tag{3.45}$$

To be specific we have here assumed $\omega_{n_e} \equiv -\omega_{e_i} > 0$. The first two roots correspond to oscillations of Alfvén type and ω_3 to the convective branch investigated in §3.1 under the assumption $k_z \gg \omega_{n_e}/c_A$ [Eq. (3.13)].

If $k_z c_A \approx \sqrt{2}\,\omega_{n_e}$, the roots ω_1 and ω_3 coincide. If the term of order z_i ignored in Eq. (3.41) is taken into account, the $\omega = \omega(k_z)$ curves near this k_z have the form shown in Fig. 3.9. If the term of order z_i is taken into account, the roots $\omega_{1,2,3}$ become complex. Oscillations with $0 < \omega/\omega_{n_e} \ll 1$ grow. In Fig. 3.9, the sections of the ω_1 and ω_3 curves lying between the abscissa and the straight line $\omega = \omega_{n_e}$ correspond to these oscillations. The oscillations of the ω_2 branch are damped for all k_z.

The maximal growth rate is attained for oscillations whose frequency lies near the point of intersection of the branches in Fig. 3.9, i.e., near

$$k_z \simeq \sqrt{2}\,\omega_{n_e}/c_A. \tag{3.46}$$

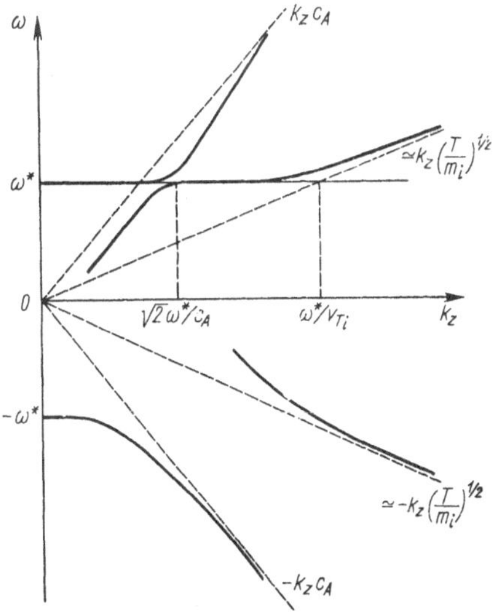

Fig. 3.9. Oscillation branches of an inhomogeneous plasma with $\beta > m_e/m_i$.

Then

$$\gamma_{max} = (\sqrt{\pi}/6)\,(z_i m_e/m_i\beta)^{1/2}\,\omega_{ne}. \qquad (3.47)$$

In accordance with (3.47), the growth rate increases with decreasing transverse wavelength, $\gamma \sim k^2$. At the limits of applicability of the long-wavelength approximation, i.e., when $z_i \simeq 1$, we have

$$\gamma_{max} \simeq (m_e/m_i\beta)^{1/2}\,v_{T_i}/a. \qquad (3.48)$$

A comparison of (3.47) and (3.48) with (3.8) and (3.17) shows that a plasma with $\beta > m_e/m_i$ is more stable against perturbations with $z_i \leq 1$ than a plasma with $\beta < m_e/m_i$.

2. <u>Short-Wavelength Perturbations.</u> If $z_i \geq 1$ and $v_{T_i} < \omega/k_z < v_{T_e}$, Eq. (3.41) reduces to

$$\left(1-\frac{\omega_{ne}}{\omega}\right)\left(1+i\sqrt{\pi}\frac{\omega}{|k_z|\,v_{Te}}\right)\left[1-\frac{\omega(\omega-\omega_{n_i})}{k_z^2 c_A^2}\frac{1-I_0 e^{-z_i}}{z_i}\right]+$$
$$+\left(1-\frac{\omega_{n_i}}{\omega}\right)(1-I_0 e^{-z_i})=0. \qquad (3.49)$$

This shows that the ratio ω/ω_{n_e} for the branch of oscillations with $\gamma > 0$ decreases (the gap between the "intersecting" curves in Fig. 3.9 widens) as z_i increases. If $z_i \ll 1$, one can assume $\omega \ll \omega_{n_e}$, and (3.49) then reduces to

$$2+\left(\frac{\omega_{ne}}{k\,dck_z}\right)^2-\frac{\omega_{ne}}{\omega\sqrt{2\pi z_i}}-\frac{i\sqrt{\pi}\,\omega_{ne}}{|k_z|\,v_{Te}}=0. \qquad (3.50)$$

This equation differs from (3.18) by a term $\sim 1/c^2$.

Assuming $\gamma \ll \mathrm{Re}\,\omega$, we find from (3.50) [cf. (3.20)]

$$\mathrm{Re}\,\omega=\frac{\omega_{ne}}{\sqrt{2\pi z_i}}\,[2+(\beta/4)(\varkappa_n/k_z)^2]^{-1},$$
$$\gamma=\sqrt{\pi}\,\omega_{ne}\,\mathrm{Re}\,\omega/|k_z|\,v_{Te}. \qquad (3.51)$$

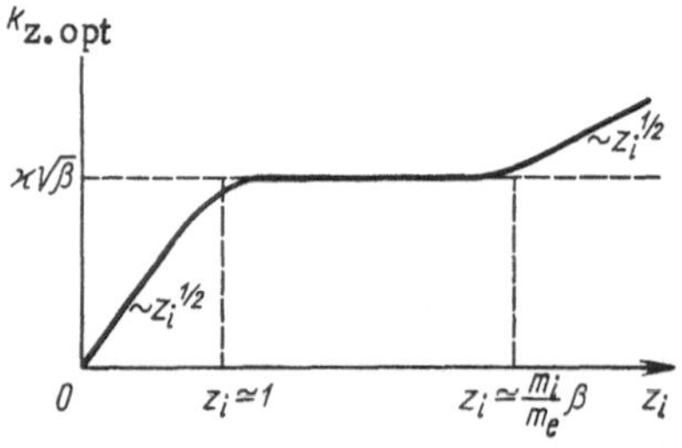

Fig. 3.10. Dependence on z_i of the longitudinal wave number $k_{z.opt}$ corresponding to the maximal growth rate for $\beta > m_e/m_i$, $\nabla T = 0$.

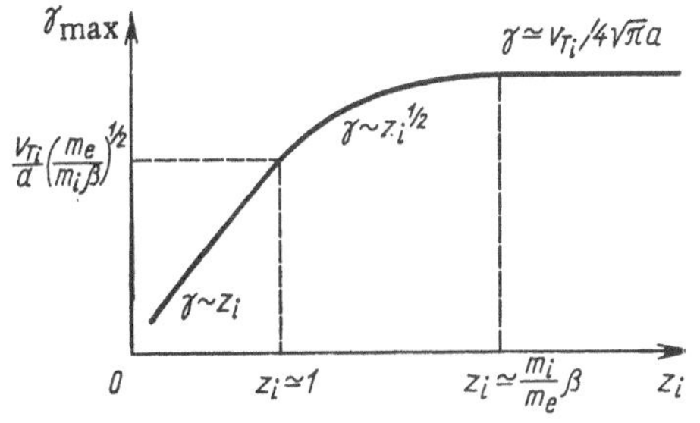

Fig. 3.11. Dependence of γ_{max} on z_i for a plasma with $\beta > m_e/m_i$, $\nabla T = 0$.

As a function of k_z, the growth rate has a maximum:

$$\gamma_{max} = \frac{\omega_{ne}}{2\sqrt{2}}\left(\frac{m_e}{m_i\beta}\right)^{1/2}. \tag{3.52}$$

This maximum is attained at

$$k_{z.opt} = \varkappa_n (\beta/8)^{1/2}. \tag{3.53}$$

The condition $\gamma_{max} \ll \mathrm{Re}\,\omega$ is violated if $z_i \geq \beta m_e/m_i$. If z_i has these large values, the fact that the perturbations are not electrostatic is not important for those with maximal growth rate, and Eq. (3.50) therefore reduces in this case to (3.18) and the maximal growth rate and optimal k_z are given by (3.20) and (3.19). The dependences $k_{z.opt} = k_{z.opt}(z_i)$ and $\gamma_{max} = \gamma_{max}(z_i)$ are shown schematically in Figs. 3.10 and 3.11.

§3.4. Plasma with Longitudinal Current

Suppose that the electrons move relatively to the ions with a velocity $V \| B_0$ (longitudinal current). At a sufficiently high V_0, $V_0 > V_{cr}$, the plasma can sustain the instabilities discussed in §12.1 in Volume 1. If $V < V_{cr}$, only gradient instabilities are possible. Those for which V_0 is not important were discussed in §§3.1-3.3. We shall now consider the gradient instabilities for which the finiteness of V_0 is important. We shall assume that the perturbations are electrostatic. If $k_\perp \simeq a^{-1}$, this is justified if $a < c/\omega_{p_e}$. Nonelectrostatic perturbations of a plasma with a longitudinal current will be considered in Chapter 8.

1. Perturbations with $\omega > k_z v_{T_e}$, $z_i \ll 1$. Hydrodynamic Current-Convective Instability. We proceed from the dispersion equation

$$\left(\frac{\omega_{p_i}}{\omega_{B_i}}\right)^2-\left(\frac{\omega_{pe}\cos\theta}{\omega}\right)^2\left[1+\frac{k_y}{k_z}\cdot\frac{\partial(n_0V_0)}{\partial x}\cdot\frac{1}{n_0\omega_{Be}}\right]=0, \qquad (3.54)$$

which holds if, in addition to the assumptions already made, $\omega \gg k_zV_0$. It can be seen that the frequency is purely imaginary if

$$[\partial(n_0V_0)/\partial x]/n_0\omega_{Be} > k_z/k_y. \qquad (3.55)$$

The growth rate is approximately

$$\gamma \simeq (m_e/m_i)^{1/2}\varkappa V_0. \qquad (3.56)$$

As in the case considered in §1.3, the instability is due to convection of the longitudinal current along x under the influence of the fields E_y and B_0.

In Eq. (3.54) we have omitted the terms of order $\omega^*/\omega \simeq k_y\varkappa \times \rho_i v_{T_i}/\omega$. In view of (3.56), this is justified if

$$V_0/v_{T_i} > (z_i m_i/m_e)^{1/2}. \qquad (3.57)$$

In the case of the perturbations with the longest wavelengths, $k_\perp \simeq \varkappa$, this means

$$V_0/v_{T_i} > (\rho_i/a)(m_i/m_e)^{1/2}. \qquad (3.58)$$

If this condition is not satisfied, the relative motion of the electrons and ions in perturbations with $\omega \gg k_z v_{T_e}$ is not important. The results of §3.1.1 then remain in force.

The transition from the instability without current to the one with current is shown in Fig. 3.12.

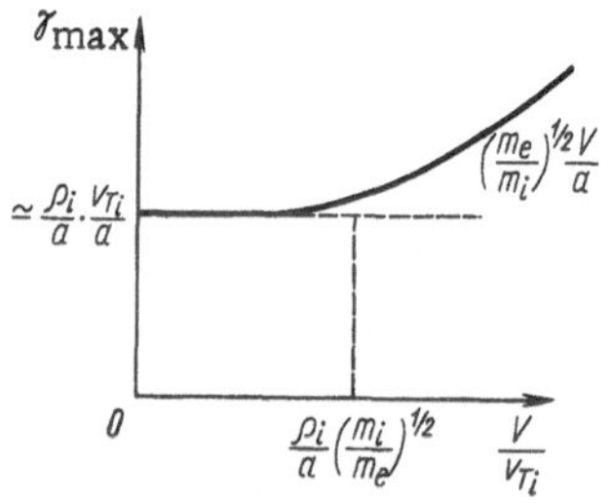

Fig. 3.12. Correspondence between the growth rates of the hydrodynamic instabilities without (see §3.1.1) and with current (see §3.4.1) when $k_\perp a \simeq 1$.

2. Perturbations with $\omega < k_z v_{T_e}$, $z_i \ll 1$. Kinetic Current-Convective Instability. In contrast to §3.4.1, we assume $v_{T_i} \ll \omega/k_z \ll v_{T_e}$. As before, we shall assume $z_i \ll 1$, $\nabla T = 0$. In this case we have the dispersion equation [cf. (3.5)]:

$$1 - \frac{\omega_{ne}}{\omega} + \frac{i\sqrt{\pi}\,\omega}{|k_z|\,v_{Te}}\left(1 - \frac{\omega_{ne}}{\omega} - \frac{k_z V_0}{\omega}\right) = 0. \tag{3.59}$$

It follows that

$$\left.\begin{aligned} &\mathrm{Re}\,\omega = \omega_{ne}, \\ &\gamma = \sqrt{\pi}\,\omega_{n_e} V_0/v_{Te}. \end{aligned}\right\} \tag{3.60}$$

The growth rate of the perturbations with the longest wavelength, $k_\perp \simeq \varkappa$, is approximately

$$\gamma \simeq (m_e/m_i)^{1/2}\,\rho_i V_0/a^2. \tag{3.61}$$

It is small compared with the hydrodynamic growth rate (3.56) as ρ_i/a but is greater than the kinetic growth rate (3.13) if

$$V_0/v_{T_i} > \rho_i/a. \tag{3.62}$$

An instability of the type (3.60) is possible only if there is a suf ficiently small temperature gradient:

$$\eta < 2V_0/v_{T_e}. \tag{3.63}$$

If the longitudinal motion of the ions is taken into account in (3.59), we find that the range of $\cos\theta$ values for which $\gamma > 0$ satisfies

$$\cos\theta \lesssim (\rho_i/a)\,(V_0/v_{T_i}). \tag{3.64}$$

This corresponds to perturbations that have a smaller scale along z than in the case of a plasma without a current [cf. (3.14)].

§3.5. Plasma Jet with Inhomogeneous Velocity Profile

Suppose the electrons and ions of the plasma move along a magnetic field with the same velocity V, this velocity having an inhomogeneous profile, $\partial V/\partial x \neq 0$. A plasma of this kind can sustain ion-electron instabilities with a mechanism similar to that of the electron instabilities considered in §1.3.

By way of an example, let us consider an instability on the ion-acoustic branch. We shall assume that the perturbations are of a small scale, $k_\perp a \gg 1$. We shall neglect the gradient of the current density. We shall be concerned with perturbations with a frequency that satisfies the conditions $k_z v_{T_i} \ll |\omega - k_z V| \ll (k_z v_{T_e}, \omega_{B_i})$. Then, proceeding as in the derivation of (1.45), we obtain the dispersion equation

$$\frac{1}{(kd_e)^2} - \left(\frac{\omega_{p_i}}{\omega'}\right)^2 \cos^2\theta \left(1 - \frac{k_y}{k_z} \cdot \frac{1}{\omega_{B_i}} \cdot \frac{dV}{dx}\right) = 0,$$
$$\omega' \equiv \omega - k_z V. \tag{3.65}$$

This shows that perturbations grow if

$$\frac{k_z}{k_y} < \frac{1}{\omega_{B_i}} \cdot \frac{dV}{dx}. \tag{3.66}$$

Numerically, this condition differs from (1.46) by the mass ratio.

The growth rate attains a maximum at

$$k_{z.\mathrm{opt}} = \frac{1}{2} \cdot \frac{k_y}{\omega_{B_i}} \cdot \frac{dV}{dx}. \tag{3.67}$$

It is equal to

$$\gamma_{\max} = \frac{1}{2} \cdot \frac{k_y}{\omega_{B_i}} \left(\frac{T_e}{m_i}\right)^{1/2} \frac{dV}{dx}. \tag{3.68}$$

The results (3.67) and (3.68) refer to a plasma with $T_e > T_i$. If $T_e \simeq T_i$, we must omit the first term in the brackets in (3.65) [since we must have $\omega' > k_z (T_e/m_i)^{1/2}$] and also remember that we now have $k_z \ll k_{z.\mathrm{opt}}$. Then the growth rate will be a few times smaller than (3.68).

If a density gradient is present, (3.65) is replaced by

$$(\omega')^2 - \omega'\omega_{ne} - \frac{T_e}{m_i}\left(k_z^2 - \frac{k_y k_z}{\omega_{B_i}} \cdot \frac{dV}{dx}\right) = 0. \tag{3.69}$$

From this there follows a stabilization condition similar to (1.49):

$$\frac{d \ln n_0}{d \ln V} > \frac{V}{\left(\frac{T_e}{m_i}\right)^{1/2}}. \tag{3.70}$$

It can be seen that an instability is impossible for $\frac{d \ln n_0}{d \ln V} > 1$ if the velocity of the jet does not exceed the velocity of the ion-acoustic oscillations.

§3.6. Plasma with an Admixture of Cold Ions

Suppose that a plasma contains a small fraction of cold ions, $v_{T_1} \ll v_{T_i}$, whose density is inhomogeneous, $\nabla n_1 \neq 0$. The existence of such ions leads to the appearance of a new branch of oscillations which can be excited by the hot ions.

In investigating an instability of this nature, we shall assume $v_{T_1} \ll \omega/k_z \ll (v_{T_i}, v_{T_e})$, $\nabla T = 0$. The dispersion equation in this case has the form

$$\frac{1}{(kd_e)^2} + \frac{1}{(kd_i)^2}\left[1 + \frac{i\sqrt{\pi}\,(\omega - \omega_{n_i})}{|k_z|\,v_{T_i}}\right] + \frac{\varkappa_1 k_y \omega_{p1}^2}{k^2 \omega \omega_{B1}} = 0. \tag{3.71}$$

Here $\omega_{p1}^2 = 4\pi e_1^2 n_1/m_1$; $\omega_{B1} = e_1 B_0/m_1 c$; $e_1 = -z_1 e_e$, where z_i is the charge of the ions; $\varkappa_1 = \partial \ln n_1/\partial x$. If $T_i = T_e$, this equation has the solution

$$\left.\begin{aligned} \operatorname{Re}\omega &= -\omega_{n1}(n_1 T/2n_0 T_1)\, z_1^2, \\ \gamma &= -\frac{\sqrt{\pi}}{2}\cdot\frac{(\operatorname{Re}\omega)^2}{|k_z|\,v_{T_i}}\left(1 + \frac{2n_0}{n_1 z_1}\cdot\frac{\varkappa}{\varkappa_1}\right). \end{aligned}\right\} \tag{3.72}$$

Perturbations grow, $\gamma > 0$, if the density gradient of the additional ions is not too great and is directed against the density gradient of the plasma:

$$\varkappa n_0/\varkappa_1 n_1 z_1 + 1/2 < 0. \tag{3.73}$$

This instability is similar to that considered in §3.1.2 [cf., for example, (3.72) and (3.13)]. The oscillation branch (3.72) is due to the convection of the additional ions; the growth rate, to the convection of resonant ions of the plasma.

Bibliography

1. L. I. Rudakov and R. Z. Sagdeev, Dokl. Akad. Nauk SSSR, 138:581 (1961) [Sov. Phys. – Doklady, 6:415 (1961)].

2. L. I. Rudakov and R. Z. Sagdeev, Nucl. Fusion. Suppl. 1962, 2:481 (1962). It is pointed out in [1, 2] that an instability can arise in a plasma with an inhomogeneous temperature, §3.2. A study is made of the instabilities that are possible if the transverse inertia of the ions and terms of order $k_\perp \rho_i$ are unimportant.
3. B. B. Kadomtsev, Zh. Eksp. Teor. Fiz., 43:1688 (1962) [Sov. Phys. – JETP, 16:1191 (1963)]. It is shown that in a plasma with a current a kinetic instability can develop even if $\nabla T_0 = 0$, (§3.4.2).
4. B. B. Kadomtsev and A. V. Timofeev, Dokl. Akad. Nauk SSSR, 146:581 (1962). [Sov. Phys. – Doklady, 7:826 (1962)]. It is shown that a plasma with an inhomogeneous density is unstable. The main results of this paper are given in §3.1. The graphical dependences shown in Figs. 3.4 and 3.5 are obtained in [4]. Kadomtsev and Timofeev [4] use the adjective "drift" for the instability they investigate.
5. A. B. Mikhailovskii, Nucl. Fusion, 2:161 (1962). Equations for nonelectrostatic perturbations (§3.3) are obtained.
6. A. B. Mikhailovskii and L. I. Rudakov, Zh. Eksp. Teor. Fiz., 44:912 (1963) [Sov. Phys. – JETP, 18:1077 (1964)]. A study is made of instabilities of a plasma with $\beta > m_e/m_i$ (§3.3) and also the nonelectrostatic instabilities of a plasma with $\beta < m_e/m_i$ and electrostatic perturbations for $\omega > k_z v_{T_i}$ (§3.2).
7. A. A. Galeev, V. N. Oraevskii, and R. Z. Sagdeev, Zh. Eksp. Teor. Fiz., 44:902 (1963) [Sov. Phys. – JETP, 17:615 (1963)]. Instabilities of a plasma with $\nabla T \neq 0$, (§3.2) are investigated, In particular, perturbations with $\omega < k_z v_{T_i}$ (§3.2) are studied.
8. L. V. Mikhailovskaya and A. B. Mikhailovskii, Sh. Tekh. Fiz., 33:1200 (1963) [Sov. Phys. – Tech. Fiz., 8:896 (1964)]. Instabilities of a plasma with $\nabla T \neq 0$, $\nabla n_0 \neq 0$ (§3.2) are analyzed numerically. It is shown that if the plasma is sufficiently short, instabilities cannot arise for any values of $\partial \ln T/\partial \ln n_0$. The graphical dependence shown in Fig. 3.8 is obtained.
9. A. B. Mikhailovskii, Reviews of Plasma Physics, Vol. 3, Consultants Bureau, New York (1967), p. 159.
10. A. A. Galeev, S. S. Moiseev, and R. Z. Sagdeev, At. Energ., 15:451 (1963).
11. B. B. Kadomtsev, Plasma Turbulence, Academic Press, London (1965). The reviews [9-11] cover the results of the stability theory of an inhomogeneous plasma available at that time.
12. A. A. Rukhadze and V. P. Silin, Usp. Fiz. Nauk, 82:499 (1964) [Sov. Phys. – Uspekhi, 7:209 (1965)]. The paper [12] gives an exposition of the results of stability theory obtained by the method of geometrical optics. It also contains a detailed bibliography of the earlier papers in this field.
13. B. Coppi, Phys. Lett., 11:226 (1964).
14. A. B. Mikhailovskii, Zh. Eksp. Teor. Fiz., 48:380 (1965) [Sov. Phys. – JETP, 21: 250 (1965)].
15. E. E. Lovetskii and A. A. Rukhadze, Nucl. Fusion, 6:9 (1966).
16. A. B. Mikhailovskii, Zh. Tekh. Fiz., 35:1933 (1965) [Sov. Phys. – Tech. Phys., 10:1490 (1966)]. The hydrodynamic current instability (§3.4.1) is discussed in [13-16].

17. A. B. Mikhailovskii, Zh. Tekh. Fiz., 37:1365 (1967) [Sov. Phys. – Tech. Phys., 12:993 (1968)]. It is shown that the electron-acoustic instability in a plasma with $\nabla T \neq 0$ (§3.2) can arise. This instability is analyzed in detail.
18. B. Coppi, H. P. Furth, M. N. Rosenbluth, and R. Z. Sagdeev, Phys. Rev. Lett., 17:377 (1966). The instability associated with admixtures of ions (§3.6) is investigated.
19. V. I. Petviashvili, Zh. Tekh. Fiz., 37:206 (1967) [Sov. Phys. – Tekh. Phys., 12:144 (1967)]. A study is made of the excitation of ion-acoustic oscillations in a plasma with an inhomogeneous velocity profile, $\nabla V \neq 0$, (§3.5).
20. N. A. Krall and M. N. Rosenbluth, Phys. Fluids, 8:1488 (1965). The instability boundary of a plasma with $\nabla T = 0$, $\nabla n_0 \neq 0$ is studied. Results similar to those of [4] are obtained.
21. M. Porkolab, Phys. Lett., 22:427 (1966).
22. M. Porkolab, Nucl. Fusion, 8:29 (1968). The instabilities of a plasma with $\nabla T_i \neq 0$ are analyzed numerically. Growth rates of perturbations with $\omega \leq k_z v_{T_i}$ are calculated.
23. E. Ya. Kogan and S. S. Moiseev, Zh. Tekh. Fiz., 37:805 (1967) [Sov. Phys. – Tech. Phys., 12:579 (1967)]. Attention is drawn to the importance of perturbations with $\omega \lessdot k_z v_{T_i}$ and $k_\perp \rho_i \gg 1$ in a plasma with $\nabla T = 0$; such perturbations cannot be well stabilized by a shear (see Chapter 8). Instabilities of a plasma with $[\nabla T, \nabla n_0] \neq 0$ are considered.
24. V. P. Silin, Zh. Tekh. Fiz., 38:945 (1968) [Sov. Phys. – Tekh. Phys., 13:715 (1968)]. The theory of admixture instabilities (cf. [18]) is developed further in this paper.
25. I. S. Baikov, Nucl. Fusion, 5:326 (1965). The influence of an admixture of cold electrons on the stability of a hot homogeneous plasma is considered.
26. Ya. I. Kolesnichenko and V. N. Oraevskii, At. Energ., 23:289 (1967). The instabilities of an inhomogeneous plasma due to products of thermonuclear reactions are discussed.
27. I. S. Baikov and A. A. Rukhadze, Zh. Tekh. Fiz., 35:1913 (1965) [Sov. Phys. – Tech. Phys., 10:1477 (1966)].
28. É. A. Pashitskii, Zh. Tekh. Fiz., 38:1020 (1968) [Sov. Phys. – Tech. Phys., 13:853 (1969)]. In [27, 28] studies are made of instabilities in colliding plasmas – interstreaming in [27] and spatially separated in [28].
29. A. B. Mikhailovskii and É. A. Pashitskii, Zh. Eksp. Teor. Fiz., 48:1787 (1965) [Sov. Phys. – JETP, 21:1197 (1965)].
30. L. S. Bogdankevich and A. A. Rukhadze, Nucl. Fusion, 6:176 (1966).
31. E. E. Lovetskii, Zh. Tekh. Fiz., 36:45 (1966) [Sov. Phys. – Tech. Phys., 11:31 (1966)].
32. S. P. Bakanov and E. E. Lovetskii, Zh. Tekh. Fiz., 36:1955 (1966) [Sov. Phys. – Tech. Phys., 11:1455 (1967)].
33. Yu. B. Ivanov and E. E. Lovetskii, Zh. Tekh. Fiz., 38:8 (1968). [Sov. Phys. – Tech. Phys., 13:5 (1968)]. The instabilities of an inhomogeneous plasma with a current are investigated in [29-33].

Chapter 4

Equations of the Oscillations of a Collisional Plasma

§4.1. Limits of Applicability of the Collisionsless Approximation and the Permittivity of an Almost Collisionless Plasma

Since the frequency of Coulomb collisions depends on the density and temperature in accordance with the law $\nu \propto nT^{-3/2}$, collisions can affect instabilities only if a plasma is sufficiently dense and cold. We now proceed to the investigation of such plasmas. In this chapter we shall obtain the basic equations that describe the instabilities of a collisional plasma. The actual types of instability will be analyzed in Chapter 5. We shall neglect the collisions of the charged particles with the neutrals. These effects, which are important for a weakly ionized plasma, will be discussed in Chapter 10.

We shall first give some conditions under which collisions can be ignored in various types of gradient instability and derive some expressions for the collisional corrections to the permittivity.

1. Perturbations with $\omega > k_z v_{T_e}$, $k_\perp \rho_i \ll 1$. Using the transport equation with the collision term $C_{\alpha\beta}$ in the Landau form (see Chapter 6 in Volume 1) and the method of successive approximation, we can show that the collisional corrections to $\varepsilon_0^{(\alpha)}$

for the given type of perturbations have the form

$$\left.\begin{aligned}\delta\varepsilon_0^{(e)} &= i\frac{\nu_{ei}}{\omega}\left(\frac{\omega_{pe}}{\omega}\right)^2\cos^2\theta\left(1-\frac{\omega_{ne}}{\omega}+\frac{\omega_{Te}}{2\omega}\right),\\ \delta\varepsilon_0^{(i)} &= \frac{7}{10}i\frac{\nu_{ii}}{\omega}z_i\left(\frac{\omega_{p_i}}{\omega_{B_i}}\right)^2\left(1-\frac{\omega_{n_i}}{\omega}+\frac{3}{28}\cdot\frac{\omega_{T_i}}{\omega}\right).\end{aligned}\right\}\qquad(4.1)$$

Here $\nu_{ei} = 0.51/\tau_e$ and $\nu_{ii} = 1/\tau_i$ are the frequencies of electron–ion and ion–ion collisions. As small expansion parameters, we have ν_{ei}/ω and ν_{ii}/ω.

In §3.1.1 we have shown that perturbations with $\gamma \simeq \omega_{ne}$ can grow in a plasma if $\omega \gg k_z v_{T_e}$, $k_\perp \rho_i \ll 1$. It follows from (4.1) that in this case collisions are not important if $\omega_{n_e} > \nu_{ei}$. If $k_\perp \simeq 1/a$, this inequality entails

$$S_e \equiv \frac{\lambda_e \rho_e}{a^2} > 1, \qquad (4.2)$$

where $\lambda_e \simeq v_{T_e}/\nu_{ei}$ is the electron mean free path.

The instability considered in §3.1 can develop only for sufficiently small k_z [the condition (3.4)]. If $k_z > k_{z.cr}$, the collisionless approximation yields $\gamma = 0$. For k_z values satisfying this inequality, collisions must be taken into account even if the condition (4.2) is satisfied, since the collisional dissipation may lead to the development of a new instability with $\gamma \ll \omega_{n_e}$ (see below, §5.2.1).

2. <u>Perturbations with $v_{T_i} < \frac{\omega}{k_z} < v_{T_e}$, $k_\perp\rho_i \ll 1$</u>. In this case the collisional correction to ε_0^e is small compared with $\text{Im}\,\varepsilon_0^{(e)}$ [the part of $\varepsilon_0^{(e)}$ due to collisionless dissipation] if $k_z v_{T_e} > \nu_{ei}$. The last inequality can be represented in the form

$$\lambda_\parallel < \lambda_e, \qquad (4.3)$$

where $\lambda_\parallel \simeq 1/k_z$ is the longitudinal wavelength.

If $\omega \gg \nu_{ii}$ and $z_i > (k_z v_{T_i})^2$, the correction to $\varepsilon_0^{(i)}$ is determined by the second formula in (4.1). Using this formula, we find that in considering the kinetic instability of §3.1.2 we can neglect ion–ion collisions if $\nu_{ii}/\omega < z_i$, i.e.,

$$\lambda_i/a > z_i^{3/2}, \qquad (4.4)$$

where $\lambda_i \simeq v_{T_i}/\nu_{ii}$ is the ion mean free path (for $T_e \simeq T_i$, $\lambda_e \simeq \lambda_i$).

At the limit of applicability of the long-wavelength approximation for the ions, $z_i \simeq 1$, the condition (4.4) has the simple meaning

$$\lambda_i > a. \tag{4.5}$$

3. Perturbations with $\omega \simeq k_z v_{T_i}$, $k_\perp \rho_i < 1$. The excitation of such perturbations occurs in a plasma with inhomogeneously heated ions (§3.2). In this case collisions are not important if $\omega_{T_i} > \nu_{ii}$. If $k_\perp \simeq 1/a$, this inequality entails

$$S_i \equiv \frac{\lambda_i \rho_i}{a^2} > 1. \tag{4.6}$$

If $k_\perp \rho_i \simeq 1$, this is replaced by (4.5).

4. Perturbations with $\omega \simeq k_z v_{T_e}$, $k_\perp \rho_i > 1$, $k_\perp \rho_e < 1$. A plasma with inhomogeneously heated electrons (§3.2.2) is unstable against such perturbations. In this case the ions are not important. Only electron collisions can therefore affect the development of the instability. These collisions can be ignored if $\omega_{T_e} > \nu_{ei}$. If $k_\perp \rho_i \simeq 1$, this inequality reduces to

$$\lambda_e > \left(\frac{m_i}{m_e}\right)^{1/2} a, \tag{4.7}$$

and if $k\rho_e \simeq 1$, to a condition similar to (4.5):

$$\lambda_e > a. \tag{4.8}$$

5. Short-Wavelength Perturbations, $k_\perp \rho_i \gg 1$, in Which the Motion of the Ions Is Important. When $z_i \gg 1$, the ion–ion collisions can be assumed weak if

$$|\omega - n\omega_{B_i}| > z_i \nu_{ii}. \tag{4.9}$$

On the right-hand side of the inequality we have the large parameter z_i and not merely unity because the collisional term contains second derivatives with respect to the velocity [see Eq. (6.2) of Volume 1] and these derivatives, applied to the exponential function $\exp[i\xi \sin(\alpha - \Psi)]$ in the perturbed distribution function [see, for example, (1.103)], lead to an additional factor in front of ν_{ii} of order $\xi^2 \simeq z_i$. The collisional correction to $\varepsilon_0^{(i)}$ for this case has been calculated by Rukhadze and Silin. The complete $\varepsilon_0^{(i)}$ has the form

$$\varepsilon_0^{(i)}=\frac{1}{(kd_i)^2}\left\{1-\sum_n\frac{1}{\sqrt{2\pi z_i}}\frac{\omega}{\omega-n\omega_{B_i}}\left[1-\frac{\omega_{n_i}}{\omega}+\right.\right.$$

$$\left.\left.+\frac{1}{2}\frac{\omega_{T_i}}{\omega}-i\frac{\nu_{ii}z_i}{\omega-n\omega_{B_i}}\cdot\frac{3(\pi+1)}{2\sqrt{2}}\left(1-\frac{\omega_{n_i}}{\omega}+\frac{3\pi+2}{4\pi+2}\cdot\frac{\omega_{T_i}}{\omega}\right)\right]\right\}. \quad (4.10)$$

A. Low-Frequency Short-Wavelength Perturbations. In §3.1.3 we considered an instability with $k_\perp\rho_i\gg 1$, $\gamma\simeq v_{T_i}/a$. It follows from (4.9) that ion–ion collisions can be ignored in this case if

$$\lambda_i/a>z_i. \quad (4.11)$$

If $z\simeq 1$, this result agrees with (4.5).

B. Cyclotron Perturbations. As we have shown in §2.1, a plasma with $\rho_i/a>2\,(m_e/m_i)^{1/2}$ can systain short-wavelength cyclotron oscillations with $\gamma\simeq(\rho_i/a)^{1/2}\omega_{B_i}$, $k_\perp\rho_i\simeq a/\rho_i$. The condition (4.9) for collisions to be negligibly weak in this case means

$$\frac{\nu_{ii}}{\omega_{B_i}}<\left(\frac{\rho_i}{a}\right)^{5/2}. \quad (4.12)$$

In particular, if $\rho_i/a\simeq(m_e/m_i)^{1/2}$, we must have

$$\nu_{ii}/\omega_{B_i}<(m_e/m_i)^{5/4}. \quad (4.13)$$

It follows from (4.11)-(4.13) that the collisionless approximation for short-wavelength ion instabilities is valid when collisions are much more infrequent than in the case $k_\perp\rho_i\leq 1$.

§4.2. Macroscopic Equations of a Collisional Plasma

1. Statement of the Problem. In this section we shall find the solution of the transport equation in the approximation of frequent collisions (we shall specify the expansion parameters below). We assume that the magnetic field is homogeneous, $\mathbf{B}_0\parallel z$, $\nabla B_0=0$, and that the electric field is electrostatic, rot $\mathbf{E}\to 0$.

We express the particle velocity v in terms of $v_z^{(a)}$, $v_\perp$, and α defined by the equation

$$\mathbf{v}=(v_z^{(a)}+V_z^{(a)})\,\mathbf{e}_z+v_\perp(\cos\alpha\mathbf{e}_x+\sin\alpha\mathbf{e}_y), \quad (4.14)$$

where $V_z^{(a)}$ is the mean velocity of the a-th component of the plasma along the magnetic field.

We represent the distribution function and the collision term in the form

$$\left.\begin{aligned} f &= \bar{f} + \tilde{f}, \\ C &= \bar{C} + \tilde{C}, \end{aligned}\right\} \tag{4.15}$$

where the bar denotes the mean over α and the tilde the oscillations with respect to α. The Boltzmann equation yields equations for $\tilde{f}$ and $\bar{f}$:

$$\hat{L}_{\parallel}\tilde{f} + \hat{L}_{\perp}\bar{f} + \hat{L}_{\perp}\tilde{f} - \langle \hat{L}_{\perp}\tilde{f}\rangle - \tilde{C} = \omega_B \frac{\partial \tilde{f}}{\partial \alpha}, \tag{4.16}$$

$$\hat{L}_{\parallel}\bar{f} + \langle \tilde{L}_{\perp}\tilde{f}\rangle = \bar{C}. \tag{4.17}$$

Here

$$\left.\begin{aligned} \hat{L}_{\parallel} &= \frac{\partial_a}{\partial t} + v_z^{(a)}\frac{\partial}{\partial z} + \\ &+ \left(\frac{e}{m}E_z - \frac{\partial_a V_z^{(a)}}{\partial t} - \frac{\partial V_z^{(a)}}{\partial z}v_z^{(a)}\right)\frac{\partial}{\partial v_z^{(a)}}; \\ \hat{L}_{\perp} &= \mathbf{v}_{\perp}\nabla + \frac{e}{m}\mathbf{E}_{\perp}\frac{\partial}{\partial \mathbf{v}_{\perp}} - (\mathbf{v}_{\perp}\nabla)V_z^{(a)}\frac{\partial}{\partial v_z^{(a)}}; \\ \frac{\partial_a}{\partial t} &= \frac{\partial}{\partial t} + V_z^{(a)}\frac{\partial}{\partial z}, \end{aligned}\right\} \tag{4.18}$$

and $\langle \ldots \rangle$ stands for averaging over α.

Equation (4.16) can be solved by making an expansion in $1/\omega_B$ and (4.17) by an expansion in $1/\nu$ if

$$\left\{\frac{\nu}{\omega_B}, \frac{\rho}{a_{\perp}}, \frac{eE_{\perp}}{m\omega_B v_T}, \frac{\omega}{\nu}, \frac{v_T}{a_{\parallel}\nu}, \frac{eE_z}{m\nu v_T}, \frac{a_{\perp}}{a_{\parallel}}\right\} \ll 1, \tag{4.19}$$

where $a_{\perp}$ and $a_{\parallel}$ are the characteristic transverse and longitudinal scales, respectively. In addition to (4.19), we assume

$$\frac{V_{ze} - V_{zi}}{v_{Te}} \ll 1. \tag{4.20}$$

This simplifies the electron–ion collision integral (cf. §6.2 of Volume 1).

To be specific we shall assume that the parameters (4.19) and (4.20) are of the same order and denote this order by $\varepsilon \ll 1$. We shall seek the functions $\bar{f}$ and $\tilde{f}$ in the form of series in powers of ε:

$$\bar{f}=\sum_{k=0}^{\infty}\bar{f}^{(k)}, \quad \tilde{f}=\sum_{k=1}^{\infty}\tilde{f}^{(k)}. \tag{4.21}$$

In the sum for $\tilde{f}$, there is no zeroth term of the series, since the solution of the transport equation in the limit $\varepsilon \to 0$ is the Maxwell function and the latter does not depend on α.

As in §6.2 in Volume 1, we shall assume that the function $\bar{f}^{(0)}$ gives the exact values of the density n, the mean longitudinal velocity V_z, and the temperature T. The transverse velocity $V_\perp \equiv \frac{1}{n}\int \tilde{f} v_\perp dv$ is a small quantity and has the form of a series analogous to (4.21).

The derivatives $\partial (n, V_z, T)/\partial t$ are series in powers of ε (as in §6.2 of Volume 1 but with the difference that $V_\perp$ is now a series as well). The products of the operator $\hat{L}_{\parallel}$ and n, V_z, and T are also series in ε.

The terms of such series are determined by the following symbolic relations:

$$\left.\begin{aligned} &\hat{L}_{\parallel}(\varepsilon^0)=\frac{\partial_a^0}{\partial t}+v_z^{(a)}\frac{\partial}{\partial z}+\\ &+\left(\frac{e}{m}E_z-\frac{\partial_a^0 V_z}{\partial t}-\frac{\partial V_z}{\partial z}v_z^{(a)}\right)\frac{\partial}{\partial v_z^{(a)}},\\ &\hat{L}_{\parallel}(\varepsilon^k)=\frac{\partial^k}{\partial t}+\frac{\partial^k V_z}{\partial t}\cdot\frac{\partial}{\partial v_z^{(a)}}, \quad k\geqslant 1, \end{aligned}\right\} \tag{4.22}$$

where

$$\frac{\partial_a^0}{\partial t}=\left(\frac{\partial_a n}{\partial t}\right)^0\frac{\partial}{\partial n}+\left(\frac{\partial_a V_z}{\partial t}\right)^0\frac{\partial}{\partial V_z}+\left(\frac{\partial_a T}{\partial t}\right)^0\frac{\partial}{\partial T}+\left(\frac{\partial_a}{\partial t}\right)_{n,\,V_z,\,T}. \tag{4.23}$$

The last term on the right-hand side of (4.23) takes into account the effect of the operator $\partial/\partial t$ on functions that do not depend explicitly on n, V_z, and T. In our adopted approximation of a homogeneous magnetic field, the only function of this kind is the electric field **E**.

We also represent the collision term in the form of series similar to (4.21). We shall also stipulate that the terms C_e and

C_i, which are small as $(m_e/m_i)^{1/2}$, are of order ε and that m_e/m_i is of order ε^2.

2. First Approximation in ε. In the zeroth approximation in ε, the function $\bar{f} \equiv \bar{f}^{(0)}$ is a Maxwell function F of the arguments $v_\perp^2$ and $(v_z^{(a)})^2$; from Eq. (4.16) we therefore find

$$\tilde{f}^{(1)} = F\mathbf{v}_\perp\left[\frac{\mathbf{e}_z}{\omega_B}, \nabla \ln F - \frac{e\mathbf{E}_\perp}{T} + \frac{mv_z^{(a)}}{T}\nabla V_z\right]. \tag{4.24}$$

Substituting this expression into the left-hand side of (4.17), we obtain an equation for $\bar{f}^{(1)}$:

$$\frac{D^0F}{Dt} + \frac{mv_z^{(a)}}{T}F\frac{D^0V_z}{Dt} + v_z^{(a)}\left(\frac{\partial F}{\partial z} - \frac{eE_z}{T}F\right) + \frac{m(v_z^{(a)})^2}{T}F\frac{\partial V_z}{\partial z} = \bar{C}^{(1)}. \tag{4.25}$$

Here $D^0/Dt = \partial^0/\partial t + V_z\partial/\partial z + V_E\nabla$; $V_E = c\,|\mathbf{E}\mathbf{e}_z|/B_0$ is the velocity of electric drift.

Integrating (4.25) with respect to the velocities, we find

$$\frac{D^0(n;\ V_z;\ T)}{Dt} = \left(-n\frac{\partial V_z}{\partial z};\ \frac{e}{m}E_z - \frac{1}{mn}\cdot\frac{\partial p}{\partial z} + \frac{R_z(\varepsilon)}{mn};\ -\frac{2}{3}T\frac{\partial V_z}{\partial z}\right). \tag{4.26}$$

Here $R_z(\varepsilon)$ is the z component of the force of friction. It does not depend on the magnetic field. An expression for this force was given in §6.2 of Volume 1.

Using (4.26) to eliminate the derivatives D^0/Dt in (4.25), we arrive at equations for $\bar{f}_a^{(1)}$ similar to the equations obtained in the absence of a magnetic field (see §6.2 of Volume 1). In particular, the equation for $\bar{f}_i^{(1)}$ has the form

$$C_{ii}[F_i,\ f_i^{(1)}] = F_i\left[\left(\frac{m_iv^2}{2T_i} - \frac{5}{2}\right)v_z\frac{\partial \ln T_i}{\partial z} + \frac{m_i}{T_i}\left(v_z^2 - \frac{v^2}{3}\right)\frac{\partial V_z^{(i)}}{\partial z}\right]. \tag{4.27}$$

For simplicity we have omitted the subscript i of v. The solution (4.27) reduces to that found by Braginskii except that $W_{zz} \equiv 2 \times (\partial V_z/\partial z - \mathrm{div}\, V^{(i)}/3)$ in the expression for $\bar{f}_i^{(1)}$ is replaced by (4/3) $\partial V_z/\partial z$. This difference will be explained in §4.4.

3. Second Approximation in ε. Substituting the results of the first approximation into (4.16), we find

$$\tilde{f}^{(2)} = \tilde{f}_1^{(2)} + \tilde{f}_2^{(2)}, \tag{4.28}$$

where

$$\left.\begin{aligned}\tilde{f}_1^{(2)} &= -\frac{1}{\omega_B}\cdot\frac{\partial}{\partial\alpha}(\hat{L}_\perp\tilde{f}_1-\tilde{C}^{(1)}),\\ \tilde{f}_2^{(2)} &= -\frac{1}{4\omega_B}\cdot\frac{\partial}{\partial\alpha}(\hat{L}_\perp\tilde{f}^{(1)}).\end{aligned}\right\} \tag{4.29}$$

The function $\tilde{f}^{(2)}$, the correction to $\tilde{f}^{(1)}$, is of order $\rho/a_\perp$ or ν/ω_B. Since $\rho_e \ll \rho_i$ and $(\nu/\omega_B)_e \ll (\nu/\omega_B)_i$, only the ion correction need be taken into account. To the same accuracy, $\tilde{C}_{ie}^{(1)}$ in (4.29) can be neglected.

In the second order in ε, Eq. (4.17) yields

$$\hat{L}_{\|}(\varepsilon^0)\bar{f}^{(1)}+\left(\frac{\partial^{(1)}}{\partial t}-\frac{\partial^{(1)}V_z}{\partial t}\cdot\frac{mv_z^{(a)}}{T}\right)F+\mathbf{V}_E\left(\nabla\bar{f}^{(1)}-\nabla V_z\frac{\partial\bar{f}^{(1)}}{\partial v_z^{(a)}}\right)+$$
$$+\frac{1}{\omega_B}\left[\mathbf{e}_z,\nabla+\frac{e\mathbf{E}}{m}\cdot\frac{\partial}{\partial v_\perp^2/2}-\nabla V_z\frac{\partial}{\partial v_z^{(a)}}\right]\langle\mathbf{v}_\perp\tilde{C}^{(1)}\rangle=\bar{C}^{(2)}. \tag{4.30}$$

Using the explicit form of the expression for $\tilde{C}^{(1)}$, we find from (4.30)

$$\left.\begin{aligned}&\frac{\partial^{(1)}n}{\partial t}=0,\quad \frac{\partial^{(1)}V_z}{\partial t}=-\frac{1}{mn}\cdot\frac{\partial\pi_{zz}}{\partial z}-\frac{1}{mn}\operatorname{div}\pi_{z\perp}^{(2)},\\ &\frac{3}{2}n\frac{\partial^{(1)}T}{\partial t}+\frac{\partial q_z}{\partial z}+\operatorname{div}\mathbf{q}_\perp^{(2)}+\pi_{zz}\frac{\partial V_z}{\partial z}+\pi_{z\perp}^{(2)}\nabla V_z=Q.\end{aligned}\right\} \tag{4.31}$$

It is assumed that π_{zz}, $\mathbf{q}_\perp^2$, and $\pi_{z\perp}^2$ occur only in the ion equations. The expressions for $q_z^{(e)}$, $q_z^{(i)}$, and Q_i have the same form as for $B_0 = 0$ (see §6.2 of Volume 1). The quantity Q_e does not contain a term with $\mathbf{R}(\mathbf{V}^{(e)}-\mathbf{V}_i^{(i)})$; it is of higher order than ε^2 and therefore $Q_e = -Q_i$. The expression for π_{zz} differs from that in the case $B_0 = 0$ in that the term with $\operatorname{div}\mathbf{V}_\perp$ is now absent. The transverse components of the ion heat flux $\mathbf{q}_\perp^{(i)}$ and the tensor $\pi_{z\perp}^{(i)}$ have the form

$$\mathbf{q}_\perp^{(i),(2)}=-\varkappa_\perp^{(i)}\nabla T_i,\quad \pi_{z\perp}^{(2)}=\eta_\perp^{(i)}\nabla_\perp V_z^{(i)}, \tag{4.32}$$

where

$$\varkappa_\perp^{(i)}=2p_i/m_i\omega_{B_i}^2\tau_i,\quad \eta_\perp^{(i)}=(6/5)\,p_i/\omega_{B_i}^2\tau_i. \tag{4.33}$$

To find $\bar{f}^{(2)}$, it is necessary to substitute (4.31) into (4.30). However, we shall not require the expression for $\bar{f}^{(2)}$.

4. Review of the Results of the First and Second Approximations. To terms of order ε^2 inclusive, the equations for n, V_z, and T are

$$\left.\begin{aligned}
&D_\alpha n_\alpha/Dt + n_\alpha\,\partial V_z^{(\alpha)}/\partial z = 0, \quad \alpha = (e, i);\\
&m_e n_e D_e V_z^{(e)}/Dt = e_e n_e E_z - \partial p_e/\partial z + R_z^{(e)};\\
&m_i n_i D_i V_z^{(i)}/Dt =\\
&= e_i n_i E_z - \partial p_i/\partial z - R_z^{(e)} - \partial\pi_{zz}^{(i)}/\partial z - \operatorname{div}\pi_{z\perp}^{(2),\,(i)};\\
&\frac{3}{2} n_e D_e T_e/Dt + p_e\,\partial V_z^{(e)}/\partial z + \partial q_z^{(e)}/\partial z = Q_e;\\
&\frac{3}{2} n_i D_i T_i/Dt + p_i\,\partial V_z^{(i)}/\partial z + \partial q_z^{(i)}/\partial z + \operatorname{div}\mathbf{q}_\perp^{(i)(2)} +\\
&+ \pi_{zz}^{(i)}\,\partial V_z^{(i)}/\partial z + \pi_{z\perp}^{(2)}\nabla V_z^{(i)} = Q_i.
\end{aligned}\right\} \tag{4.34}$$

Here $D_\alpha/Dt = \partial/\partial t + V_z^{(\alpha)}\partial/\partial z + V_E\nabla$; the expressions for $\mathbf{q}_\perp$ and $\pi_{z\perp}$ are determined by (4.32) and (4.33) and the expressions for $R_z^{(e)}$, $\pi_{zz}^{(i)}$, $q_z^{(e)}$, $q_z^{(i)}$, Q_e, Q_i are

$$\left.\begin{aligned}
&R_z^{(e)} = -0.51\,\frac{m_e n_e}{\tau_e}\left(V_z^{(e)} - V_z^{(i)}\right) - 0.71 n_e \frac{\partial T_e}{\partial z};\\
&\pi_{zz}^{(i)} = -0.96 p_i \tau_i \frac{4}{3}\frac{\partial V_z^{(i)}}{\partial z}, \quad q_z^{(i)} = -3.9\,\frac{p_i\tau_i}{m_i}\cdot\frac{\partial T_i}{\partial z};\\
&q_z^{(e)} = -3.16\,\frac{p_e\tau_e}{m_e}\cdot\frac{\partial T_e}{\partial z};\\
&Q_i = -Q_e = 3\,\frac{m_e}{m_i}\cdot\frac{n_e}{\tau_e}(T_e - T_i);\\
&\tau_e = 3\sqrt{m_e}\,T_e^{3/2}/(4\sqrt{2\pi}\,\lambda e^2 e_i^2 n_e);\\
&\tau_i = 3\sqrt{m_i}\,T_i^{3/2}/(4\pi\lambda e_i^4 n_i).
\end{aligned}\right\} \tag{4.35}$$

In deriving the macroscopic equations (4.34), we have assumed that the parameters (4.13) are small, as $\varepsilon \ll 1$. It follows, in particular, that

$$\omega/\omega_{B_i} \simeq (\rho_i/a)^2 \simeq \varepsilon^2. \tag{4.36}$$

This corresponds to processes whose frequency is of the order of the frequency of the gradient instabilities, $\omega \simeq \omega^*$ (see Chapter 3).

§4.3. Macroscopic Equations with Allowance for Transverse Inertia and Transverse Viscosity of the Ions

For very small $k_z/k_\perp$, the total contribution of the electrons and ions to the electrical neutrality equation is relatively small [see the continuity equation of the system (4.34)]. In this connection, the ion continuity equation in the limit $k_z/k_\perp \to 0$ must be augmented by the contribution of terms of order ε^3, which correspond, as will be shown in §4.4, to the transverse inertia and the transverse magnetic viscosity of the ions. However, if one wishes to make an investigation of dissipative effects for such small values of $k_z/k_\perp$, one must also take into account the terms of order ε^4 corresponding to the transverse collisional viscosity of the ions.

To take into account these effects, it is necessary to calculate $\partial^{(2)}n/\partial t$ and $\partial^{(3)}n/\partial t$. From (4.11) we find that these quantities satisfy the equations

$$\frac{\partial^{(2)}n}{\partial t}+\int d\mathbf{v}\hat{L}_\perp\tilde{f}_1^{(3)}=0, \tag{4.37}$$

$$\frac{\partial^{(3)}n}{\partial t}+\int d\mathbf{v}\hat{L}_\perp\tilde{f}_1^{(4)}=0. \tag{4.38}$$

The subscript 1 in $\tilde{f}^{(n)}$ denotes the part of $\tilde{f}^{(n)}$ proportional to $\cos\alpha$ and $\sin\alpha$ [as also in Eq. (4.28)].

We find the expressions for $\tilde{f}_1^{(3)}$ and $\tilde{f}_1^{(4)}$ by means of (4.16):

$$\tilde{f}_1^{(3)}=-\frac{1}{\omega_B}\cdot\frac{\partial}{\partial\alpha}\left\{\frac{\partial^0\tilde{f}^{(1)}}{\partial t}+\hat{L}_\perp\bar{f}^{(2)}+(\hat{L}_\perp\tilde{f}^{(2)})_1-\tilde{C}_1^{(2)}\right\}, \tag{4.39}$$

$$\tilde{f}_1^{(4)}=-\frac{1}{\omega_B}\cdot\frac{\partial}{\partial\alpha}\left\{\frac{\partial^0\tilde{f}_1^{(2)}}{\partial t}+\frac{\partial^{(1)}\tilde{f}^{(1)}}{\partial t}+\hat{L}_\perp\bar{f}^{(3)}+(\hat{L}_\perp\tilde{f}^{(3)})_1-\tilde{C}_1^{(3)}\right\}. \tag{4.40}$$

1. Terms of Order ε^3. Substituting (4.39) into (4.28), we obtain

$$\frac{\partial^{(2)}n}{\partial t}-\frac{1}{m\omega_B^2}\operatorname{div}\frac{\partial^{(0)}}{\partial t}(\nabla_\perp p-en\mathbf{E}_\perp)-\frac{1}{\omega_B}\operatorname{rot}_z\int d\mathbf{v}\mathbf{v}_\perp(\hat{L}_\perp\tilde{f}_2^{(2)})_1=0. \tag{4.41}$$

In calculating the integral in (4.41), we use formula (4.29) for $\tilde{f}_2^{(2)}$. As a result we find

$$\operatorname{rot}_z\int d\mathbf{v}\mathbf{v}_\perp(\hat{L}_\perp\tilde{f}_2^{(2)})_1=\operatorname{div}[\mathbf{e}_z,(\mathbf{V}_E+\mathbf{V}_L)(\mathbf{V}_E\nabla)n-n(\mathbf{V}_E\nabla)(\mathbf{V}_E+\mathbf{V}_L)]. \tag{4.42}$$

Substituting (4.42) into (4.41), we finally obtain

$$\frac{\partial^2 n}{\partial t}+\operatorname{div}\left[n\frac{\mathbf{e}_z}{\omega_B},\left(\frac{\partial}{\partial t}+\mathbf{V}_E\nabla\right)(\mathbf{V}_E+\mathbf{V}_L)\right]=0. \tag{4.43}$$

Here, $\mathbf{V}_L \equiv [\mathbf{e}_z, \nabla p]/mn\omega_B$ is the mean velocity of the Larmor currents (cf. §1.6).

Using (4.43), we find that if terms of order ε^2 are included, the continuity equation for the ions contained in the system (4.34) must be replaced by

$$\left(\frac{\partial}{\partial t}+\mathbf{V}_E\nabla\right) n+\frac{\partial (nV_z)}{\partial z}+\operatorname{div}\left[n\frac{\mathbf{e}_z}{\omega_B},\ \left(\frac{\partial}{\partial t}+\mathbf{V}_E\nabla\right)(\mathbf{V}_E+\mathbf{V}_L)\right]=0. \qquad (4.44)$$

2. <u>Terms of Order ε^4.</u> Substituting (4.40) into (4.38) and using (4.31), we obtain

$$\frac{\partial^3 n}{\partial t}+\frac{2}{3}\cdot\frac{1}{m\omega_B^2}\Delta(\operatorname{div}\mathbf{q}_\perp-Q)-\frac{1}{\omega_B}\operatorname{rot}_z\int d\mathbf{v}\mathbf{v}_\perp(\hat{L}_\perp\tilde{f}^{(3)})_1=0. \qquad (4.45)$$

The expression we require for $\tilde{f}_2^{(3)}$ has the form

$$\tilde{f}_2^{(3)}=-\frac{1}{4\omega_B^2}\cdot\frac{\partial}{\partial\alpha}\hat{L}_\perp\frac{\partial\tilde{C}^{(1)}}{\partial\alpha}+\frac{1}{4\omega_B}\cdot\frac{\partial\tilde{C}_2^{(2)}}{\partial\alpha}, \qquad (4.46)$$

where

$$\left.\begin{aligned}&\tilde{C}^{(1)}=C_{ii}[F_i,\tilde{f}_i^{(1)}];\\&\tilde{C}_2^{(2)}=C_{ii}[\tilde{f}_i^{(1)},\tilde{f}_i^{(1)}]-\langle C_{ii}[\tilde{f}_i^{(1)},\tilde{f}_i^{(1)}]\rangle+C_{ii}[F_i,\tilde{f}_{2i}^{(2)}].\end{aligned}\right\} \qquad (4.47)$$

Below, we shall restrict ourselves to a calculation of the integral in (4.45) under the assumption $\nabla T_i = 0$. It then follows from (4.24) that $\tilde{f}^{(1)}$ has the form $\tilde{f}^{(1)} = (\mathbf{bv})F$, and as a result $\tilde{C}^{(1)} = 0$ and the contribution to (4.45) of the term with $C_{ii}[\tilde{f}^{(1)}, \tilde{f}^{(1)}]$ vanishes. For simplicity we shall assume that the collisions are not too frequent and we shall neglect the term with Q. As a result, we write (4.45) in the form

$$\frac{\partial^{(3)} n}{\partial t}-\frac{1}{4\omega_B^2}\operatorname{rot}_z\int d\mathbf{v}\mathbf{v}_\perp(\mathbf{v}_\perp\nabla)\frac{\partial}{\partial\alpha}C_{ii}[F_i,f_{i2}^{(2)}]=0. \qquad (4.48)$$

The integral in this equation reduces to that calculated by Braginskii. It is convenient to express the result in terms of the transverse velocity $\mathbf{V}_\perp^1$, which for $\nabla T = 0$ is equal to

$$\mathbf{V}_\perp^{(1)}\equiv\frac{1}{n}\int\mathbf{v}_\perp f^{(1)}\,d\mathbf{v}=\frac{cT}{eB_0}[\mathbf{e}_z,\nabla\ln n]+\mathbf{V}_E. \qquad (4.49)$$

Finally

$$\frac{\partial^{(3)} n}{\partial t}+\frac{1}{m\omega_B}\left\{\operatorname{rot}_z\left(\frac{\partial}{\partial x_k}\eta_1\frac{\partial\mathbf{V}_\perp^{(1)}}{\partial x_k}\right)+\left[\nabla\frac{\partial\eta_1}{\partial x_k},\ \nabla V_{\perp k}^{(1)}\right]_z\right\}=0, \qquad (4.50)$$

where $\eta_1 = 0.3p_i/\omega_{Bi}^2\tau_i$.

As a result, the system of macroscopic equations takes the form

$$\left.\begin{aligned}
&\partial n_e/\partial t + \mathbf{V}_E \nabla n_e + \partial\,(n_e V_z^{(e)})/\partial z = 0;\\
&\partial n_i/\partial t + \mathbf{V}_E \nabla n_i + \partial\,(n_i V_z^{(i)})/\partial z +\\
&+ \operatorname{div}\left[n_i \frac{\mathbf{e}_z}{\omega_{B_i}},\ \left(\frac{\partial}{\partial t} + \mathbf{V}_E \nabla\right) \mathbf{V}_\perp^{(i)}\right] +\\
&+ \frac{1}{m_i \omega_{B_i}} \left\{\operatorname{rot}_z \left(\frac{\partial}{\partial x_k} \eta_1 \frac{\partial \mathbf{V}_\perp^{(1)}}{\partial x_k}\right) + \left[\nabla \frac{\partial \eta_1}{\partial x_k},\ \nabla \mathbf{V}_k^{(1)}\right]_z\right\} = 0;\\
&m_e n_e D_e V_z^{(e)}/Dt = e_e n_e E_z - \partial p_e/\partial z + R_z^{(e)};\\
&(3/2)\, n_e D_e T_e/Dt + p_e \partial V_z^{(e)}/\partial z + \partial q_z^{(e)}/\partial z = 0.
\end{aligned}\right\} \qquad (4.51)$$

In this system we have not assumed $T_e = T_i$, so that the linearized equations take into account the perturbation of the electron temperature even if $T_{0e} \equiv T_{0i}$.

§4.4. Hydrodynamic Treatment of the Macroscopic Equations of §§4.2 and 4.3

In §§4.2 and 4.3 we have solved the transport equation under the assumption that the mean transverse velocity $\mathbf{V}_\perp^{(a)}$ is small. Instead of this one could use the Chapman–Enskog approach (§6.2 of Volume 1), in which $\mathbf{V}_\perp^{(a)}$ is regarded as a quantity of zeroth order. One would then obtain the following system of equations for n, V, and T:

$$\left.\begin{aligned}
&\frac{\partial n}{\partial t} + \operatorname{div}(n\mathbf{V}) = 0;\\
&mn\left(\frac{\partial \mathbf{V}}{\partial t} + (\mathbf{V}\nabla)\,\mathbf{V}\right) =\\
&= en\left(\mathbf{E} + \frac{[\mathbf{V},\,\mathbf{B}]}{c}\right) - \nabla p - \operatorname{div} \overleftrightarrow{\pi} + \mathbf{R};\\
&\frac{3}{2} n \left[\frac{\partial T}{\partial t} + (\mathbf{V}\nabla)\,T\right] + p \operatorname{div} \mathbf{V} =\\
&= -\operatorname{div} \mathbf{q} - \pi_{\alpha\beta} \frac{\partial V_\alpha}{\partial x_\beta} + Q,
\end{aligned}\right\} \qquad (4.52)$$

where the viscosity $\overleftrightarrow{\pi}$, the heat flux **q**, the frictional force **R**, and the heat transfer Q are functions of n, V, and T.

We shall now show how we can obtain from (4.52) the macroscopic equations given in §§4.2 and 4.3 containing $\partial\,(n,\ V_z,\ T)/\partial x_i$ and with what accuracy it is necessary to calculate the fluxes $\overleftrightarrow{\pi}$

q, etc., in order that (4.52) contain all the terms allowed for in the equations of §§4.2 and 4.3.

1. Equations of Order ε. In this approximation we have the system of equations (4.26) for n, V_z, and T and the expression (4.24) for the part of the distribution function $\tilde{f}^{(1)}$ that oscillates with respect to the angle α. The following macroscopic parameters of the plasma are connected with this part of the distribution

A. The velocity of the plasma components at right angles to the magnetic field:

$$\mathbf{V}_{\perp}^{(1)} = \frac{1}{\omega_B}\left[\mathbf{e}_z, \frac{\nabla p}{mn} - \frac{e}{m}\mathbf{E}\right]. \tag{4.53}$$

B. The heat flux vector at right angles to the magnetic field:

$$\mathbf{q}_{\perp}^{(1)} = \frac{5}{2}\cdot\frac{p}{m\omega_B}[\mathbf{e}_z, \nabla T]. \tag{4.54}$$

C. The components of the viscosity tensor with subscripts (x, y) and (y, z):

$$\pi_{\perp z}^{(1)} \equiv (\pi_{xz}^{(1)}, \pi_{yz}^{(1)}, 0) = \frac{p}{\omega_B}[\mathbf{e}_z, \nabla V_z]. \tag{4.55}$$

Using (4.53)-(4.55), we can represent the system (4.26) in the form

$$\left.\begin{aligned} &\frac{d_a n}{dt} + n\,\mathrm{div}\,\mathbf{V} = 0;\\ &mn\frac{d_a V_z}{dt} = enE_z - \frac{\partial p}{\partial z} - (\nabla\overleftrightarrow{\pi})_z + R_z;\\ &en\left(\mathbf{E}_{\perp} + \frac{[\mathbf{V}, \mathbf{B}]}{c}\right) = 0;\\ &\frac{3}{2}n\frac{d_a T}{dt} + p\,\mathrm{div}\,\mathbf{V} = -\mathrm{div}\,\mathbf{q}. \end{aligned}\right\} \tag{4.56}$$

Here, as in §6.2 of Volume 1, $\frac{d_a}{dt} = \frac{\partial}{\partial t} + (\mathbf{V}\nabla)$.

Thus, the equations of order ε are a consequence of the approximation of hydrodynamics in which one takes into account the heat flux q and the viscosity tensor $\overleftrightarrow{\pi}$ of order ε. In the determination of the corrections to the distribution function by the Chapman–Enskog method, these terms in q and $\overleftrightarrow{\pi}$ appear in the first-order in the expansion parameters, i.e., for this it is sufficient to calculate only f_1 in accordance with the Chapman–Enskog procedure. The function f_1 and the corresponding terms in q and $\overleftrightarrow{\pi}$ have been calculated by Braginskii, so that Eqs. (4.26) can be obtained as a consequence of the system of hydrodynamic equations given in his paper.

A comparison of (4.56) and (4.26) shows that some of the terms in the operators d (n, V_z, T)/dt cancel against other terms of the hydrodynamic equations:

$$\left.\begin{aligned} \mathbf{V}_\perp\nabla n + n\,\mathrm{div}\,\mathbf{V}_\perp &= \mathbf{V}_E\nabla n;\\ \mathbf{V}_\perp\nabla V_z + \frac{(\nabla\overleftrightarrow{\pi})_z}{mn} &= \mathbf{V}_E\nabla V_z;\\ \frac{3}{2}n\mathbf{V}_\perp\nabla T + p\,\mathrm{div}\,\mathbf{V}_\perp + \mathrm{div}\,\mathbf{q} &= \frac{3}{2}n\mathbf{V}_E\nabla T. \end{aligned}\right\} \tag{4.57}$$

This canceling of terms shows that in the phenomena in which we are interested (see §4.2) physical meaning does not attach to the total velocity V but only that part of it $\mathbf{V}_{\mathrm{eff}}$ that does not contain a term with the gradient of the pressure [i.e., the term with the Larmor currents; (see §1.6)]:

$$\mathbf{V}_{\mathrm{eff}} = \mathbf{e}_z V_z + \mathbf{V}_E. \tag{4.58}$$

2. Equations of Order ε^2. The terms of order ε^2 in the equations of §4.2 take into account dissipative effects such as longitudinal heat conduction, the collisional part of the transverse heat conduction, and the collisional viscosity.

In the Chapman–Enskog procedure, all these effects are determined by the correction to the distribution function of first and second orders. If only terms of first order are taken into account, we obtain, in particular,

$$\pi_{zz} = -\eta_0\left(\frac{4}{3}\cdot\frac{\partial V_z}{\partial z} - \frac{2}{3}\,\mathrm{div}\,\mathbf{V}_\perp\right). \tag{4.59}$$

This expression differs from that found in §4.2. If we take into account the correction to the distribution function of second order in the Chapman–Enskog procedure, we obtain the following correction to (4.59):

$$\pi_{zz}^{(2)} = \frac{2}{3}\eta_0\frac{1}{mn\omega_B}[\nabla T, \nabla n]_z. \tag{4.60}$$

Adding (4.59) and (4.60) and noting that, in accordance with (4.53),

$$\operatorname{div} \mathbf{V}_{\perp} = -\frac{1}{mn\omega_B}[\nabla T, \nabla n]_z, \tag{4.61}$$

we arrive at the expression for π_{zz} obtained in §4.2.

Note that in the approximation adopted in §4.2 of a homogeneous magnetic field and electrostatic electric field, it is sufficient to calculate only the corrections of first-order to obtain a correct description by the Chapman-Enskog method of all the remaining effects (expect viscosity) of order ε^2.

3. Equations of Order ε^3. The ε^3 approximation corresponds to allowance in the ion equation (4.52) for transverse inertia (the left-hand side of the equation) and the "collisionless" part of the viscosity tensor. If these small terms are included, the transverse part of Eq. (4.52) can be represented in the form

$$\mathbf{V}_{\perp} = \mathbf{V}_{\perp}^{(1)} + \mathbf{V}_{\perp}^{(3)}, \tag{4.62}$$

where $\mathbf{V}_{\perp}^{(1)}$ is determined by Eq. (4.53) and

$$\mathbf{V}_{\perp}^{(3)} = \frac{1}{\omega_B}\left[\mathbf{e}_z, \frac{\partial \mathbf{V}}{\partial t} + (\mathbf{V}\nabla)\mathbf{V} + \frac{(\nabla \overleftrightarrow{\pi})_{\perp}}{mn}\right]\Bigg|_{\mathbf{V}=\mathbf{V}_{\perp}^{(1)}}. \tag{4.63}$$

The tensor $\overleftrightarrow{\pi}$ is determined by the distribution function $\tilde{f}_2^{(2)}$ [see Eq. (4.29)], which contains double oscillations with respect to α ($\cos 2\alpha$, $\sin 2\alpha$), and by the function $\bar{f}^{(2)}$, which does not depend on α [the latter is determined by Eq. (4.30) and in the limit $V_z \to 0$ depends on the temperature gradient], so that

$$\overleftrightarrow{\pi} = \overleftrightarrow{\pi}\{\tilde{f}_2^{(2)}\} + \overleftrightarrow{\pi}\{\bar{f}^{(2)}\}. \tag{4.64}$$

For our purposes we do not require $\mathbf{V}_{\perp}^{(3)}$ itself but its contribution to the continuity equation, i.e., the expression

$$\frac{\partial^{(2)} n}{\partial t} \equiv -\operatorname{div}(n\mathbf{V}^{(3)}) = \frac{1}{\omega_B}\operatorname{rot}_z\left\{n\left(\frac{\partial \mathbf{V}^{(1)}}{\partial t} + (\mathbf{V}\nabla)\mathbf{V} + (\nabla\overleftrightarrow{\pi})_{\perp}\right\}\right.. \tag{4.65}$$

It turns out that

$$\operatorname{rot}_z(\nabla\overleftrightarrow{\pi}\{\bar{f}^{(2)}\})_{\perp} = 0, \tag{4.66}$$

and therefore the part of the viscosity tensor associated with $\bar{\mathbf{f}}^{(2)}$ need not be calculated. The part of the tensor $\overleftrightarrow{\pi}$ associated with $\tilde{\mathbf{f}}_2^{(2)}$ can be represented in the form

$$\left.\begin{aligned} &\pi_{xx} = -\pi_{yy} = \\ &= -\frac{p}{2\omega_B}\left(\frac{\partial V_x^{(1)}}{\partial y} + \frac{\partial V_y^{(1)}}{\partial x}\right) - \frac{1}{5\omega_B}\left(\frac{\partial q_x^{(1)}}{\partial y} + \frac{\partial q_y^{(1)}}{\partial x}\right), \\ &\pi_{xy} = \pi_{yx} = \\ &= \frac{p}{2\omega_B}\left(\frac{\partial V_x^{(1)}}{\partial x} - \frac{\partial V_y^{(1)}}{\partial y}\right) + \frac{1}{5\omega_B}\left(\frac{\partial q_x^{(1)}}{\partial x} - \frac{\partial q_y^{(1)}}{\partial y}\right). \end{aligned}\right\} \quad (4.67)$$

In the first approximation of the Chapman–Enskog method, one obtains only the part of the tensor $\overleftrightarrow{\pi}$ that depends on the velocity derivatives. The derivatives of the heat flux are important if $\nabla T \neq 0$.

4. Equations of Order ε^4. In this approximation we take into account the transverse collisional viscosity of the ions. We have considered this effect under the assumption $\nabla T = 0$, when $\mathbf{q} = 0$ and the viscosity is determined solely by the velocity derivatives [cf. (4.67)]. The velocity derivatives are taken into account in the first approximation of the Chapman–Enskog method. Equation (4.38) expressed in terms of the collisional viscosity tensor has the form

$$\frac{\partial^{(3)} n}{\partial t} = \frac{1}{\omega_B} \operatorname{rot}_z (\nabla \overleftrightarrow{\pi}^{\mathrm{cl}}), \quad (4.68)$$

where

$$\left.\begin{aligned} \pi_{xx}^{\mathrm{cl}} &= -\pi_{yy}^{\mathrm{cl}} = -\eta_1\left(\frac{\partial V_x}{\partial x} - \frac{\partial V_y}{\partial y}\right), \\ \pi_{xy}^{\mathrm{cl}} &= \pi_{yx}^{\mathrm{cl}} = -\eta_1\left(\frac{\partial V_x}{\partial y} + \frac{\partial V_y}{\partial x}\right). \end{aligned}\right\} \quad (4.69)$$

If $\nabla T \neq 0$, the right-hand sides of (4.69) would also contain derivatives of the heat flux $\mathbf{q}$.

5. Comparison with Other Methods. The assumptions concerning the orders of magnitude adopted in §§4.2 and 4.3 do not accord with the assumptions on which the Chapman–Enskog method is based. The point is that in this method the time and spatial variations of the macroscopic parameters are taken into

account in the same order:

$$\partial/\partial t \sim \partial/\partial x, \tag{4.70}$$

whereas, in accordance with (4.19),

$$\frac{\partial}{\partial t} \sim \left(\frac{\partial}{\partial x}\right)^2 \tag{4.71}$$

in the gradient effects in which we are interested. The velocity of the transverse motion of a plasma in processes with the frequency (4.19) is, in accordance with (4.53), a small quantity, whereas in the Chapman–Enskog method it is assumed to be of zeroth order. The leading terms in the heat flux and the viscosity tensor in the method of §4.3 are assumed to be of the same order as the transverse velocity, but in the Chapman–Enskog method **q** and $\overleftrightarrow{\pi}$ are assumed to be small.

If it assumed that $\mathbf{V}_\perp$ is of zeroth order, the results of §§4.2 and 4.3 can be obtained by two methods:

(1) in the Chapman-Enskog method one takes into account the second-order corrections;

(2) one assumes that, besides n, **V**, and T, the quantities **q** and $\overleftrightarrow{\pi}$ are of zeroth order. This approach has been developed, in particular, by Herdan and Liley and Kogan, Moiseev, and Oraevskii.

§4.5. Dispersion Equations when Small Dissipative Effects Are Ignored

This and the following section consist essentially of a compilation of the dispersion equations that follow in the different limiting cases from the macroscopic equations of §§4.2 and 4.3. In deriving these equations we neglect an equilibrium electric field, $\mathbf{E}_{\perp 0} = 0$, $E_{z0} = 0$, and the equilibrium motion of the plasma components along the magnetic field, $V_{z0}^{(e)} = V_{z0}^{(i)} = 0$, and we also set $T_{0e} = T_{0i} = T_0$. In this section we also neglect the small dissipative effects of order $(k_z v_{T_i})^2/\omega\nu_{ii}$, $(k\rho_i)^2$, $(k\rho_i)^2\nu_{ii}/\omega$. The dispersion equations that include these effects will be given in §4.6.

Under these assumptions the system of equations (4.34), in which the ion continuity equation is replaced by (4.44) and allowance is made for the quasineutrality condition $n'_e = n'_i = n'$, yields

$$\left.\begin{aligned}
&-i\omega n' + \mathbf{V}_E\nabla n_0 + ik_z V_z^{(e)} n_0 = 0;\\
&-i\omega n' + \mathbf{V}_E\nabla n_0 + ik_z V_z^{(i)} n_0 -\\
&\qquad -\frac{ik_\perp^2 e_i n_0}{m_i\omega_{Bi}^2}(\omega - \omega_{p_i}^*)\psi = 0;\\
&0 = e_e n_0 E_z - ik_z T_0 n' - 1{,}71\, ik_z n_0 T'_e -\\
&\qquad - \nu_{ei} n_0 m_e (V_z^{(e)} - V_z^{(i)});\\
&-i\omega n_0 m_i V_z^{(i)} = -ik_z(p'_i + p'_e);\\
&-\frac{3}{2} i\omega n_0 T'_e + \frac{3}{2} n_0 \mathbf{V}_E \nabla T_0 + ik_z p_0 V_z^{(e)} =\\
&\qquad = -3{,}16\frac{k_z^2 p_0}{m_e}\tau_e T'_e - \frac{3}{2}\nu_{ie} n_0 (T'_e - T'_i);\\
&-\frac{3}{2} i\omega n_0 T'_i + \frac{3}{2} n_0 \mathbf{V}_E \nabla T_0 + ik_z p_0 V_z^{(i)} =\\
&\qquad = \frac{3}{2}\nu_{ie} n_0 (T'_e - T'_i).
\end{aligned}\right\} \tag{4.72}$$

Here $\nu_{ei} \equiv 0.51/\tau_e$, $\nu_{ie} \equiv {}^2/_3\, m_e/m_i\tau_e$. In the equation of motion of the electrons we have ignored their inertia, which is small, $\sim\omega/\nu_{ei}$. The equation of motion of the ions is written down with allowance for the corresponding electron equation. In this system $V_z^{(e)}$ and $V_z^{(i)}$ stand for the perturbations of the longitudinal electron and ion velocities; $\omega_{p_i}^* = (ck_y/e_i\, B_0 n_0)\, \partial p_{0i}/\partial x$.

The last term of the left-hand side of the ion continuity equation obtained in the ε^3 approximation is small compared with the first as $z_i \equiv (k_\perp \rho_i)^2$. It can be seen, however, that if k_z is sufficiently small, the sum of the first two terms of the given equation vanishes. This is a consequence of the electron continuity equation. Under these conditions the terms of order ε^3 are the principal terms.

Formally, the large terms of the second equation in (4.72) can be eliminated by subtracting the first equation from the second. Then

$$k_z(V_z^{(i)} - V_z^{(e)}) - z_i(\omega - \omega_{p_i}^*)\, e_i\psi/T_0 = 0. \tag{4.73}$$

We can now estimate the relative importance of the terms of order ε^3 by eliminating from this equation the difference $V_z^{(i)} - V_z^{(e)}$ by means of the third equation of (4.72):

$$-i\,(k_z^2 T_0/m_e\nu_{ei})\,(e_e\psi + n'T_0/n_0 + 1.71 T'_e) + e_i z_i(\omega - \omega_{p_i}^*)\,\psi = 0. \tag{4.74}$$

It can be seen that the terms of order ε^3 (transverse inertia of the ions and their transverse magnetic viscosity) are important if if

$$\omega_s \lesssim \omega^*, \tag{4.75}$$

where $\omega^* = \max(\omega_n, \omega_T, \omega_p^*)$,

$$\omega_s \equiv k_z^2 T_0 / m_e \nu_{ei} z_i. \tag{4.76}$$

1. Perturbations with Small k_z, $\omega_s \le \omega^*$. Let us now consider what simplifications can be made in the system (4.72) if the condition (4.75) holds. From the first two equations of (4.72) we obtain approximately

$$n' = -(k_y e\psi / m\omega\omega_B)\, \partial n_0/\partial x. \tag{4.77}$$

From the last three

$$T_e' = -(k_y e\psi / m\omega\omega_B)\, \partial T_0/\partial x. \tag{4.78}$$

As a result, (4.74), (4.77), and (4.78) yield the dispersion equation

$$1 - \frac{\omega_{pi}^*}{\omega} + i\frac{\omega_s}{\omega}\left(1 - \frac{\omega_{ne}}{\omega} - 1.71\frac{\omega_{Te}}{\omega}\right) = 0. \tag{4.79}$$

This is the collisional analog of Eq. (3.2).

2. Perturbations with Large cos θ, $\omega_s \gg \omega^*$. In this case Eq. (4.78) entails

$$V_z^{(e)} = V_z^{(i)} \equiv V_z. \tag{4.80}$$

Equation (4.74) is now replaced by

$$e_e\psi + n' T_0/n_0 + 1.71 T_e' = 0. \tag{4.81}$$

We find the expression for V_z from the ion equation of motion:

$$V_z = k_z (p_e' + p_i')/m_i n_0 \omega. \tag{4.82}$$

Substituting (4.82) into the first equation of (4.72), we obtain

$$-i\omega n' - i(e\psi k_y / m\omega_B)\, \partial n_0/\partial x + i k_z^2 (p_e' + p_i')/m_i\omega = 0. \tag{4.83}$$

It can be seen that V_z must be included if

$$k_z \gtrsim \omega^*/v_{T_i}. \tag{4.84}$$

Therefore, we must consider the following limiting cases.

A. Perturbations with $k_z < \omega^*/v_{T_i}$. In this case $V_z = 0$, so that (4.77) remains in force. From the last two equations of the system (4.72) we find T'_e. If $\omega^* > \nu_{ie}$,

$$T'_e = -(e\psi k_y/m\omega_B)(\partial T_0/\partial x)(1 + ik_z^2 T_0\tau_e/3c_0\omega m_e), \tag{4.85}$$

where $c_0 \equiv (2\cdot 3.16)^{-1}$. From (4.74), (4.81), and (4.85) we obtain the dispersion equation

$$1 - \frac{\omega_{ne}}{\omega} - 1.71\frac{\omega_{Te}}{\omega + 2.1ik_z^2 T_0\tau_e/m_e} = 0. \tag{4.86}$$

Here, the imaginary terms are of the same order as the real terms if

$$\omega_s \simeq \omega^*/z_i. \tag{4.87}$$

B. Perturbations with $k_z \simeq \omega^*/v_{T_i}$ in the Case $\omega^* > \nu_{ie}$. Under these conditions the electron thermal conductivity is high, so that approximately

$$T'_e = 0. \tag{4.88}$$

We then obtain from (4.72)

$$1 - \frac{\omega_{ne}}{\omega} - \frac{k_z^2 T_0}{m_i\omega^2}\left(\frac{8}{3} - \frac{2}{3}\cdot\frac{\omega_{ne}}{\omega} + \frac{\omega_{Te}}{\omega}\right) = 0. \tag{4.89}$$

C. Perturbations with $k_z \simeq \omega^*/v_{T_i}$ in the Case $\omega^* < \nu_{ie}$. In accordance with the last inequality, heat exchange is important under these conditions, and therefore

$$T'_e = T'_i \equiv T'. \tag{4.90}$$

The last two equations of (4.72) reduce to the single equation

$$-i\omega T' + \mathbf{V}_E\nabla T_0 + \frac{2}{3}ik_z T_0 V' = 0, \tag{4.91}$$

and Eqs. (4.82) and (4.83) then entail

$$V' = 2k_z p'/m_i n_0 \omega, \quad -i\omega n' + \mathbf{V}_E \nabla n_0 + 2ik_z p'/m_i \omega = 0. \tag{4.92}$$

Equations (4.81), (4.91), and (4.92) yield the dispersion equation

$$1 - \frac{\omega_{ne} + 1.71\omega_{T_e}}{\omega} - \frac{k_z^2 T_0}{m_i \omega^2}\left[\frac{10}{3} - \frac{0.71}{\omega}\left(\frac{2}{3} - \eta\right)\omega_{ne}\right] = 0. \tag{4.93}$$

In contrast to (4.89), it describes perturbations of a strongly collisional plasma, $\omega^* < \nu_{ie}$.

§4.6. Dispersion Equations Which Take into Account Small Dissipative Effects

1. Inclusion of Terms of Order $(k_\perp \rho_i)^2$. Considering perturbations with $\omega_s \simeq \omega^*/z_i$ in §4.5, we neglected the terms of order z_i, and this led us to the dispersion equation (4.86). If $\nabla T_0 = 0$, the latter has real solutions. Therefore, in the determination of the growth rate of perturbations of a plasma with $\nabla T_0 = 0$, the small terms of order z_i are important.

Making the same assumptions as in the derivation of (4.86) but retaining terms of order z_i, we find that (4.80) and (4.82) are replaced by

$$V_z^{(e)} = -(e_i \psi / k_z T_0)\, z_i (\omega - \omega_{ni}), \tag{4.94}$$

$$V_z^{(i)} = 0. \tag{4.95}$$

The perturbations of the density and the electron temperature are then given by $(\nabla T_0 = 0)$:

$$n' = -(e\psi k_y / m\omega\omega_B)\, \partial n_0/\partial x - (e_i \psi / T)\, z_i (1 - \omega_{ni}/\omega), \tag{4.96}$$

$$T_e' = -\frac{2}{3} e_i \psi z_i (1 - \omega_{ni}/\omega)(1 + 2.1 i k_z^2 T_0 \tau_e / m_e \omega)^{-1}. \tag{4.97}$$

Substituting these expressions into (4.74), we arrive at the dispersion equation

$$1 - \frac{\omega_{ne}}{\omega} + \left(1 - \frac{\omega_{ni}}{\omega}\right) z_i \left(1 - i0.51 \frac{m_e \omega}{k_z^2 T_0 \tau_e} + \frac{2}{3}\,\frac{1.71}{1 + i2.1 k_z^2 T_0 \tau_e / m_e \omega}\right) = 0. \tag{4.98}$$

In particular, if $\omega_s \gg \omega^*/z_i$, this can be written in a form similar to (3.12):

$$(1 + i\nu_{\text{eff}}\, \omega / k_z^2 v_{Te}^2)(1 - \omega_{ne}/\omega) + z_i (1 - \omega_{ni}/\omega) = 0, \tag{4.99}$$

where

$$\nu_{\text{eff}} = 2(0.51 + 1.71/3.16)/\tau_e.$$

Equation (4.99) ceases to hold if $k_z v_{T_i}/\omega^* \simeq z_i^{1/2}$. For such values of k_z, it must be augmented by terms due to the longitudinal ion inertia. Then (4.99) is replaced by

$$(1 + i\nu_{\text{eff}}\omega/k_z^2 v_{T_e}^2)(1 - \omega_{ne}/\omega) + (z_i - k_z^2 T/m_i\omega^2)(1 - \omega_{ni}/\omega) = 0. \tag{4.100}$$

2. Inclusion of Transverse Collisional Viscosity. The terms corresponding to transverse collisional viscosity are contained in the equations of the system (4.51). This system yields the following generalization of the dispersion equation (4.98):

$$\left(1 - \frac{\omega_{ne}}{\omega}\right)\left(1 - i\,\frac{0.51\omega m_e}{k_z^2 T_0 \tau_e} + \frac{2}{3}\,\frac{1.71}{1 + i2.1k_z^2 T_0 \tau_e/m_e\omega}\right)^{-1} +$$

$$+ z_i\left(1 - \frac{\omega_{ni}}{\omega}\right)\left(1 + 0.3\,\frac{iz_i}{\omega\tau_i}\right) = 0. \tag{4.101}$$

The additional terms are important if $z_i > \omega/\nu_{ii}$.

3. Inclusion of Dissipative Effects in Perturbations with $k_z \simeq \omega^*/v_{T_i}$. In this case, Eqs. (4.72) must be augmented by the longitudinal ion viscosity and the ion thermal conductivity. On the other hand, one can neglect the transverse inertia and the magnetic viscosity of the ions. As a result, the dispersion equation takes the form

$$D_0 + i(\Delta_1/\omega_{T_e})a_1 + i(\nu_{ie}/\omega_{T_e})a_2 = 0, \tag{4.102}$$

where

$$\left.\begin{aligned}
D_0 &= x^2(x - 1/\eta) + b(8x/3 + 1 - 2/3\eta);\\
a_1 &= (c_1 + 1)(x - 1/\eta)x - 2c_1 b;\\
a_2 &= x(x - 1/\eta) - 2b - (c_0 x/b)[x^3 - x^2 \times\\
&\times (1 + s + 1/\eta) - (10/3)bx + 2b(1 - 2/3\eta)];\\
x &= \omega/\omega_T,\ b = k_z^2 T_0/m_i\omega_T^2;\ \omega_T \equiv \omega_{Te};\\
\Delta_1 &= (4/3)\cdot 0.96k_z^2 T_0 \tau_i/m_i,\ c_1 = 3.9/2\cdot 0.96.
\end{aligned}\right\} \tag{4.103}$$

§4.7. Application of a Model Collision Integral

In §4.1 we considered the approximation of weak collisions; in §§4.2-4.6, the approximation of strong collisions. The equations of the oscillations in these two limiting cases were obtained by essentially different methods. The reason for this is the complexity of the collision integral (also known as the collision term of the transport equation). It is for precisely this reason that one cannot use a single method to solve the transport equation and obtain a unified picture of the oscillations of a collisional plasma.

If accuracy is sacrificed for the sake of a unified picture, the problem can be solved. This is achieved by replacing the exact collision integral by an approximate integral. On the one hand, this approximation must take into account correctly the main features of the collision process and, on the other hand, must enable one to find the solution of the transport equation by the same method as in the absence of collisions.

In the case of a plasma with a homogeneous temperature, $\nabla T_{0i} = \nabla T_{0e} = 0$, and vanishing longitudinal current, $V_{0e} = V_0$, these requirements are satisfied by a collision integral of the form (the Bhatnagar–Gross–Krook model)

$$C_\alpha = -\sum_\beta \nu_{\alpha\beta} \left\{ f^{(\alpha)} - \left(\frac{n^{(\alpha)}}{n_0} + \frac{m_\alpha}{T_\alpha} \mathbf{v}\mathbf{V}^{(\alpha)} \right) f^{(\alpha)} \right\}, \tag{4.104}$$

where $\nu_{\alpha\beta}$ is the frequency of collisions of particles of the species α with particles of the species β; and $f^{(\alpha)}$, $n^{(\alpha)}$, and $V^{(\alpha)}$ are the perturbations of the distribution function, the density, and the velocity of the α-th component of the plasma.

The collisional term of the form (4.104) satisfies the laws of conservation of the particle number and momentum. However, it does not take into account the perturbations of the temperature and effects such as heat conduction. Nor is allowance made for effects associated with the detailed mechanism of the Coulomb interaction such as the anisotropy of the frictional force, the thermal force, and the increase in the effective collision frequency in short-wavelength perturbations (see §4.1). A collisional term of the form (4.104) can therefore give qualitatively correct results only if all these effects are unimportant.

It follows from the foregoing analysis that such a situation obtains in the case of low-frequency, $\omega \ll \omega_{B_i}$, perturbations of a plasma with an inhomogeneous density, $\nabla n_0 \neq 0$, $\nabla T = 0$, at a not too high frequency of ion-electron collisions, $\nu_{ie} < \omega^*$, and not too large wave numbers, $k_\perp \rho_i \leq 1$.

Despite these restrictions, the transport equation with collision term of the form (4.104) enables one to obtain more extensive information than the equation obtained by rigorous methods in the limits of weak and strong collisions. As regards the electrons, one can use this method to take into account their mutual collisions and the force of friction against the ions for arbitrary relationships between the parameters ω, $k_z v_{T_e}$, and ν_{ee}. The ion transport equation with C_α of the form (4.104) takes into account ion viscosity, first, for an arbitrary relationship between ω and ν_{ii} and, secondly, in a wide range of transverse wave numbers right up to $k_\perp \rho_i \simeq 1$ inclusive. In this manner one can also describe correctly the friction of the ions against the electrons provided k_z is not too large and a thermal force does not play a role.

The desired distribution function on the right-hand side of (4.104) is not differentiated (as in the case of the exact collision integral) but is merely multiplied by a number. As a result, the method of trajectory integrals can be used. For $f^{(\alpha)}$ we obtain

$$f^{(\alpha)} = -\int_{-\infty}^{t} dt' \exp\left[\nu_\alpha (t'-t)\right] \left\{ \frac{e_\alpha}{m_\alpha} \nabla\psi \frac{\partial f_0^{(\alpha)}}{\partial \mathbf{v}} + \right.$$

$$\left. + \sum_\beta \nu_{\alpha\beta} f_0^{(\alpha)} \left[\frac{n^{(\alpha)}}{n_0} + \frac{m_\alpha}{T_\alpha} \mathbf{v}\mathbf{V}^{(\beta)} \right] \right\}. \tag{4.105}$$

Here, $\nu_\alpha = \sum_\beta \nu_{\alpha\beta}$. The integral is taken along trajectories which have the same form as in the absence of collisions (cf. §1.7). The integral is calculated in the same way as in §1.7 and leads to the result

$$f^{(\alpha)} = -\frac{e f_0^{(\alpha)}}{T_\alpha} \psi \left\{ 1 - \frac{(\omega + i\nu_\alpha - \omega_{n\alpha}) J_0(\xi) \exp\left[i\xi \sin(\alpha - \Psi)\right]}{\omega + i\nu_\alpha - k_z v_z} \right\} +$$

$$+ i \sum_\beta \nu_{\alpha\beta} \frac{\exp\left[i\xi \sin(\alpha-\Psi)\right]}{\omega + i\nu_\alpha - k_z v_z} \left\{ \left[\frac{n^{(\alpha)}}{n_0} + \frac{m_\alpha}{T_\alpha} V_z^{(\beta)} \right] \times \right.$$

$$\left. \times J_0(\xi) + i \frac{m_\alpha}{T_\alpha} \frac{v_\perp J_0'(\xi)}{k_\perp} [\mathbf{k}, \mathbf{V}^{(\beta)}]_z \right\}. \tag{4.106}$$

Using (4.106), we obtain a system of equations for $n^{(\alpha)}$, $V_z^{(\alpha)}$, $[\mathbf{k}, \mathbf{V}^{(\alpha)}]_z$, ($\alpha$ = i, e). It can readily be solved if allowance is made for the smallness of the parameters ν_{ii}/ν_{ei} and ν_{ie}/ν_{ii}. Calculating the charge densities $n^{(\alpha)}$, we can find the permittivity $\varepsilon_0^{(\alpha)}$ (see §1.1) and obtain the dispersion equation with allowance for the collisions. We shall now give some limiting cases of such a dispersion equation.

If $z_i \ll 1$, $\nu_e \gg (\omega, k_z v_{T_e})$, and small terms are ignored,

$$1-\frac{\omega_{ne}}{\omega}+\left(1-2i\frac{\nu_{ei}\omega}{k_z^2 v_{T_e}^2}\right) z_i \left(1-\frac{\omega_{ni}}{\omega}\right)=0. \tag{4.107}$$

This agrees qualitatively with (4.99). Note that in deriving (4.99) we assumed that both the electrons and the ions are collisional. In reality this result also remains in force in the case of collisionless ions. This is a natural result, since a contribution to (4.99) arises only from the transverse ion inertia and the transverse magnetic viscosity, i.e., from effects that are insensitive to collisions.

Using (4.106), we can obtain a generalization of Eq. (4.107) to the case of arbitrary ω/ν_e:

$$1-\frac{\omega_{ne}}{\omega}+z_i\left[1-i\sqrt{\pi}\frac{\omega}{|k_z|v_{T_e}}\left(\frac{W(x)}{1+i\sqrt{\pi}\,xW(x)}+\right.\right.$$
$$\left.\left.+2i\frac{x}{\sqrt{\pi}}\right)\left(1-\frac{\omega_{ni}}{\omega}\right)\right]=0, \tag{4.108}$$

where $x=(\omega+i\nu_e)/|k_z|v_{T_e}$.

If small terms of order z_i are included, (4.107) is replaced by an equation similar to (4.101):

$$1-\frac{\omega_{ne}}{\omega}+\left(1-2i\frac{\nu_{ei}\omega}{k_z^2 v_{T_e}^2}\right) z_i\left(1-\frac{\omega_{ni}}{\omega}\right)\left[1+z_i\frac{i\nu_{ii}}{\omega}\left(\frac{9}{4}-\frac{\omega}{\omega+i\nu_{ii}}\right)\right]=0. \tag{4.109}$$

This equation takes into account the collisional viscosity of the ions – both when $\nu_{ii} > \omega$ and $\nu_{ii} < \omega$.

In Eq. (4.107) one could also include the terms corresponding to the longitudinal motion of the ions. In particular, one would then obtain an equation similar to (4.100).

Thus, the use of the model collision integral leads to qualitatively correct results in the problems we have discussed.

Bibliography

1. S. Chapman and T. G. Cowling, Mathematical Theory of Non-uniform Gases, Cambridge (1952). Contains an exposition of the manner in which the Boltzmann equation is solved and a derivation of the transport equations known as the Chapman-Enskog method.
2. S. I. Braginskii, Zh. Eksp. Teor. Fiz., 33:459 (1957). The Chapman–Enskog method is used to obtain the transport equations of a plasma in a magnetic field for an arbitrary value of the ratio ν/ω_B. The case of a strong magnetic field is considered. The main results of this paper are given in Braginskii's review in: Reviews of Plasma Physics, Vol. 1, Consultants Bureau, New York (1965), p. 205.
3. A. N. Kaufman, "Plasma Transport Theory" in: La Theorie des Gaz Neutres et Ionizés, Hermann, Paris (1960). An expansion in 1/B yields the transport equations of a plasma. Kaufman's results agree qualitatively with those of Braginskii.
4. A. B. Mikhailovskii, Zh. Eksp. Teor. Fiz., 52:943 (1967) [Sov. Phys. – JETP, 25:623 (1967)]. Equations are obtained for slow processes in a collisional plasma of vanishing pressure. The main results of this paper are given in §§4.2-4.4.
5. R. Herdan and B. S. Liley, Rev. Mod. Phys., 32:731 (1960). Transport equations that take into account the derivatives of the fluxes are obtained.
6. E. Ya. Kogan, S. S. Moiseev, and V. N. Oraevskii, Zh. Prikl. Mekh. Tekh. Fiz., No. 6, p. 41 (1965) [J. Appl. Mech. Tech. Phys., No. 6, p. 25 (1965)]. Equations with derivatives of the fluxes are derived for the case of a collisionless plasma.
7. P. L. Bhatnagar, E. P. Gross, and M. Krook, Phys. Rev., 94:511 (1954). The use of a transport equation with a model collision integral (the Bhatnagar–Gross–Krook model; §4.7) is proposed for problems of plasma oscillations.
8. L. P. Pitaevskii, Zh. Eksp. Teor. Fiz., 44:969 (1963) [Sov. Phys. – JETP, 17:658 (1963)]. The effect of the increase of the collision frequency in short-wavelength perturbations (§4.1) is noted.
9. A. A. Rukhadze and V. P. Silin, Dokl. Akad. Nauk SSSR, 169:558 (1966) [Sov. Phys. – Doklady, 11:606 (1967)]. The collisional corrections to the permittivity of an almost collisionless plasma (§4.1) are obtained.
10. A. B. Mikhailovskii and O. P. Pogutse, Zh. Tekh, Fiz., 36: 205 (1966) [Sov. Phys. – Tech. Phys., 11:153 1966)]. Expressions are given for the permittivity obtained with the use of the Bhatnagar–Gross–Krook model (§4.7).

Chapter 5

Instabilities of a Collisional Plasma

§5.1. Instabilities of an Almost Collisionless Plasma

1. Inertial-Dissipative Instability of a Plasma with $\nabla T = 0$. The long-wavelength instabilities of a collisionless plasma with $\nabla T = 0$ are shown in Fig. 3.1. This figure shows that there is an interval of wave numbers k_z at which neither a hydrodynamic nor a kinetic instability can develop. Let us consider the influence of weak collisions on perturbations with k_z in this interval.

Using (4.1), we obtain the following dispersion relation, which replaces (3.2):

$$1-\frac{\omega_{ni}}{\omega}-\frac{m_e}{m_i}\left(\frac{\omega_{B_i}}{\omega}\right)^2\cos^2\theta\left(1-\frac{\omega_{ne}}{\omega}\right)\left(1-\frac{i\nu_{ei}}{\omega}\right)=0. \tag{5.1}$$

The results that follow from this equation are shown in Fig. 5.1, which also includes the kinetic instability.

A comparison of Fig. 5.1 and Fig. 3.1 shows that in the region between the hydrodynamic and the kinetic instabilities an inertial-dissipative instability can arise which is due to collisions between the electrons and ions.

Near the boundary of the hydrodynamic instability

$$\gamma_{\text{diss}} \simeq \nu_{ei}. \tag{5.2}$$

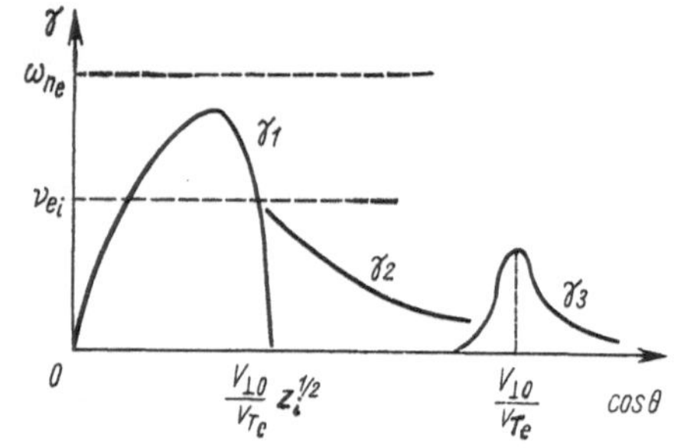

Fig. 5.1. Instabilities of a plasma with $\nu_{ei} < \omega_{n_e}$, $z_i < 1$, $\nabla T = 0$ (γ_1 is the growth rate of the hydrodynamic instability; γ_2, that of the dissipative instability; γ_3, that of the kinetic instability; $V_{\perp 0} \approx v_{T_i}\rho_i/a$).

As k_z increases, the growth rate of the inertial-dissipative instability decreases in accordance with the law

$$\gamma_{\text{diss}} = 2\frac{m_e}{m_i \cos^2\theta}\nu_{ei}\left(\frac{\omega_{ne}}{\omega_{B_i}}\right)^2. \tag{5.3}$$

It follows from a comparison of (5.2) and (3.13) that $\gamma_{\text{diss}} > \gamma_{\text{kin}}$ if

$$\nu_{ei}/\omega_{ne} > z_i. \tag{5.4}$$

<u>2. Excitation of Perturbations with $k_z = 0$, $k_\perp \rho_i \ll 1$, in a Plasma with $\eta \equiv \partial \ln T/\partial \ln n < 0$.</u> If collisions are unimportant, such perturbations do not grow (see §3.2). If we include the collisions [see Eq. (4.1)], we obtain

$$\operatorname{Re}\omega = \omega^*_{p_i}, \qquad \gamma = -\frac{3}{10}\nu_{ii}z_i\eta/(1+\eta). \tag{5.5}$$

It can be seen that collisions give rise to an instability if

$$-1 < \eta < 0. \tag{5.6}$$

If $k_\perp \rho_i \simeq 1$, the growth rate is $\gamma \simeq \nu_{ii}$.

<u>3. Suppression of the Short-Wavelength Kinetic Instability of a Plasma with $\nabla T = 0$ by Ion–Ion Collisions.</u> If ion–ion collisions are taken into account [see Eq. (4.10)], the dispersion equation (3.18) is replaced by

$$2 - \frac{\omega_{ne}}{\omega\sqrt{2\pi z_i}}\left(1 - \frac{3(\pi+1)}{2\sqrt{2}}\, i\,\frac{\nu_{ii}}{\omega} z_i\right) - \frac{i\sqrt{\pi}\omega_{ne}}{|k_z|\, v_{T_e}} = 0. \tag{5.7}$$

Assuming that the growth rate is small, we find

$$\left.\begin{aligned} \operatorname{Re}\omega &= \omega_{ne}/2\sqrt{2\pi z_i},\\ \gamma &= \frac{\sqrt{\pi}\,(\omega_{ne})^2}{2\,|k_z|\, v_{T_e}\sqrt{2\pi z_i}} - \frac{3(\pi+1)}{2\sqrt{2}}\nu_{ii}z_i. \end{aligned}\right\} \tag{5.8}$$

It can be seen that the ion–ion collisions lead to damping of the oscillations and suppression of the kinetic instability discussed in §3.1.

§5.2. Collisional Plasma with Inhomogeneous Density

1. Inertial-Dissipative Instability. Suppose that there is no temperature gradient, $\nabla T = 0$, and that only the density is inhomogeneous, $\nabla n_0 \neq 0$; suppose also that the collision frequency is not too high, so that $\omega^* > \nu_{ie}$. In this case, Eq. (4.78) is the only one of the dispersion equations obtained in §4.5 that describes perturbations with $\gamma > 0$. The maximal growth rate of these perturbations is approximately

$$\gamma_{max} \simeq \omega^*. \tag{5.9}$$

It is attained when

$$\omega_s \simeq \omega^*. \tag{5.10}$$

As the ratio ω_s/ω^* increases, the growth rate decreases, and if $\omega_s \gg \omega^*$ the solution (4.78) has the form

$$\left.\begin{aligned} \mathrm{Re}\,\omega &= \omega_{ne}, \\ \gamma &\simeq \omega_{ne}^2/\omega_s. \end{aligned}\right\} \tag{5.11}$$

Equation (4.79) is valid only if $\omega_{ne}/\omega_s \ll z_i$. If this is not the case, the perturbations of this branch are described by Eq. (4.98). The latter shows that the result (5.11) is also approximately true for $\omega_{ne}/\omega_s > z_i$ until the values of k_z are reached at which the longitudinal inertia of the ions begins to play a role. Equation (4.98) is then replaced by Eq. (4.100). From the latter, we find the instability boundary

$$k_z/k_\perp < \varkappa k_\perp \rho_i^2, \tag{5.12}$$

which agrees with the boundary of the kinetic instability of a collisionless plasma [cf. Eq. (3.14)].

The above treatment refers to perturbations with sufficiently small $k_\perp \rho_i$, $z_i < \omega/\nu_{ii}$. If this condition is not satisfied, we must take into account the transverse collisional viscosity of the ions. From Eq. (4.101) we find that viscosity plays a stabilizing role.

2. Thermal-Force Instability. Assuming, as above, $\omega^* > \nu_{ie}$, let us consider perturbations with $\omega \leq \nu_{ie}$. Using (4.103), we obtain the dispersion equation for such perturbations:

$$\omega\left(\omega+4i\frac{m_e}{m_i\tau_e}\right)\left(\omega+i3.16k_z^2\frac{T}{m_e}\tau_e\right)+$$
$$+\frac{8}{3}0.71\frac{k_z^2T}{m_i}\left(\omega+4i\frac{m_e}{m_i\tau_e}-i\frac{3.16}{2}\cdot\frac{k_z^2T}{m_e}\tau_e\right)=0. \quad (5.13)$$

If $k_zv_{T_i}\simeq\nu_{ie}$, we obtain approximately

$$\mathrm{Re}\,\omega\approx 0;$$
$$\gamma\simeq 0.71\nu_{ie}\left[1-\frac{3.16}{4}\cdot k_z^2T\frac{m_i}{m_e^2}\tau_e^2\right]. \quad (5.14)$$

Instability occurs if

$$k_zv_{T_i}\leqslant\nu_{ie}. \quad (5.15)$$

Approximately

$$\gamma\simeq\nu_{ie}. \quad (5.16)$$

The growth rate, as follows from (5.14), is proportional to the parameter s = 0.71, i.e., the instability is due to the thermal force. If $k_zv_{T_i}\ll\nu_{ie}$, the growth rate decreases with decreasing k_z:

$$\gamma\simeq k_zv_{T_i}. \quad (5.17)$$

The correspondence between the instabilities considered in §5.2.1 and §5.2.2 is shown in Fig. 5.2.

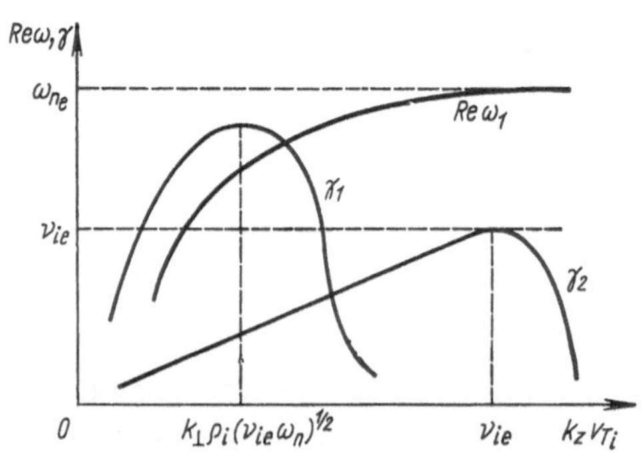

Fig. 5.2. Correspondence between the inertial-dissipative (subscript 1) and thermal-force (subscript 2) instabilities of a plasma with $\nabla n_0\neq 0$, $\omega_n\gg\nu_{ie}$.

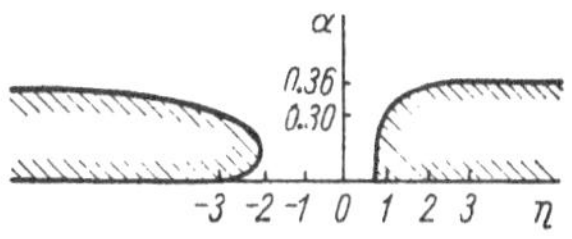

Fig. 5.3. Boundary of the hydrodynamic instability of a collisional plasma with an inhomogeneous temperature on the plane of $\alpha = (k_z v_{T_i}/\omega_T)^2$ and $\eta = \partial \ln T/\partial \ln n_0$.

§5.3. Collisional Plasma with Inhomogeneous Temperature

As in §5.2, we assume $\omega^* > \nu_{ie}$. In contrast to §5.2, we shall assume a nonvanishing temperature gradient, $\nabla T \neq 0$. Under these conditions, the following types of instability can arise.

1. Hydrodynamic Instability at $k_z \simeq \omega^*/v_{T_i}$. The perturbations of a plasma with $\nabla T \neq 0$ that are least extended along z are described by Eq. (4.89). This equation shows that when $|\eta| \gg 1$ these perturbations are unstable if

$$k_z < 0.6 \varkappa_T k_\perp \rho_i. \tag{5.18}$$

As a function of k_z, the growth rate attains a maximum at $k_{z.opt} = 0.37 \varkappa k_\perp \rho_i$ equal to

$$\gamma_{max} \simeq 0.21 \omega_T. \tag{5.19}$$

At the same time, $\mathrm{Re}\,\omega \ll \gamma$, i.e., the instability is almost aperiodic.

For finite η, the instability boundary of perturbations of the type (4.89) is shown in Fig. 5.3. These perturbations do not grow if

$$\eta > 2/3 \quad \text{or} \quad \eta < -2. \tag{5.20}$$

2. Viscous Instability. If Eq. (4.89) does not have solutions that grow, i.e., the hydrodynamic instability considered above is absent, it is necessary to take into account small dissipative effects. They are contained in Eq. (4.102). If $\nu_{ii} < (m_i/m_e)^{1/4}\omega^*$, the principal dissipative effects are the longitudinal ion viscosity

and longitudinal ion heat conduction. Retaining in (4.102) only these effects, we find that they can lead to instability if

$$k_z < \varkappa_T \rho_i (3\sqrt{3}/8)(1 - 4/9\eta^2)^{1/2}. \tag{5.21}$$

The frequency and growth rate of the unstable perturbations are approximately

$$\mathrm{Re}\,\omega \simeq \omega_T, \qquad \gamma \simeq \omega_T^2/\nu_{ii}. \tag{5.22}$$

It follows from a comparison of (5.21) and (5.20) that the viscous instability can play a role only if

$$-2 < \eta < -2/3. \tag{5.23}$$

3. Heat-Exchange Instability. In contrast to §5.3.2, we shall now assume that the collisions are fairly frequent, $\nu_{ii} > (m_i/m_e)^{1/4}\omega^*$. Under these conditions the principal dissipative effects are the heat exchange between the ions and electrons and the perturbation of the electron temperature as a result of the insufficiently high thermal conductivity. If $|\eta| \gg 1$, these effects give rise to an instability if

$$k_z < 1.1\varkappa k_\perp \rho_i. \tag{5.24}$$

The frequency and growth rate of the unstable perturbations are approximately

$$\mathrm{Re}\,\omega \simeq \omega_T, \qquad \gamma \simeq \nu_{ie}. \tag{5.25}$$

In a plasma with finite η, the heat-exchange instability arises if $\eta > 2/3$ or $\eta < 0$. Since the growth rate (5.25) is small compared with (5.19), this instability can play a role only if

$$-2 < \eta < 0. \tag{5.26}$$

4. Heat-Conduction Instability. It follows from (4.102) that with decreasing k_z the heat exchange becomes unimportant and the principal dissipative effect is the finite electron thermal conductivity. The imaginary terms of the dispersion equation are then of the same order as the real terms. As a result, Eq. (4.102) reduces to Eq. (4.86). From the latter we find that the

perturbations grow if

$$\eta < 0. \tag{5.27}$$

At $k_z v_{T_i} \simeq (\omega_T \nu_{ei})^{1/2}$, the growth rate attains a maximum given approximately by

$$\gamma_{\max} \simeq |\omega_T \omega_n|^{1/2}. \tag{5.28}$$

5. Thermal-Force Instability. Let us consider perturbations with $\omega < \nu_{ie}$. From (4.102) we obtain

$$\Gamma(\Gamma+1)(\Gamma+\theta) + 0.22\left(\eta - \frac{2}{3}\right)(\Gamma + 1 - \theta/2) = 0, \tag{5.29}$$

where

$$\Gamma \equiv -i\omega\tau_e m_i/4m_e, \qquad \theta = 3.16 k_z^2 T m_i \tau_e^2/4m_e^2.$$

Regarding 0.22 as a small parameter and taking $\theta \simeq 1$, we find the roots of (5.29). Two of them correspond to damped perturbations, $\Gamma_1 = -1$, $\Gamma_2 = -\theta$. The third root is [cf. (5.14)]

$$\Gamma_3 = 0.22\left(\frac{2}{3} - \eta\right)\left(1 - \frac{\theta}{2}\right). \tag{5.30}$$

It follows that the thermal-force instability discussed in §5.2.1 arises if

$$\eta < 2/3. \tag{5.31}$$

The instability condition and the growth rate are given by (5.15) and (5.16).

6. Inertial-Dissipative Instability. Let us now consider perturbations that are very extended along $\mathbf{B}_0$, i.e., $\omega_s/\omega^* < z_i^{-1}$. They are described by the dispersion equation (4.79). For $\eta = 0$, this equation was discussed in §5.2.1. If $\eta \neq 0$, there are no fundamental changes, so that an instability of this type takes place for arbitrary η. The growth rate and the characteristic wave number are determined by Eqs. (5.9) and (5.10) with ω_{n_e} replaced by max $(\omega_{n_e}, \omega_{T_e})$.

§5.4. Strongly Collisional Plasma

In §§5.2 and 5.3 we have assumed that the collisions are not too frequent and that the time of heat transfer between the electrons and ions is long compared with the characteristic gradient frequency, ν_{ie}/ω^*. We shall now make the opposite assumption, $\nu_{ie} > \omega^*$. We shall consider separately the cases $\nabla n_0 \neq 0$, $\nabla T = 0$ and $\nabla T \neq 0$, $\nabla n_0 \neq 0$.

1. Plasma with an Inhomogeneous Density. It can be shown that the inertial-dissipative instability discussed in §5.2.1 can develop in a strongly collisional plasma with an inhomogeneous density; for this instability is insensitive to the value of the parameter ν_{ie}/ω^*. However, it is associated with perturbations that are very extended along the magnetic field, $\omega_s \simeq \omega^*$, and cannot therefore be manifested in short systems.

In addition to this instability, the thermal-force instability described by Eq. (4.93) must play an important role when $\nu_{ie} > \omega^*$. The maximal growth rate of this instability is attained when $k_z \simeq \omega^*/v_{Tji}$. It is of the same order as the gradient frequency:

$$\gamma_{max} \simeq 0.2\,|\,\omega_n\,|. \tag{5.32}$$

At the same time $\mathrm{Re}\,\omega \simeq \gamma_{max}$. The boundary of the thermal-force instability is characterized by the inequality

$$k_z \leqslant 1.3 k_\perp \varkappa \rho_i. \tag{5.33}$$

The correspondence between the inertial-dissipative and thermal-force instabilities of a strongly collisional plasma is shown in Fig. 5.4.

2. Plasma with Inhomogeneous Temperature. If $\nabla T \neq 0$, the following types of instability can be important.

A. Thermal-Force Instability. Assuming $\omega \ll \omega_{ne}$ in (4.93), we find

$$\omega^2 = -0.71\left(\frac{2}{3} - \eta\right)(1 + 1.71\eta)^{-1} k_z^2 T/m_i. \tag{5.34}$$

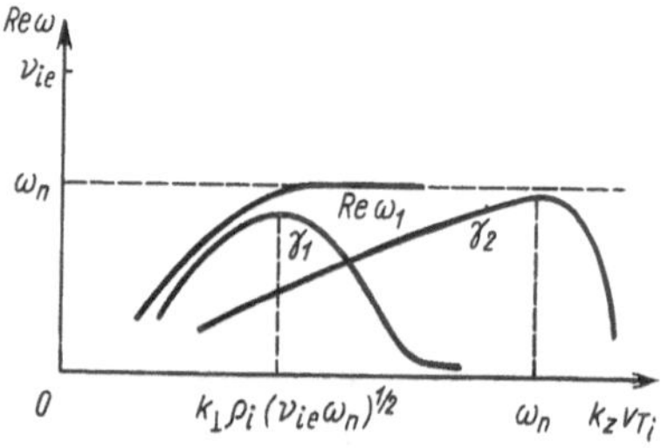

Fig. 5.4. Correspondence between the inertial-dissipative (subscript 1) and thermal-force instabilities (subscript 2) of a strongly collisional plasma (for $\omega_n < \nu_{ie}$).

It follows that the thermal-force instability arises if

$$-1/1.71 < \eta < 2/3. \tag{5.35}$$

Thus, the thermal-force instability is possible only in a plasma with a temperature gradient that is not too strong.

B. Dissipative Instability Due to the Finite Electron Thermal Conductivity. Suppose η does not satisfy the conditions (5.35). Then in (4.93) it is necessary to take into account the small terms of order ω^*/ν_{ie}. Using (4.102), we obtain the dispersion equation

$$x^3 - x^2(1+1.71\eta) - \frac{10}{3}\xi x + 1.42\xi\left(\eta - \frac{2}{3}\right) + 3.16i\,(\xi/\sigma)\,(x^2 - x - 2\xi) = 0. \tag{5.36}$$

Here

$$x \equiv \omega/\omega_{ne}, \qquad \xi \equiv k_z^2 T/m_i\omega_{ne}^2,$$
$$\sigma = 2m_e/m_i\tau_\epsilon\omega_{ne}.$$

Equating the real and imaginary parts of (5.36) to zero, we obtain the equation for the instability boundary:

$$\frac{16}{9}\xi_0^2 + \xi_0\left[2\left(\frac{2}{3} - \eta\right) - \frac{8}{9}1.71\eta\right] + 0.71\cdot 1.71\eta\left(\eta - \frac{2}{3}\right) = 0, \tag{5.37}$$

where $\xi_0 - \xi(k_{z.cr})$. In contrast to (5.35), it follows from this equation that the plasma is unstable even if $|\eta| \gg 1$. In this case the growth rate is approximately

$$\gamma \simeq \omega_T^2/\nu_{ie}, \tag{5.38}$$

and the instability boundary is determined by the inequality

$$k_z < \frac{3}{4} k_\perp \varkappa_T \rho_i. \tag{5.39}$$

At the limit of applicability of the approximation of frequent collisions, when $\omega^* \simeq \nu_{ie}$, the growth rate is $\sim\omega^*$.

§5.5. Review of Results Obtained in §§5.2-5.4

The picture we obtain for the long-wavelength instabilities of a collisional plasma is somewhat more complicated than that

TABLE 1. Classification of Types of Unstable Plasma

Class of plasma	Conditions satisfied by the plasma parameters	Type of instability	Based on the results of
I $(k_z \simeq k_{z.max},\ \gamma \simeq \gamma_{max})$	a) $S \geqslant \mu,\ \eta > 2/3,\ \eta < -2$ b) $S \leqslant \mu,\ -0.6 < \eta < 2/3$	Hydrodynamic due to temperature gradient Thermal-force	§5.3.1 § 5.4
II $(k_z \simeq k_{z.max},\ \gamma \ll \gamma_{max})$	a) $S > \mu^{1/2},\ -2 < \eta < 2/3$ b) $\mu^{1/2} > S > \mu,\ -2 < \eta \leqslant 0$ c) $S \ll \mu,\ \eta > 2/3,\ \eta < -0.6$	Viscous Heat-exchange Dissipative	§ 5.3.2 § 5.3.3 § 5.4.2B
III $(k_z \ll k_{z.max},\ \gamma \simeq \gamma_{max})$	$S > \mu,\ \eta < 0$	Due to finite electron thermal conductivity	§ 5.3.4
IV $(k_z \ll k_{z.max},\ \gamma \ll \gamma_{max})$	$S > \mu,\ 0 \leqslant \eta < 2/3$	Thermal-force	§ 5.3.5 § 5.2.2
V $(k_z \lll k_{z.max},\ \gamma \simeq \gamma_{max})$	S and η arbitrary	Inertial-dissipative	§ 5.2.1 § 5.3.6

Notation: $\eta \equiv \partial \ln T/\partial \ln n_0$, $\mu = (m_e/m_i)^{1/2}$, $S = \rho_i \lambda_e/a^2$; see §4.1. It is assumed that $k_\perp \simeq 1/a$.

for a collisionless plasma [cf. §§ 3.1 and 3.2]. This is due to the appearance in the problem of a new parameter – the collision frequency. However, there are aspects that are characteristic of a plasma with an arbitrary collision frequency:

A. The frequency and growth rate of unstable perturbations never exceed the gradient frequency:

$$\max(\gamma, \operatorname{Re}\omega) \leqslant k_y v_T \rho / a. \tag{5.40}$$

B. The longitudinal wave number of unstable perturbations is always bounded above:

$$k_z \leqslant k_y \rho_i / a. \tag{5.41}$$

The most unstable plasma is one in which there is growth of perturbations with

$$\gamma \simeq \gamma_{max} \simeq k_y v_{T_i} \rho_i / a; \tag{5.42}$$

$$k_z \simeq k_{z.max} \simeq k_y \rho_i / a. \tag{5.43}$$

Not every plasma belongs to this class, but only one for which the steady-state parameters η and S lie in definite intervals. These intervals have been already determined. They are given in Table 1, which summarizes the results of the foregoing subsections that apply to perturbations with $k_\perp \simeq 1/a$. The most unstable plasmas belong to class I in Table I.

A plasma in which only instabilities with $\gamma \ll \gamma_{max}$ can develop when $k_z \simeq k_{z.max}$ belongs to class II (see Table 1). A plasma with parameters η and S that do not lie in the limits corresponding to the classes I and II can be unstable provided its longitudinal dimension is sufficient great, $k_z \ll k_{z.max}$. A fairly long plasma in which instabilities can arise with $\gamma \simeq \gamma_{max}$ belongs to class III; with $\gamma \ll \gamma_{max}$, to class IV.

It is assumed that the ratio $k_z/k_{z.max}$ corresponding to the plasma classes III and IV is nevertheless not as small as ρ_i/a. Plasmas for which this is not the case, i.e., very long plasmas, $k_z/k_{z.max} \leq \rho_i/a$, belong to class V. Very long plasmas are unstable with $\gamma \simeq \gamma_{max}$ for arbitary S and η.

Table I also indicates which type of perturbation is responsible for the instability of a given class (or subclass) of plasma and gives references to the corresponding formulas.

§5.6. Collisional Plasma with a Current

A homogeneous collisional plasma with a current is unstable if the current velocity V_0 is a few times greater than the velocity of ion-acoustic oscillations (see §6.3 in Volume 1). Here we shall assume that $V_0 \ll v_{T_i} \lesssim (T_e/m_i)^{1/2}$ and take into account the plasma inhomogeneity. Our aim is to establish which additional instabilities (apart from those considered in §§5.2 and 5.3) can arise under these conditions (cf. the similar statement of the problem in §3.4).

1. Inertial Instability. Like the small terms of order ε^3, the terms with V_0 can play an appreciable role only if $\cos\theta$ is relatively small. We shall therefore assume $\omega \gg k_z V_0$. It is then sufficient to include the terms with V_0 only in the electron equations of continuity and motion, i.e., in the first and third equations of the system (4.72). These equations now have the form

$$\begin{gathered}
-\mathrm{i}(\omega-k_zV_0)\,n'+\mathbf{V}_E\nabla n_0+\mathrm{i}k_zV'_{ze}n_0=0,\\
0=e_en_0E_z-\mathrm{i}k_zT_0n'-1.71\,\mathrm{i}k_zn_0T'_e-\\
-\nu_{ei}n_0m_e(V'_{ze}-V'_{zi})-\nu_{ei}n'm_eV_0+\frac{2}{3}\nu_{ei}\frac{T'_e}{T_0}n_0m_eV_0,\\
(\nu_{ei}\equiv 0.51/\tau_{e0}).
\end{gathered}\tag{5.44}$$

The new term with T'_e is due to the temperature dependence of $\tau_e (\tau_e \simeq T_e^{3/2})$. We shall neglect the weak dependence of the Coulomb logarithm on the density. In linearizing the equation of motion, we have taken into account the fact that in the steady state

$$e_eE_z^{(0)}-m_e\nu_eV_0=0.\tag{5.45}$$

Using (5.44), we find that (4.74) is replaced by

$$-\mathrm{i}\frac{\cos^2\theta}{\nu_e}\left(e_e\psi+\frac{n'}{n_0}T_0+1.71T'_e\right)+\frac{3}{2}\cdot\frac{m_e}{T_0}\cdot\frac{k_zV_0}{k_\perp^2}T'_e+\frac{m_e}{m_i}\frac{e_i(\omega-\omega^*_{p_i})}{\omega^2_{B_i}}\psi=0.\tag{5.46}$$

Setting $\omega_s \simeq \omega^*$ in this equation, we find that the term with V_0 is important if

$$\frac{V_0}{v_{T_i}}>k_\perp\rho_i\left(\frac{\omega^*}{\nu_{ie}}\right)^{1/2}.\tag{5.47}$$

However, this refers only to a plasma with $\nabla T \neq 0$, since, in accordance with (4.78), it is precisely in this case that $T_e' \neq 0$.

We shall now assume that condition (5.47) is satisfied and that the plasma temperature is inhomogeneous, $\nabla T_0 \neq 0$. If $\omega \gg \omega^*$, Eq. (4.79) is replaced by the dispersion equation [we use Eqs. (4.77), (4.78), and (5.46)]

$$1 - i\frac{3}{2}\cdot\frac{\omega_{T_e}}{\omega}\cdot\frac{m_e V_0}{k_z T_0}\nu_{ei} - i\frac{m_e}{m_i}\cdot\frac{\omega\nu_{ei}}{\omega_{B_i}^2\cos^2\theta} = 0. \quad (5.48)$$

The maximum of the growth rate is attained at

$$\cos\theta \simeq \frac{(\varkappa_T V_0 \nu_{ie}^2)^{1/3}}{\omega_{B_i}}. \quad (5.49)$$

Approximately

$$\gamma_{\max} \simeq (\nu_{ie}\varkappa_T^2 V_0^2)^{1/3}. \quad (5.50)$$

Using the condition (5.47), we see that $\gamma_{\max} > \omega_T$. Thus, the passage of a longitudinal current in an inhomogeneously heated collisional plasma leads to the excitation of perturbations whose growth rate is large compared with ω^* for small $k_\perp$.

The expressions (5.49) and (5.50) depend essentially on the transverse ion inertia [the last term on the left-hand side of (5.48)]. For this reason, this instability is frequently called the inertial current-convective instability.

2. Noninertial Instability. As $\cos\theta$ increases, the growth rate of the perturbations described by (5.48) decreases,

$$\gamma = \frac{3}{2}\cdot\frac{\omega_{T_e} m_e V_0 \nu_{ei}}{k_z T_0}. \quad (5.51)$$

This is the noninertial current-convective instability.

To find the limits of applicability of Eq. (5.51) at large k_z, we must augment Eq. (5.48) with the omitted terms of order ω^*/ω and also take into account the longitudinal electron thermal conductivity. In (5.46) we neglect the transverse ion inertia (the last term on the left-hand side), and we take the expression for T_e' from (4.85) and not (4.78), assuming at the same time $\omega \gg \nu_{ie}$. Then instead of (5.48) we obtain

$$1 - \frac{\omega_{ne}}{\omega} - \frac{\omega_{T_e}\left(1.71 + \frac{3}{2}i\frac{m_e V_0 \nu_{ei}}{k_z T_0}\right)}{\omega + 2.1 i k_z^2 T_0/\nu_{ei} m_e} = 0. \quad (5.52)$$

From this we find the conditions at the instability boundary:

$$\left.\begin{aligned}\omega_{\mathrm{cr}} &= \omega_{ne} + 1.71\,\omega_{Te},\\ k_{z,\mathrm{cr}} &= \left(\frac{\nu_{ei}^2 \omega_{\mathrm{cr}} V_0 m_e^2}{2.8 T_0^2}\right)^{1/3}.\end{aligned}\right\} \tag{5.53}$$

The second equation entails the approximate relation [cf. (5.49)]

$$(\cos\theta)_{\mathrm{cr}} \simeq \frac{(\varkappa_T V_0 \nu_{ie}^2)^{1/3}}{\omega_{B_i}} \left(\frac{1}{k_\perp \rho_i}\right)^{2/3}. \tag{5.54}$$

Bibliography

1. B. B. Kadomtsev, Zh. Tekh. Fiz., 31:1209 (1961) [Sov. Phys. – Tech. Phys., 6:882 (1962)]. The instability of an inhomogeneous plasma with a current (§5.6) is investigated.
2. A. A. Galeev, V. N. Oraevskii, and R. Z. Sagdeev, Zh. Eksp. Teor. Fiz., 44:903 (1963) [Sov. Phys. – JETP, 17:615 (1963)]. The heat conduction instability of a plasma with $\partial \ln T/\partial \ln n_0 < 0$ (§5.3) is found.
3. S. S. Moiseev and R. Z. Sagdeev, Zh. Eksp. Teor. Fiz., 44:763 (1963) [Sov. Phys. – JETP, 17:515 (1963)].
4. S. S. Moiseev and R. Z. Sagdeev, Zh. Tekh. Fiz., 34:249 (1964) [Sov. Phys. – Tech. Phys., 9:196 (1964)].
5. F. F. Chen, Phys. Fluids, 7:949 (1964). The papers [3-5] demonstrate the possibility of development of an instability in a plasma without a current but with an inhomogeneous density (§5.2.1).
6. G. M. Zaslavskii and S. S. Moiseev, Zh. Tekh. Fiz., 34:410 (1964) [Sov. Phys. – Tech., 9:324 (1964)]. A study is made of the effect on the stability of a plasma of transverse viscosity, heat conduction, a finite conductivity, and a longitudinal current. It is shown, in particular, that transverse ion viscosity has a stabilizing effect (§§5.2 and 5.3).
7. A. A. Galeev, S. S. Moiseev, and R. Z. Sagdeev, At. Energ., 15:451 (1963).
8. B. B. Kadomtsev, Plasma Turbulence, Academic Press, London (1965). Reviews of results on the instabilities of a collisional plasma are given in [7, 8].
9. A. B. Mikhailovskii and O. P. Pogutse, Dokl. Akad. Nauk SSSR, 156:64 (1964) [Sov. Phys. – Doklady, 9:379 (1964)].
10. A. B. Mikhailovskii and O. P. Pogutse, Zh. Tekh. Fiz., 36:205 (1966) [Sov. Phys. – Tech. Phys., 11:153 (1966)]. In [9-10] a model collision integral is used to study the instabilities of a plasma with $\nabla n_0 \neq 0$, (§5.2). In [10] a study is also made of the instability of nonelectrostatic perturbations.
11. F. F. Chen, Phys. Fluids, 8:912 (1965). The instability mechanism of a plasma with an inhomogeneous density is established.
12. F. F. Chen, Phys. Fluids 8:1323 (1965). The instability of a plasma with $\nabla n_0 \neq 0$ (§5.2) is discussed.

13. F. C. Hoh, Phys. Fluids, 7:956 (1964). The current instability (§ 5.6) is discussed. It is shown that the maximal growth rate is $\sim \nu_e^{1/3}$.
14. A. A. Rukhadze and V. P. Silin, Dokl. Akad. Nauk SSSR, 169:558 (1966) [Sov. Phys. – Doklady, 11:606 (1967)].
15. L. S. Bogdankevich and A. A. Rukhadze, Zh. Eksp. Teor. Fiz., 51:628 (1966) [Sov. Phys. – JETP, 24:418 (1967)].
16. L. S. Bogdankevich, B. Milich, and A. A. Rukhadze, Zh. Tekh. Fiz., 37:1936 (1967) [Sov. Phys. – Tech. Phys., 12:1424 (1968)].
17. A. A. Rukhadze and V. P. Silin, Usp. Fiz. Nauk, 96:87 (1968) [Sov. Phys. – Uspekhi, 11:659 (1969)]. In [14-17] the influence of weak collisions on the instability is investigated. All the results given in § 5.1 are obtained in these papers.
18. S. S. Moiseev, ZhETF Pis. Red., 4:81 (1966) [JETP Letters, 4:55 (1966)]. The thermal-force instability of a strongly collisional plasma (§ 5.4) is found. It is shown that if collisions are frequent the temperature perturbations must be taken into account even if $\nabla T_0 = 0$ and that the longitudinal motion of the ions does not always play a stabilizing role.
19. I. S. Baikov, ZhETF Pis. Red., 4:299 (1966) [JETP Letters, 4:201 (1966)].
20. J. D. Jukes, Phys. Fluids, 10:1107 (1967).
21. A. B. Mikhailovskii, Zh. Eksp. Teor. Fiz., 52:1251 (1967) [Sov. Phys. – JETP, 25:831 (1967)]. The papers [19-21] are devoted to instabilities of a plasma with an inhomogeneous temperature (§§ 5.3 and 5.4).
22. H. W. Hendel et al., Phys. Rev. Lett., 18:439 (1967).
23. H. W. Hendel, T. K. Chu, and P. A. Politzer, Phys. Fluids, 11:2426 (1968). In [22, 23] the instability of a plasma with $\nabla n_0 \neq 0$ is discussed with allowance for the effect of transverse ion viscosity (§ 5.2).
24. F. F. Chen, Phys. Fluids, 8:752 (1965).
25. F. F. Chen, J. Nucl. Energy, Pt. C. 7:399 (1965).
26. B. B. Kadomtsev, Proceedings of the 7th Intern. Conf. on Phen. in Ioniz. Gases, 2:610 (1966).
27. A. M. Levine and A. F. Kuckes, Phys. Fluids, 9:2263 (1966).
28. A. V. Shut'ko, Zh. Tekh. Fiz., 38:1431 (1968) [Sov. Phys. – Tech. Phys., 13:1174 (1969)]. In [24-28] the effect of end electrodes and processes near the electrodes on the instability of a plasma are discussed.

Chapter 6

Plasmas in a Gravitational Field

§6.1. Flute Instability of a Dense Plasma

A gravitational field is introduced into the theory of the instabilities of an inhomogeneous plasma to simulate the effects of curvature of the magnetic lines of force. Curvature of the lines of force leads to drift of the particles at right angles to the magnetic field with a velocity of the order $V_{dr} \simeq v_T \rho / R$, where R is the radius of curvature. Because of the presence of a force of gravity, the particles (ions) drift with a velocity $V_{dr} \simeq g/\omega_{Bi}$. These two velocities are of the same order if $g \simeq T/m_i R$. The use of the last relationship enables one to obtain simple estimates of the effect of curvature on instabilities without considering the actual geometry of the magnetic field if the instability problem is solved with allowance for the force of gravity. We shall have this idea in mind in the following exposition (§§6.1-6.6) and it will be justified subsequently (see, for example, Chapter 11).

The introduction of a gravitational field is also convenient for simulating various other effects in a real plasma, for example, the effect on the plasma stability of an averaged high-frequency force. This assertion will be justified in §6.7.

In the present section we shall show that a force of gravity directed against a density gradient leads to a growth of perturbations with $k_z = 0$; this is the flute instability.

Suppose that the plasma is situated in a magnetic field $\mathbf{B}_0 \parallel z$ and a gravitational field $\mathbf{g} \parallel x$. Under the influence of these fields the particles will move at right angles to $\mathbf{B}_0$ and $\mathbf{g}$ with the velocity

$$\mathbf{V}_g = -\frac{g}{\omega_B}\mathbf{e}_y. \tag{6.1}$$

In contrast to the velocity $\mathbf{V}_E$ of electric drift, this velocity is not the same for the ions and the electrons. It depends on the mass and the charge of the particles and $|\mathbf{V}_{gi}| \gg |\mathbf{V}_{ge}|$. Therefore, if a force g is present, there is a relative motion of the ions and electrons at right angles to the magnetic field, and this may be the cause of new types of instability. We now proceed to their investigation.

We shall be interested in low-frequency perturbations, $\omega \ll \omega_{B_i}$, with $k_z = 0$. In investigating them, we shall neglect effects of order ω^*/ω and the fact that the perturbed electric field is not electrostatic. We find the perturbed density of each species of particles by means of the hydrodynamic equations of a cold plasma (as in §1.1). These equations have the form

$$\left.\begin{aligned} &\frac{\partial n}{\partial t} + \operatorname{div}(n\mathbf{V}) = 0 \quad \text{for the ions and electrons} \\ &\frac{d\mathbf{V}^{(i)}}{dt} = \frac{e_i}{m_i}\mathbf{E} + [\mathbf{V}^{(i)}, \boldsymbol{\omega}_{B_i}] + \mathbf{g}; \\ &0 = \frac{e_e}{m_e}\mathbf{E} + [\mathbf{V}^{(e)}, \boldsymbol{\omega}_{B_e}]. \end{aligned}\right\} \tag{6.2}$$

From this we find that the expression for the perturbed density of the ion charge differs from (1.11) with $k_z = 0$ by the substitution $\omega \to \omega + k_y g/\omega_B$. Substituting this result into Poisson's equation, we arrive at the following equation for the perturbed potential ψ:

$$\frac{\partial}{\partial x}\left(\varepsilon_\perp \frac{\partial \psi}{\partial x}\right) - k_y^2 \varepsilon_\perp \psi - \frac{k_y}{\omega_{B_i}} \psi \frac{\partial \omega_{p_i}^2}{\partial x}\left(\frac{1}{\omega + \frac{k_y g}{\omega_{B_i}}} - \frac{1}{\omega}\right) = 0. \tag{6.3}$$

Here

$$\varepsilon_\perp \equiv 1 + (\omega_{p_i}/\omega_{B_i})^2.$$

The physical meaning of the individual terms of this equation is as follows. The terms containing $\varepsilon_\perp$ are due to the fact that the perturbations do not satisfy electrical neutrality (the unity in $\varepsilon_\perp$) and the transverse inertia of the ions [the term with $(\omega_{p_i}/\omega_{B_i})^2$]. The gradient terms are due to the perturbed convection ot the ions and electrons (see §1.1). We have taken into account the relation $\omega_{p_i}^2/\omega_{B_i} = -\omega_{p_e}^2/\omega_{B_e}$. The convection of the particles occurs in a

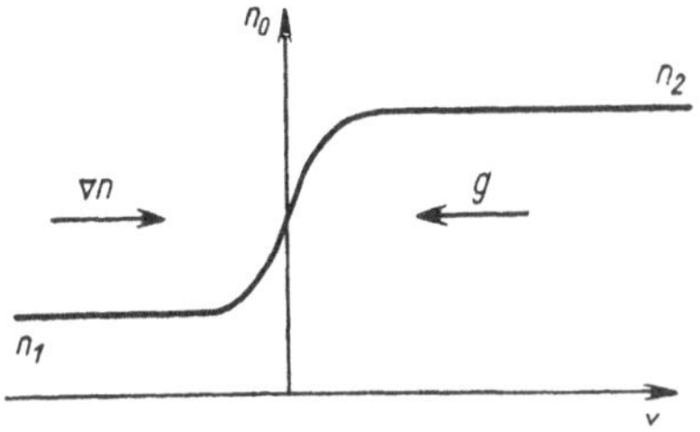

Fig. 6.1. Plasma with an abrupt boundary.

field of effective frequency $\omega_{eff} = \omega + kg_y/\omega_B$. The effect of the gravitational force on the motion of the electrons is ignored.

In the present section we shall assume that the plasma is fairly dense, $\omega_{p_i} \gg \omega_{B_i}$. Then, in accordance with (6.3), the basic equation is

$$\frac{\partial}{\partial x}\left(n_0 \frac{\partial \psi}{\partial x}\right) - k_y^2 \psi n_0 + \frac{k_y^2 g}{\omega^2} \psi \frac{\partial n_0}{\partial x} = 0. \tag{6.4}$$

We assume $\omega \gg k_y g/\omega_{B_i}$.

1. Instability of a Plasma with an Abrupt Boundary. Suppose that the plasma density changes abruptly over a distance $a \ll 1/k_y$. Then the solution of Eq. (6.4) can be found in the approximation of surface waves (see §1.7 of Volume 1). To be specific we shall assume that the density varies from the value n_1 to $n_2 \gg n_1$ (Fig. 6.1). We then have the dispersion equation

$$\omega^2 - |k_y| g = 0. \tag{6.5}$$

This equation shows that if the gravitational force is directed against the density gradient, i.e., $g < 0$ (see Fig. 6.1), the plasma boundary is unstable with the growth rate

$$\gamma = \sqrt{|k_y g|}. \tag{6.6}$$

The perturbations have the form of flutes oriented along the magnetic field, and this instability of a plasma in a gravitational field is therefore called the flute instability. We shall use this terminology in what follows.

We shall also mention some other names given to the flute instability. Since the instability mechanism is due to the convection of charges (see §1.1), it is also called a convective instability. The flute instability of a plasma with an abrupt boundary is occasionally called the Kruskal–Schwarzschild instability [Kruskal and Schwarzschild first obtained Eq. (6.6) for a plasma]. However, the possibility of the excitation of perturbations with the growth rate (6.6) was already known in the theory of an ordinary liquid (Rayleigh and Taylor). In this connection the flute instability of a plasma in a gravitational field is sometimes called a Rayleigh–Taylor instability. The same instability in the case of a plasma with a smooth boundary is frequently called the interchange instability. This term will be explained a little later.

2. Small-Scale Perturbations of a Plasma with a Smooth Boundary. In the quasiclassical approximation, Eq. (6.4) yields the local dispersion equation

$$\omega^2 - g\varkappa\left(\frac{k_y}{k_\perp}\right)^2 = 0. \tag{6.7}$$

As in the approximation of an abrupt boundary (§6.1.1), we find that the plasma is unstable if the gravitational force is directed against the density gradient, $g\varkappa < 0$. The growth rate is

$$\gamma = \sqrt{|g\varkappa|}\,|k_y|/k_\perp. \tag{6.8}$$

The growth rate predicted by (6.8) and (6.6) is shown schematically as a function of k_y in Fig. 6.2.

3. Investigation of the Flute Instability by the Energy Method. It follows approximately from the equations of motion (6.2) that the electrons and ions move at right angles to the magnetic field with the same perturbed velocity

$$\mathbf{V}' = c\,[\mathbf{E}, \mathbf{e}_z]/B_0. \tag{6.9}$$

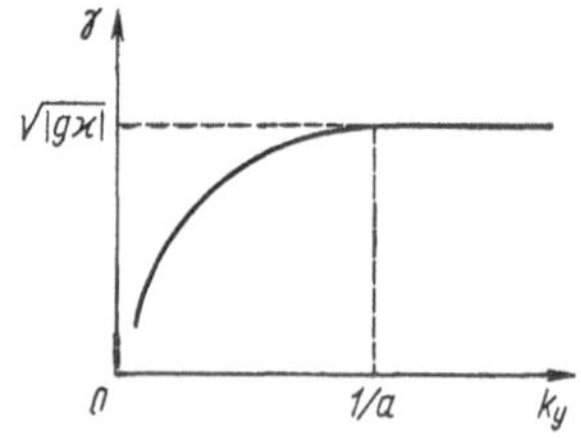

Fig. 6.2. Qualitative dependence of the growth rate of the flute instability on the wave number k_y.

Using this relation and the electrical neutrality condition, we can represent the linearized system of equations (6.2) in the form

$$\frac{\partial \rho'}{\partial t} + \operatorname{div}(\rho_0 \mathbf{V}') = 0,$$
$$\rho_0 \frac{\partial \mathbf{V}'}{\partial t} = \rho' \mathbf{g}, \tag{6.10}$$

where $\rho' \equiv m_i n'$ and $\rho_0 = m_i n_0$ are the perturbed and equilibrium mass densities. Replacing the perturbed velocity $\mathbf{V}'$ by the displacement ξ of the plasma defined by the equation

$$\mathbf{V}' = \partial \xi / \partial t, \tag{6.11}$$

we obtain from the first equation of (6.10)

$$\rho' = -\operatorname{div}(\rho_0 \xi). \tag{6.12}$$

Substituting (6.11) and (6.12) into the second equation of (6.10), we arrive at the following equation for ξ:

$$\rho_0 \frac{\partial^2 \xi}{\partial t^2} = -\hat{K}\xi, \tag{6.13}$$

where

$$\hat{K}\xi = \mathbf{g} \operatorname{div}(\rho_0 \xi). \tag{6.14}$$

It is important for us to know whether the perturbations described by this second-order differential equation grow in time. It follows from the theory of differential equations that in the case of a self-adjoint operator $\hat{K}$ this depends on the sign of the potential energy of the perturbations, which, by definition, is equal to

$$W_{pot} = \frac{1}{2} \int \xi \hat{K} \xi \, dr. \tag{6.15}$$

Here the integration is extended over the volume of the plasma with appropriate boundary conditions. Using (6.13) with (6.14) and assuming perturbations with a time dependence $\exp(-i\omega t)$, we obtain

$$\omega^2 = W_{pot} \Big/ \int \rho_0 |\xi|^2 \, dr. \tag{6.16}$$

There is therefore stability if for all ξ

$$W_{\text{pot}} > 0. \tag{6.17}$$

Thus, a necessary and sufficient condition for a plasma to be stable against the flute instability can be obtained by first proving that the operator $\hat{K}$ is self-adjoint and then investigating the sign of the potential energy W_{pot}.

The operator $\hat{K}$ is self-adjoint if for all ξ and η

$$\int \eta \hat{K} \xi \, d\mathbf{r} = \int \xi \hat{K} \eta \, d\mathbf{r}. \tag{6.18}$$

If $\hat{K}$ has the form (6.14),

$$\int \eta \hat{K} \xi \, d\mathbf{r} = - \int \eta \mathbf{g} \operatorname{div} (\rho_0 \xi) \, d\mathbf{r} = - \int \{(\eta \mathbf{g}) \, \xi \nabla \rho_0 + \eta \mathbf{g} \rho_0 \operatorname{div} \xi\} \, d\mathbf{r}. \tag{6.19}$$

For the irrotational plasma perturbations considered here, $\operatorname{div} \xi = 0$ in a rectilinear magnetic field. Noting also that $\mathbf{g} \parallel \nabla \rho_0 \parallel \mathbf{x}$, we obtain

$$\int \eta \hat{K} \xi \, d\mathbf{r} = - \int \eta_x \xi_x (\mathbf{g} \nabla \rho_0) \, d\mathbf{r}. \tag{6.20}$$

From this we immediately obtain the self-adjointness condition (6.18).

In this example the potential energy of the perturbations is, in accordance with (6.15),

$$W_{\text{pot}} = \frac{1}{2} \int (\xi \mathbf{g}) (\xi \nabla \rho_0) \, d\mathbf{r}. \tag{6.21}$$

Therefore, the necessary and sufficient condition of stability that follows from (6.17) and (6.21) is

$$\mathbf{g} \nabla \rho_0 > 0. \tag{6.22}$$

This agrees with the results of §§6.1.1 and 6.1.2, in which a condition of the type (6.22) was obtained by the dispersion equation method. It should be noted that in these subsections we had to consider the limiting cases of large- or small-scale perturbations. In the present derivation of the condition (6.22), these additional assumptions were not required.

The energy method also enables one to estimate the order of magnitude of the square of the frequency of the perturbations. In particular, if $\hat{K}$ has the form (6.14), then Eq. (6.16) yields

$$\omega^2 \simeq g\varkappa, \tag{6.23}$$

which agrees with the estimate for the frequency obtain in §6.1.2.

§6.2. Effect of a Finite Ion Larmor Radius

In §6.1 we have ignored terms of order ω^*/ω. If $\omega \simeq \sqrt{g\varkappa}$ this corresponds to the approximation

$$\omega^{*2} \ll g\varkappa. \tag{6.24}$$

If we recall that the gravitational field is introduced to simulate the effect of curvature of the lines of force and we set $g \simeq v_{T_i}^2/R$ (R is the radius of curvature), the inequality (6.24) for $k_\perp a \approx 1$ entails

$$(\rho_i/a)^2 \ll a/R. \tag{6.25}$$

A plasma whose parameters satisfy this relation is generally known in the theory of the flute instability as a plasma with a vanishing ion Larmor radius. If this condition is not satisfied, one speaks of a plasma with a finite ion Larmor radius. In what follows we shall consider the flute instability of such a plasma, assuming $\beta \ll 1$, $\omega_{p_i}^2 \gg \omega_{B_i}^2$. As in §6.1, we shall be concerned only with perturbations with $k_z = 0$.

1. Small-Scale Perturbations, $k_\perp a \gg 1$.

If allowance is made for the finite temperature of the particles but the gravitational field is ignored, the dispersion equation of small-scale perturbations with $k_z = 0$ and $\omega \ll \omega_{B_i}$ can be obtain by using the results of §1.7. If the gravitational field is taken into account, the substitution $\omega \to \omega + \frac{gk_y}{\omega_B}$ must be made in the ion terms. Thus, if the plasma temperature is included,

$$1 + \frac{1}{(kd_i)^2}\left[1 - I_0(z_i)\, e^{-z_i}\right] + \frac{\varkappa \omega_{p_i}^2 k_y}{k^2 \omega_{B_i}} \left(\frac{I_0(z_i)\, e^{-z_i}}{\omega + \frac{gk_y}{\omega_B}} - \frac{1}{\omega} \right) = 0. \tag{6.26}$$

Here it is also assumed that $\nabla T = 0$.

In (6.26) we assume $z_i \ll 1$, $\omega_{p_i}^2 \gg \omega_{B_i}^2$. It then yields

$$\omega^2 - \omega\omega_{ni} - g\varkappa = 0. \tag{6.27}$$

The same result follows if the plasma is described by the equations of the collisional approximation (§4.3) if the collisional viscosity (which is small as $(\nu_{ii}/\omega)\, z_i$) is ignored (the role of collisional viscosity is discussed in §6.6).

The term of Eq. (6.27) that is linear in ω is due to the inclusion in (6.26) of the gradient terms of order z. In accordance with §1.7, the latter are due to two effects – the perturbed convection of the charge density at right angles to **E'** and $\mathbf{B}_0$ and the "beam" transport of the perturbed charge density in the direction of **E'** as a result of the steady-state Larmor currents. The perturbed convection is also responsible for terms of zeroth order in z_i which cancel to within terms of order g (cf. §6.1.1) on the summation over the ions and electrons. Using the results of §1.7, we can represent the genealogy of the term ω_{ni} in (6.7) as follows:

Total contribution to the dispersion equation		Perturbed convection of order $(k_\perp \rho)^2$		"Beam" transport
ω_{ni}	=	$\frac{3}{2}\omega_{ni}$	−	$\frac{1}{2}\omega_{ni}$.

(6.28)

The term in (6.27) that is linear in ω can also be obtained on the basis of the following model representation. One includes only convection and assumes that, due to the finite value of the ion Larmor radius, the ions are subjected to a field $\mathrm{E}'\left(\mathrm{r}+\frac{[\mathrm{v},\,\mathrm{e}_z]}{\omega_B}\right)$ that differs from $\mathrm{E}'(\mathrm{r})$. Then the effective field averaged over the Larmor oscillations and the equilibrium distribution function is

$$\left\langle \mathrm{E}'\left(\mathrm{r}+\frac{[\mathrm{v},\,\mathrm{e}_z]}{\omega_B}\right)\right\rangle = \mathrm{E}'(\mathrm{r})\left(1-\frac{k^2T_i}{m_i\omega_{B_i}^2}\right). \tag{6.29}$$

The convective contribution of the ions to Eq. (6.2) (in the quasiclassical approximation) will also differ by the factor $1-\frac{k^2T_i}{m_i\omega_{B_i}^2}$. In this nonrigorous approach, we obtain (6.27) from the modified equation (6.2).

From (6.27) we find that the flute instability does not develop if

$$\omega_{ni}^2 > 4\,|g\varkappa|. \tag{6.30}$$

This condition describes the effect of stabilization of the flute instability due to the finite value of the ion Larmor radius. The meaning of this terminology is clear from the foregoing comments – the stabilization is due to effects of order z_i. The parameter ρ_i/a also appears explicitly in the following form of expression of (6.30) for perturbations with $k_y \simeq 1/a$ [cf. (6.25)]:

$$\left(\frac{\rho_i}{a}\right)^2 > \frac{\gamma_0}{\omega_{B_i}}, \tag{6.31}$$

where γ_0 is the growth rate in the limit $z \to 0$ [Eq. (6.7)].

2. Large-Scale Perturbations with $k_z = 0$. A. Plasma with Planar Symmetry. In Eq. (6.3) we take into account terms with ω_{ni}/ω. This can be done by proceeding from Eqs. (4.34) and (4.44). If $k_z = 0$, these equations are applicable for the description of both collisional and collisionless plasmas. We introduce a force of gravity in (4.34) and (4.44) by making the substitution $V_E \to V_E + V_g$ in the ion equations; here, V_g is determined by Eq. (6.1). As a result

$$\frac{\partial}{\partial x}\left[n_0\left(1-\frac{\omega_{ni}}{\omega}\right)\frac{\partial\psi}{\partial x}\right]-k_{\parallel}^2 n_0\left(1-\frac{\omega_{ni}}{\omega}-\frac{g\varkappa}{\omega^2}\right)\psi=0 \tag{6.32}$$

(we assume $\nabla T_i = 0$).

In the approximation of a thin transition layer ($k_y a \ll 1$), we obtain from this equation the dispersion equation [cf. (6.5)]:

$$\omega^2 - k_y^2\rho_i^2\omega\omega_{B_i} - g\,|k_y| = 0. \tag{6.33}$$

It can be seen from (6.33) that for very small g, when

$$(k_y\rho_i)^2 > 2\,(|gk_y|)^{1/2}/\omega_{B_i}, \tag{6.34}$$

the flute instability of the type (6.5) does not develop. At the limit of applicability of the approximation of a thin transition layer ($k_y \simeq 1/a$) the stabilization condition (6.34) is qualitatively the same as (6.28).

Thus, in the case of a plasma of planar symmetry the stabilization effect is operative in both small- and large-scale flute perturbations.

B. Dense Plasma with Cylindrical Symmetry. We shall assume that the plasma is cylindrically symmetric, $n_0 = n_0(r)$, and that the force g has only a radial component, $\mathbf{g} = (g, 0, 0)$. Using (4.34) and (4.44), we now obtain instead of (6.32)

$$\nabla_\perp \left[\nabla_\perp \psi \left(n_0 - \frac{T_i}{m_i \omega_{B_i} \omega} \frac{l}{r} \cdot \frac{\partial n_0}{\partial r} \right) \right] +$$

$$+ \psi \left[\frac{T_i}{m_i \omega_{B_i} \omega} \cdot \frac{l}{r} \cdot \frac{\partial}{\partial r} \left(\frac{1}{r} \cdot \frac{\partial n_0}{\partial r} \right) + \frac{l^2}{r^2} \frac{\partial n_0}{\partial r} \frac{g}{\omega^2} \right] = 0. \qquad (6.35)$$

Here, l is the number of the azimuthal mode ($\psi \sim e^{il\varphi}$).

For $l = 2, 3, \ldots$, Eq. (6.35) yields qualitatively the same results as in the case of a plasma of planar symmetry. This can be seen as follows. Integrate Eq. (6.35) with weight ψ^* over the plasma cross section assuming that $n_0 = 0$ at the integration boundary. The result is

$$A\omega^2 + B\omega + C = 0, \qquad (6.36)$$

where

$$\left.\begin{aligned} A &= \int n_0 |\nabla\psi|^2 \, dr, \\ B &= -\frac{lT_i}{m_i \omega_{B_i}} \int dr \frac{1}{r} \cdot \frac{\partial n_0}{\partial r} \left\{ (l^2 - 1) \frac{|\psi|^2}{r^2} + \right. \\ &\qquad \left. + \left| \frac{\psi}{r} - \frac{\partial \psi}{\partial r} \right|^2 \right\}, \\ C &= -l^2 \int |\psi|^2 \frac{g}{r^2} \cdot \frac{\partial n_0}{\partial r} \, dr. \end{aligned}\right\} \qquad (6.37)$$

It can be seen that, provided $l \neq 1$, B does not vanish for any dependence $\psi(r)$, and the stabilization condition (6.30) then remains true in order of magnitude.

Perturbations with $l = 1$ present a special case. Equation (6.35) in this case can be written in the form

$$\frac{d}{dr} \left(rG \frac{d\psi}{dr} \right) - \frac{\psi}{r} \cdot \frac{d}{dr} (rG) = -\psi \frac{dn_0}{dr} \left(1 + \frac{g}{r\omega^2} \right), \qquad (6.38)$$

where

$$G = n_0 - \frac{T_i}{m_i \omega_{B_i} \omega} \frac{1}{r} \cdot \frac{\partial n_0}{\partial r}.$$

If $\omega \ll \omega_{ni}$, Eq. (6.38) belongs to the class of equations with a large parameter of the highest derivative. In the zeroth approximation in this parameter the solution of (6.38) is

$$\psi_0 = r. \tag{6.39}$$

This function makes the left-hand side of (6.37) vanish for any form of $n_0(r)$. If the metallic casing is not situated too near the plasma, the solution of (6.38) can be fitted with a decreasing solution outside the plasma. Taking this into account and writing down the condition that the right-hand side of (6.38) be orthogonal to the solution of the zeroth approximation $\psi_0(r)$, we obtain

$$\omega^2 = \frac{\int_0^a r\,dr\,\frac{\partial n_0}{\partial r}g}{2\int_0^a n_0 r\,dr}. \tag{6.40}$$

It can be seen that perturbations with $l = 1$ and $\psi(r) \approx r$ are not stabilized even for large ω_{ni}/ω [this also follows from (6.36) and (6.37)]. This result can be readily understood by noting that a potential of the form $\psi = re^{i\varphi}$ corresponds to a coordinate-independent electric field:

$$\mathbf{E}' = \text{const}. \tag{6.41}$$

In such a field the ions move in the same way as in the limit $\rho_i \to 0$, and the fact that ρ_i/a is finite is not significant.

Perturbations with $l = 1$ are stabilized if the plasma is bounded by a conducting vessel. In this case, the eigenfunctions differ appreciably from $\psi(r)$ and, then, in accordance with (6.37), $B \neq 0$.

C. Gaussian Density Distribution and $g \sim r$. If the density decreases along the radius in accordance with the law $n_0 = \bar{n}e^{-(r/a)^2}$, and the gravitational force that simulates curvature of the lines of force is proportional to the radius, $g = g_0 r/a$, Eq. (6.35) can be reduced to

$$\psi'' + \frac{\psi'}{r} - \frac{l^2}{r^2}\psi - \frac{2}{a^2}(r\psi' - \nu\psi) = 0, \tag{6.42}$$

where

$$\nu = -\frac{\frac{\omega_i^*}{\omega} + \frac{l^2 g_0}{a\omega^2}}{1 - \frac{\omega^*}{\omega}}, \qquad \omega_i^* \equiv -2l\frac{T_i}{a^2\omega_{B_i}}. \tag{6.43}$$

Making the substitutions $(r/a)^2 = x$, $(1+\nu)/2 = \mu$, $l/2 = s$, $\psi = y(x)e^{x/2}/\sqrt{x}$, we obtain Whittaker's equation:

$$y'' + \left\{ -\frac{1}{4} + \frac{\mu}{x} + \frac{\frac{1}{4} - s^2}{x^2} \right\} y = 0. \tag{6.44}$$

The solutions of this equation are the Whittaker functions $W_{\mu s}(x)$ and $W_{-\mu s}(-x)$. From these functions we construct an eigenfunction, assuming that the plasma is surrounded by a conducting vessel situated at $r = r_0 \gg a$, i.e., for $x \gg 1$. Using the asymptotic representation of the Whittaker functions, we conclude that if the boundary condition is chosen in this manner an approximate eigenfunction is $W_{\mu s}(x)$ with subscript $\mu = 1/2 + s + n$, where n is an integer. This equality yields the following equation for the eigenfrequencies of the oscillations:

$$\nu = l + 2n, \qquad n = 0, 1, 2 \ldots, \tag{6.45}$$

where ν is determined by Eq. (6.43). Using (6.43) and (6.45), we find

$$\omega^2 - \omega_i^* \omega \left(1 - \frac{1}{l+2n}\right) + \frac{l^2 g_0}{a(l+2n)} = 0. \tag{6.46}$$

It can be seen that the perturbations of the lower level of the first mode ($n = 0$, $l = 1$) are insensitive to the value of ω_i^* and have the same growth rate as in the case $\omega_i^* = 0$. This is in agreement with the general results of §6.2.2B. At sufficiently large ω_i^*, all the other perturbations must have a real frequency.

§6.3. Flute Instability of a Low-Density Plasma

In contrast to §§6.1 and 6.2, we shall now assume $\omega_{p_i} \ll \omega_{B_i}$. We shall restrict the treatment to small-scale perturbations, $k_\perp a \gg 1$. Initially we shall neglect the terms of order ω_i^*/ω, but then take them into account in §6.3.2.

1. Plasma with Vanishing Ion Larmor Radius.

We shall proceed from Eq. (6.3). In §6.1 we have investigated this equation for $\omega_{p_i} \gg \omega_{B_i}$. We now make the opposite assumption, $\omega_{p_i} \ll \omega_{B_i}$. It then follows from (6.3) that

$$\omega\left(\omega + \frac{g k_y}{\omega_B}\right) - \frac{g \varkappa \omega_{p_i}^2 k_y^2}{\omega_{B_i}^2 k^2} = 0. \tag{6.47}$$

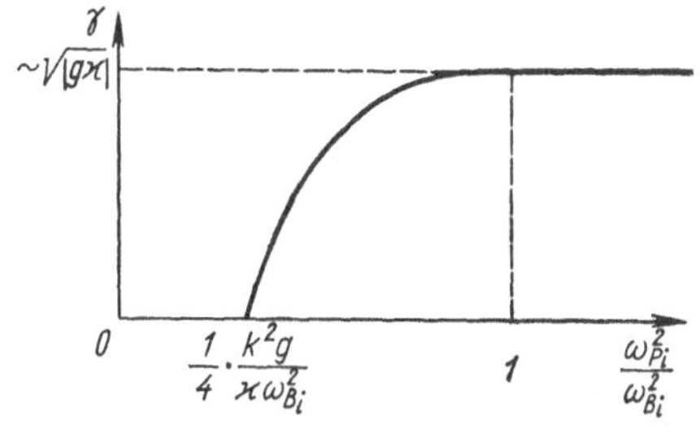

Fig. 6.3. Growth rate of the flute instability as a function of the density.

From this we find that the flute instability is impossible if the plasma density is very low:

$$\omega_{p_i}^2 < \frac{1}{4} k_y^2 \left|\frac{g}{\varkappa}\right| . \tag{6.48}$$

If the opposite condition holds, the plasma is unstable. The growth rate is then

$$\gamma = \frac{\omega_{p_i}}{\omega_{B_i}} \sqrt{|g\varkappa|}\, \frac{|k_y|}{k_\perp} . \tag{6.49}$$

This is small compared with the growth rate of the perturbations of a dense plasma (6.7). The growth rate of the flute instability is shown schematically as a function of the density in Fig. 6.3.

2. Plasma with Finite Ion Larmor Radius. Using (6.26), we find that if the density is reduced the condition for the stabilization of the flute instability is more stringent than (6.30):

$$\omega_{ni}^2 > 4\,|g\varkappa| \left(1 + \frac{\omega_{B_i}^2}{\omega_{p_i}^2}\right) . \tag{6.50}$$

At a sufficiently low density, $(\omega_{p_i}/\omega_{B_i})^2 < |g\varkappa|/v_{T_i}^2$, it follows from (6.26) that only perturbations with $k_\perp \rho_i \geq 1$ are not excited. At the same time, however, we must take into account all the powers

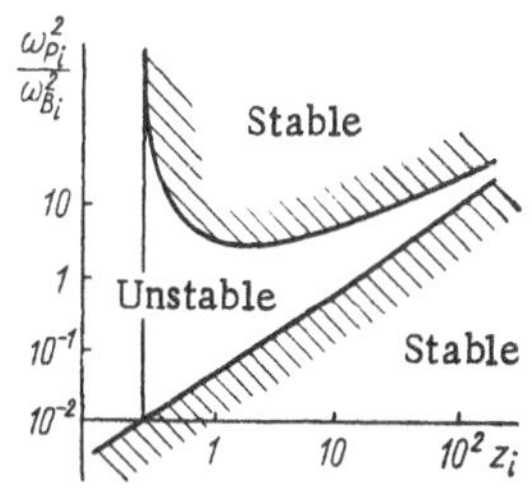

Fig. 6.4. Boundary of the flute instability on the plane $(\omega_{p_i}^2/\omega_{B_i}^2,\ z_i \equiv k_\perp^2 \rho^2)$.

of the parameter z_i in (6.26). The stability boundary obtained by means of the complete equation (6.26) is shown in Fig. 6.4.

§6.4. Stabilizing Influence of Conducting Ends

1. Plasma with a Small Value of $\Pi_e \equiv (\omega_{pe} a/c)^2$. We shall consider perturbations with $k_z \neq 0$, assuming that they have an electrostatic field. In the case of a collisionless plasma with $\omega > k_z v_{T_e}$, this assumption is justified if $\Pi_e \equiv \left(\frac{\omega_{p_e} a}{c}\right)^2 \ll 1$ [we shall call Π_e the dimensionless (electron) density number]. If the plasma is collisional and $\omega > (k_z v_{T_e})^2/\nu_{ei}$, the fact that the perturbations are not electrostatic can be ignored if $\Pi_e \ll \nu_{ei}/\omega$. If k_z is greater than in the above inequalities, this assumption is also valid at larger values of Π_e. Suppose $\sqrt{g\varkappa} \gg \omega^*$. Then, as in the case $k_z = 0$, oblique perturbations with $\omega > k_z v_{T_e}$ can be described on the basis of the hydrodynamic equations of a cold plasma of the type (6.2). We then arrive at Eq. (6.3), to which the term $(k_z \omega_{p_e}/\omega)^2 \psi$, due to the longitudinal motion of the electrons, is added on the left-hand side. For small-scale perturbations the dispersion equation is

$$1+\left(\frac{\omega_{p_i}}{\omega_{B_i}}\right)^2-\left(\frac{\omega_{p_e}}{\omega}\cos\theta\right)^2+\frac{\varkappa k_y \omega_{p_i}^2}{\omega_{B_i} k^2}\left(\frac{1}{\omega+\frac{k_y g}{\omega_{B_i}}}-\frac{1}{\omega}\right)=0. \quad (6.51)$$

A. Plasma with $\omega_{p_i} > \omega_{B_i}$. In this case (6.51) yields [cf. (6.7)]

$$\omega^2-\omega_{B_i}^2\frac{m_i}{m_e}\cos^2\theta-g\varkappa\left(\frac{k_y}{k_\perp}\right)^2=0. \quad (6.52)$$

From this equation we find that the flute instability occurs only for sufficiently small k_z:

$$\frac{k_z}{k}<\left(\frac{m_e}{m_i}\right)^{1/2}\frac{\sqrt{|g\varkappa|}}{\omega_{B_i}}. \quad (6.53)$$

If the plasma is bounded by metal at its ends, k_z takes a discrete series of values:

$$k_z=\frac{\pi n}{L}, \qquad n=1,2\ldots \quad (6.54)$$

It can be seen from (6.53) that the instability condition when $k_\perp a \simeq 1$ cannot be satisfied if ($g \simeq v_{Ti}^2/R$)

$$\frac{L}{a}<\pi\left(\frac{m_i}{m_e}\cdot\frac{aR}{\rho_i^2}\right)^{1/2}. \quad (6.55)$$

Thus, contact at the ends can lead to stabilization of the plasma even if the condition (6.31) is not satisfied. Moreover, the instabilities of all modes, including $l = 1$, can be suppressed.

B. Plasma with $\omega_{p_i} < \omega_{B_i}$. Instead of (6.52) we now have [cf. (6.47)]

$$\omega^2 - \omega_{p_e}^2 \cos^2\theta - g\varkappa\left(\frac{\omega_{p_i}}{\omega_{B_i}}\right)^2\left(\frac{k_y}{k}\right)^2 = 0. \tag{6.56}$$

The stability conditions (6.53) and (6.55) remain unaltered.

2. Plasma with a large Π_e. In §6.4.1 we considered oblique waves with $g \neq 0$ under conditions when the nonelectrostatic nature of the perturbations is not important. In the opposite limiting case of essentially nonelectrostatic perturbations, the analysis can be carried through relatively easily by noting that the flute instability is associated with the excitation of Alfvén perturbations. For $g = 0$ and $k_z \neq 0$ such perturbations were discussed in §3.3; for $k_z = 0$ and $g \neq 0$, in §6.2. Combining the results of these two sections, we arrive at the following local dispersion equation for the case $k_z \neq 0$ and $g \neq 0$:

$$\omega(\omega - \omega_{ni}) - k_z^2 c_A^2 - g\varkappa\frac{k_y^2}{k^2} = 0. \tag{6.57}$$

In the case of a vanishing ion Larmor radius, $|g\varkappa| \gg \omega_{ni}^2$, we then obtain the following stabilization condition for oblique waves:

$$k_z^2 c_A^2 > |g\varkappa|. \tag{6.58}$$

For a plasma in contact with conducting ends, this means [cf. (6.55)]:

$$\frac{L}{a} < \pi\left(\frac{R}{\beta a}\right)^{1/2}. \tag{6.59}$$

Equation (6.57) describes small-scale perturbations, $k_\perp a \approx 1$, and the condition (6.59) is obtained by extrapolating the results that follow from (6.57) to the region of wave numbers satisfying $k_\perp a \approx 1$. A more correct treatment of large-scale perturbations can be made in the approximation $\sqrt{|g\varkappa|} > \omega_{ni}$ by the energy method expounded in §6.1.3. Let us consider this in more detail.

As the original system of equations, we shall use equations of the type (6.10), adding, however, a term with a magnetic force to the right-hand side of the second equation in (6.10):

$$\left.\begin{aligned}&\frac{\partial\rho'}{\partial t}+\operatorname{div}(\rho_0\mathbf{V}')=0;\\&\rho_0\frac{\partial\mathbf{V}'}{\partial t}=\rho'\mathbf{g}+\frac{1}{4\pi}[\operatorname{rot}\mathbf{B},\mathbf{B}]'.\end{aligned}\right\}\tag{6.60}$$

The prime denotes the part that is linear in the perturbations. Using (6.9), (6.11), and the equation $\frac{\partial\mathbf{B}'}{\partial t}=-c\operatorname{rot}\mathbf{E}$, we express $\mathbf{B}'$ in terms of ξ:

$$\mathbf{B}'=\operatorname{rot}[\xi,\mathbf{B}_0].\tag{6.61}$$

From (6.60) we then obtain an equation of the type (6.13) but with a different operator $\hat{K}$:

$$\hat{K}\xi=\mathbf{g}\operatorname{div}(\rho_0\xi)+\frac{k_z^2B_0^2}{4\pi}\xi.\tag{6.62}$$

In this case, the potential energy (6.15) is

$$W=\frac{1}{2}\int\left\{(\xi\mathbf{g})(\xi\nabla\rho_0)+k_z^2\frac{B_0^2}{4\pi}\xi^2\right\}d\mathbf{r}.\tag{6.63}$$

It is not negative for any ξ, if for all $\mathbf{r}$ the condition (6.58) is satisfied. This means that the stabilizing effect of the conducting ends described by the condition (6.59) obtains for all perturbations (including perturbations of the first mode, l = 1).

§6.5. Stabilization of a Collisionless Plasma by a Negative g. Gravitational-Kinetic Instability for g > 0

We have shown above that a gravitational force with $g\varkappa > 0$ gives rise to a flute instability and with $g\varkappa > 0$ leads to the appearance of new branches of oscillations with a real frequency. This applied to perturbations with $k_z = 0$. If $k_z \neq 0$, the plasma can sustain other oscillation branches not associated with the presence of g. Let us consider how a gravitational force affects these oscillation branches.

We shall first introduce the concept of positive and negative g. An investigated plane layer of a plasma in a gravitational field is a model of a real plasma in a curved magnetic field with a radially decreasing density, $\partial n_0/\partial r < 0$. Therefore, the case $g\nabla n_0 < 0$ corresponds to a positive direction of g, g > 0; the case

$g\nabla n_0 > 0$, to a negative g, $g < 0$. We may therefore refer to these two situations as plasmas with positive or negative g.

The flute instability arises if $g > 0$. We shall show below that if g has this sign the gravitational force exerts a destabilizing influence on other oscillation branches as well. In contrast, a plasma with $g < 0$ is more stable than one with $g = 0$.

1. Perturbations with $\omega \gg k_z v_{T_e}$. In accordance with §3.1.1, perturbations with $\omega > k_z v_{T_e}$ and sufficiently small $\cos\theta$ grow in time if there is no gravitational force. Let us consider the effect of a gravitational force on this instability. We shall proceed from the dispersion equation

$$1 - \frac{\omega_{ni}}{\omega} - \left(\frac{\omega_{B_i}}{\omega}\right)^2 \frac{m_i}{m_e} \cos^2\theta \left(1 - \frac{\omega_{ne}}{\omega}\right) - \frac{g\varkappa}{\omega^2} = 0. \tag{6.64}$$

It follows from this equation that if g is directed against the density gradient, $g\varkappa < 0$ (positive g), the growth rate of oblique perturbations is only increased. In the opposite case, $g\varkappa > 0$, there is stabilization of these perturbations if in order of magnitude

$$g\varkappa > \omega_{ne}^2. \tag{6.65}$$

Expressed in terms of the curvature of the lines of force, this stabilization condition in the case $T_e \simeq T_i$ entails

$$a/R > (k_\perp \rho_i)^2. \tag{6.66}$$

It can be seen that long-wavelength perturbations, $k_\perp \simeq 1/a$, can be stabilized even for small values of $a/R \simeq (\rho_i/a)^2$ [cf. (6.25)].

2. Perturbations with $v_{T_i} < \omega/k_z < (v_{T_e}, \omega_{B_i})$, $z_i \ll 1$ in a Plasma with $\nabla T = 0$. If $g = 0$, perturbations of this type are described by Eq. (3.12). Including a gravitational force in this equation by making the substitution $\omega \to \omega + gk_y/\omega_{B_i}$ in the ion terms, we obtain

$$1 + z_i \left(1 + \frac{\omega_{ni}}{\omega + gk_y/\omega_{B_i}}\right) - \frac{\omega_{ne}}{\omega + gk_y/\omega_{B_i}} + \frac{i\sqrt{\pi}(\omega - \omega_{ne})}{|k_z| v_{T_e}} = 0. \tag{6.67}$$

Hence

$$\left.\begin{aligned} \operatorname{Re}\omega &= \omega_{ne}(1 - 2z_i) - gk_y/\omega_{B_i}, \\ \gamma &= \sqrt{\pi}\,(\omega_{ne}^2/k_z v_{T_e})(2z_i - gm_i/\varkappa T). \end{aligned}\right\}. \tag{6.68}$$

If $g\varkappa > 0$, the force of gravity leads to stabilization of the perturbations if the condition (6.65) is satisfied or, equivalently, (6.66) holds. This is the effect of stabilization by a negative g of the kinetically excited perturbations of a plasma with an inhomogeneous density.

If $g\varkappa < 0$ and the condition (6.65) is satisfied, we have the gravitational-kinetic instability with growth rate

$$\gamma \simeq z_i \gamma_0^2 / |k_z| v_{T_e}, \tag{6.69}$$

where $\gamma_0 \equiv \sqrt{|g\varkappa|}$ is the growth rate of the flute instability.

§6.6. Stabilization of a Collisional Plasma by a Negative g. Gravitational-Dissipative Instability for g > 0

The perturbations of a collisional plasma with $\nabla n_0 \neq 0$, $\nabla T = 0$ and $g = 0$ can be described by the dispersion equation (4.79). If we include a gravitational force by making the substitution $\mathbf{E} \to \mathbf{E} + \frac{m}{e}\mathbf{g}$ in the ion equations (4.34) and (4.44), Eq. (4.79) is replaced by

$$1 - \frac{\omega_{ni}}{\omega} + i\frac{\omega_s}{\omega}\left(1 - \frac{\omega_{ne}}{\omega}\right) - \frac{g\varkappa}{\omega^2}\left(\frac{k_y}{k}\right)^2 = 0. \tag{6.70}$$

From this equation we see that the gravitational force affects the inertial-dissipative instability considered in §5.2 if the condition (6.65) is satisfied. If this condition is satisfied, the instability is suppresed – by the negative g effect – if $g\varkappa > 0$.

If $k_z = 0$, $g\varkappa < 0$, and $\omega_{ni}^2 < |g\varkappa|$, Eq. (6.70) describes a flute instability with the growth rate $\gamma = \gamma_0 \equiv |g\varkappa|^{1/2}$. Collisions begin to affect this instability when

$$\cos\theta \gtrsim (\nu_{ie}\gamma_0)^{1/2}/\omega_{B_i}. \tag{6.71}$$

At the same time the growth rate is

$$\gamma \simeq z_i \gamma_0^2 \nu_{ei} / (k_z v_{T_e})^2. \tag{6.72}$$

This is the gravitational-dissipative instability. Its growth rate is greater than that of the inertial-dissipative instability (if the condition $\omega_n^2 < g\varkappa$ stipulated above holds). It should be noted that at the limits of applicability of the collisional and collisionless approximations, i.e., at $\nu_{ei} \simeq k_z v_{T_e}$, Eqs. (6.72) and (6.69) are qualitatively the same. If the model description of collisions is used [see §6.7], both these results can be obtained as limiting cases of the same expression.

Taking into account ion viscosity, we find that a gravitational force with $g > 0$ also leads to excitation of flute-type oscillations, $k_z = 0$, even under conditions when the stabilization effect of the finite ion Larmor radius is operative. In this case, the given type of perturbations is described by the dispersion equation

$$\left(1-\frac{\omega_{ni}}{\omega}\right)\left(1-iz_i\frac{\nu_{i.\text{eff}}}{\omega}\right)-\frac{g\varkappa}{\omega^2}\left(\frac{k_y}{k}\right)^2=0. \tag{6.73}$$

Here, $\nu_{i.\text{eff}} = 0.7\nu_{ii}$ if $\omega \gg \nu_{ii}$ [see §4.1] and $\nu_{i.\text{eff}} = 0.3\nu_{ii}$ if $\omega \ll \nu_{ii}$ [see §4.4]. In the Bhatnagar-Gross-Krook model, $\nu_{i.\text{eff}} = \nu_{ii}\left[\frac{9}{4}-\frac{\omega}{\omega+i\nu_{ii}}\right]$ [see §4.7].

Let us consider Eq. (6.73) under the assumptions $g\varkappa < 0$ but $\omega_{ni}^2 > 4|g\varkappa|$; then the usual flute instability does not arise [see §6.2]. For simplicity we shall assume $\omega_{ni}^2 \gg |g\varkappa|$, $\omega \gg \nu_{i.\text{eff}} z_i$. From (6.73) we then obtain

$$\left.\begin{aligned} \operatorname{Re}\omega &= \omega_{ni}+g\varkappa/\omega_{ni},\\ \gamma &= (\gamma_0/\omega_{ni})^2 z_i \nu_{i.\text{eff}}. \end{aligned}\right\} \tag{6.74}$$

It can be seen that the flute instability develops because of the ion viscosity despite the stabilizing effect of the finite ion Larmor radius.

§6.7. Plasma in a High-Frequency Field

1. Averaged High-Frequency Force. Suppose that in addition to the constant field $\mathbf{B}_0$ there is an alternating electromagnetic field (high-frequency field) in the plasma with the components

$$\left.\begin{aligned} \tilde{\mathbf{E}} &= (0,\ 0,\ \tilde{E}_z(x)\,e^{-i\Omega t}),\\ \tilde{\mathbf{B}} &= (0,\ \tilde{B}_y(x)\,e^{-i\Omega t},\ 0). \end{aligned}\right\} \tag{6.75}$$

In such a field the particles acquire a longitudinal velocity equal to

$$\tilde{v}_z = \frac{ie}{m\Omega}\tilde{E}_z. \tag{6.76}$$

The high-frequency field also affects the transverse motion of the particles. The latter are subjected to the additional transverse force

$$F_x = -\frac{e}{c}\,\tilde{v}_z\tilde{B}_y. \tag{6.77}$$

Averaging this result over a period of the high-frequency field and expressing $\tilde{v}_z$ and $\tilde{B}_y$ in terms of $\tilde{E}_z$, we find that the effect of the high-frequency field on the plasma is similar to the effect of a certain effective gravitational force that acts basically on the electrons:

$$g_{\text{eff}}^{(e)} = -\left(\frac{e}{m\Omega}\right)^2 \frac{1}{2}\cdot\frac{\partial}{\partial x}\,|\tilde{E}_z|^2. \tag{6.78}$$

2. Effect of a High-Frequency Force on the Plasma Stability. A polarization field of the type (6.75) with $\Omega < \omega_{p_e}$ penetrates into the plasma a distance of the order of c/ω_{p_e} (this can be found from Maxwell's equations). It will have an appreciable effect on the plasma if

$$\frac{c}{\omega_{pe}} \gg a. \tag{6.79}$$

A. Low-Frequency Instabilities. For a plasma with $\rho_i/a < (m_e/m_i)^{1/2}$ the condition (6.79) entails

$$\beta < (m_e/m_i)^2. \tag{6.80}$$

In this case, the low-frequency instabilities (see §3.1) are the most important. Since g_{eff} is in the direction of increasing density, the high-frequency field must, in accordance with §6.1, have a stabilizing effect. In particular, a long-wavelength instability (§3.1.1) cannot arise if

$$\frac{|\tilde{v}_{ze}|}{v_{T_i}} \gg \frac{\rho_i}{a} \simeq \left(\frac{m_i}{m_e}\beta\right)^{1/2}. \tag{6.81}$$

B. Instabilities with $\omega \geq \omega_{B_i}$. If the plasma pressure is not too low,

$$\left(\frac{m_e}{m_i}\right)^2 < \beta < \frac{m_e}{m_i}, \tag{6.82}$$

the condition (6.79) can be satisfied for $\left(\frac{m_e}{m_i}\right)^{1/2} < \frac{\rho_i}{a} < 1$. If this is the case and $g_{\mathrm{eff}} > 0$, instabilities with frequencies near the ion cyclotron harmonics can develop (§2.1). We shall find the effect of a high-frequency force on these oscillations by making the substitution

$$\omega \rightarrow \omega + k_y \frac{g_{\mathrm{eff}}}{\omega_{B_e}} \tag{6.83}$$

in the electron contribution to Eq. (2.6). The latter is then replaced by

$$1 + k_\perp^2 \left(d_i^2 + \frac{m_e}{m_i} \rho_i^2\right) - \frac{\omega_{ni}}{\omega + k_y \frac{g_{\mathrm{eff}}}{\omega_{B_e}}} - \frac{\omega - \omega_{ni}}{\sqrt{2\pi z_i}\,(\omega - n\omega_{B_i})} = 0. \tag{6.84}$$

It follows that the cyclotron instability does not develop if

$$\frac{g_{\mathrm{eff}}}{\omega_{B_e}} \gtrsim \frac{\rho_i}{a} v_{T_i}, \tag{6.85}$$

i.e., if

$$m_e |\tilde{v}_{ze}|^2 \gtrsim T. \tag{6.86}$$

This means that to stabilize the cyclotron perturbations we need $\tilde{E}$ values that are so large that the plasma is confined by the gradient of the high-frequency field and not the magnetic field gradient.

It follows from §6.7.1 and §6.7.2 that the use of a high-frequency field of the above polarization to stabilize a plasma confined by a constant magnetic field can be effective only if the condition (6.80) is satisfied. If this is not the case, a high-frequency field either does not penetrate to a sufficient depth into the transition layer between the plasma and the vacuum or – if this layer is sufficiently narrow – does not suppress the most rapidly developing instabilities.

Bibliography

1. S. Chandrasekhar, Hydrodynamic and Hydromagnetic Stability, Clarendon Press, Oxford (1961). In Chapter 10 of this monograph the Rayleigh-Taylor instability in an ordinary liquid is discussed in detail. A bibliography on the subject is also included.
2. M. Kruskal and M. Schwarzchild, Proc. Roy. Soc., A233:348 (1954). The instability of the boundary of a plasma in a gravitational field is considered (§6.1.1).

3. C. L. Longmire and M. N. Rosenbluth, Ann. Phys., 1:120 (1957). The mechanism of the flute instability is discussed.
4. I. B. Bernstein et al., Proc. Roy. Soc., A244:17 (1958). The energy method used in §6.1 is developed. Earlier, similar methods had also been used in the instability theory of an ordinary liquid (see, for example, [1]).
5. E. P. Velikhov, Zh. Tekh. Fiz., 31:180 (1961) [Sov. Phys. – Tech. Phys., 6:130 (1961)].
6. M. Vuillemin, Nucl. Fusion, Suppl., 1:341 (1962). In [5, 6] the stabilizing effect of conducting ends in the case of a dense plasma (§ 6.4.2) is considered.
7. B. Lehnert, Phys. Fluids, 4:847 (1961). The stabilizing effect of conducting ends in the case of a low-density plasma (§ 6.4.1) is considered.
8. B. B. Kadomtsev, Zh. Eksp. Teor. Fiz., 40:328 (1961) [Sov. Phys. – JETP, 13:223 (1961)].
9. B. B. Kadomtsev, Nucl. Fusion, 1:286 (1961). In [8, 9] the flute instability of a low-density plasma is considered in the approximation $k_\perp \rho_i \to 0$.
10. M. N. Rosenbluth, N. A. Krall, and N. Rostoker, Nucl. Fusion, Suppl., 1:143 (1962). It is pointed out that a finite ion Larmor radius can lead to stabilization (§ 6.2). A study is made of the case of a plane layer of a plasma and a cylindrical plasma with a Gaussian density distribution.
11. L. I. Rudakov, Nucl. Fusion, 2:107 (1962).
12. K. V. Roberts and J. B. Taylor, Phys. Rev. Lett., 8:197 (1962). It is pointed out in [11, 12] that the effect of a finite Larmor radius is contained in the hydrodynamic equations with magnetic viscosity.
13. F. C. Hoh, Phys. Fluids, 6:1359 (1963). A lucid explanation is given of the stabilizing effect of a finite ion Larmor radius.
14. A. B. Mikhailovskii, Zh. Eksp. Teor. Fiz., 43:509 (1962) [Sov. Phys. – JETP, 16:364 (1963)].
15. C. C. Damm et al., Phys. Rev. Lett., 10:323 (1963). In [14, 15] the flute instability of a low-density plasma with a finite ion Larmor radius (§ 6.3) is investigated.
16. A. B. Mikhailovskii, in: Reviews of Plasma Physics, Vol. 3, Consultants Bureau, New York (1967), p. 159. A condition is obtained for stabilization due to a finite ion Larmor radius of a plasma with an abrupt boundary (§ 6.2.2A).
17. V. F. Kuleshov and A. A. Rukhadze, Nucl. Fusion, 4:169 (1964).
18. J. D. Jukes, Phys. Fluids, 7:1468 (1964).
19. N. A. Krall and M. N. Rosenbluth, Phys. Fluids, 8:1488 (1965). In [17-19] a study is made of the effect of a gravitational force on the gradient instabilities of a collisionless plasma (§ 6.5).
20. J. L. Johnson, J. M. Greene, and B. Coppi, Phys. Fluids, 6:1169 (1963).
21. J. D. Jukes, Phys. Fluids, 7:52 (1964).
22. F. F. Chen, Phys. Fluids, 8:1323 (1965). In [20-22] the gravitational-dissipative instability is disucssed (§ 6.6).
23. B. Coppi and M. N. Rosenbluth, Plasma Physics and Controlled Nuclear Fusion Research, Vol. 1, IAEA, Vienna (1966), p. 617. The instability due to ion viscosity (§ 6.6) is considered.

24. A. B. Mikhailovskii, Nucl. Fusion, 4:108 (1964); L. V. Mikhailovskaya and A. B. Mikhailovskii, Nucl. Fusion, 5:249 (1965).
25. K. Jungwirth and M. Seidl, J. Nucl. Energy, C7:563 (1965).
26. M. N. Rosenbluth and A. Simon, Phys. Fluids, 8:1300 (1965). The papers [24-26] contain a discussion of the first mode of the flute instability in a plasma with a finite ion Larmor radius (§6.2).
27. M. V. Babykin et al., Zh. Eksp. Teor. Fiz., 47:1631 (1964) [Sov. Phys. – JETP, 20:1096 (1965)]. The stabilizing influence of plasma beyond the stopper (mirror) is discussed.
28. N. A. Krall, J. Nucl. Energy, C7:283 (1965).
29. N. A. Krall, Phys. Fluids, 9:820 (1966).
30. A. F. Kuckes, Phys. Fluids, 9:2239 (1966). In [28-30] some stabilizing effects that are important in a plasma with hot electrons, $T_e \gg T_i$, are discussed.
31. L. G. Kuo et al., Phys. Fluids, 7:988 (1964).
32. L. G. Kuo et al., J. Nucl. Energy, C6:505 (1964). In [31-32] the flute instability of a low-density plasma of cylindrical symmetry is discussed.
33. B. Lehnert, Phys. Fluids, 9:1367 (1966). The stabilizing role of a finite a/R and a dense cold plasma sheath is discussed.
34. F. F. Chen, J. Nucl. Energy, C7:399 (1965). The instabilities of a collisional plasma when a gravitational force is present are reviewed. The effect of conducting and nonconducting ends on the flute instability is also discussed.

Papers devoted to the high-frequency stabilization of a plasma with low β:

35. T. F. Volkov and B. B. Kadomtsev, At. Energ., 13:429 (1962). The suppression of the flute instability is discussed.
36. A. B. Mikhailovskii and V. P. Sidorov, Zh. Tekh. Fiz., 37:1630 (1967) [Sov. Phys. – Tech. Phys., 12:1192 (1968)]. It is shown that in a collisional plasma, $\Omega < \nu_{ei}$, the effect of the gradient of the high-frequency field is not manifested.
37. V. Kopecky, Plasma Phys., 10:609 (1968). The influence of the gradient of a high-frequency field on the ion-cyclotron oscillations is investigated.
38. R. A. Demirkhanov, T. I. Gutkin, and S. N. Lozovskii, Zh. Eksp. Teor. Fiz., 55:2195 (1968) [Sov. Phys. – JETP, 28:1164 (1968)]. A study is made of the effects of a high-frequency field on the low-frequency instability of a plasma with inhomogenous density and temperature.

Chapter 7

Plasmas in Crossed Electric and Magnetic Fields

§7.1. Low-Frequency Centrifugal Instability

In the present chapter we shall consider the instabilities of a plasma in which there is, besides a magnetic field $\mathbf{B}_0 \| z$, an electric field $E_0 \perp z$. An electric field can lead to instabilities for two reasons. In a cylindrical plasma in which the electric field is directed along the radius, instabilities may arise because of a centrifugal effect – a difference between the azimuthal drift velocities of the ions and electrons due to a centrifugal force acting on the ions. The second excitation mechanism can be manifested even when the cylindrical geometry of the plasma is unimportant; this can happen say, when $E_0 \| x$ and the charges of both signs drift with the same velocity $V_0 = -cE_0/B_0$. If $\nabla E_0 \neq 0$, this velocity varies in space so that in this case the plasma constitutes a system of spatially separated streams, and such a system (see, for example, §1.7 in Volume 1) may be unstable.

1. Equilibrium State. Suppose that in a cylindrical plasma with $T = 0$ there is an electric field $E_0 \| e_r$. This field gives rise to a rotation of each of the components of the plasma with a velocity determined by the equation of motion

$$-\frac{V_{0\varphi}^2}{r} = \frac{e}{m} E_0 + V_{0\varphi}\omega_B. \tag{7.1}$$

For the electrons this yields

$$V_{0\varphi}^{(e)} = -cE_0/B_0. \tag{7.2}$$

For the ions and with the inclusion of the centrifugal force [the left-hand side of (7.1)]

$$V_{0\varphi}^{(i)}=\frac{r\omega_{B_i}}{2}\left[-1+\sqrt{1-\frac{4e_iE_0}{m_i\omega_{B_i}^2 r}}\right]. \tag{7.3}$$

It is assumed that for $E_0 = 0$ the ions do not rotate, and the plus sign is therefore taken in front of the radical in (7.3). If the electric force e_iE_0 acting on the ions is directed from the center of the cylinder, a steady state is possible only if

$$E_0<\frac{m_i\omega_{B_i}^2}{4e_i}\,r. \tag{7.4}$$

2. General Equation for Electrostatic Perturbations of a Cold Plasma in an Electric Field. We shall proceed from the hydrodynamic equations of the type (1.1) and (1.6). In the case of a plasma with cylindrical symmetry, the equations of motion are

$$\left.\begin{aligned}
&-i\Omega V_r'-\frac{2V_0V_\varphi'}{r}=\frac{e}{m}E_r'+V_\varphi'\omega_B,\\
&-i\Omega V_\varphi'+\frac{V_0V_r'}{r}+\frac{\partial V_0}{\partial r}V_r'=\frac{e}{m}E_\varphi'-V_r'\omega_B,\\
&-i\Omega V_z'=\frac{e}{m}E_z.
\end{aligned}\right\} \tag{7.5}$$

Here $\Omega = \omega-(l/r)V_0$, and the azimuthal dependence of the perturbations is taken in the form $\exp(il\varphi)$.

We substitute $\mathbf{V}'$ as found from this system into the continuity equation,

$$-i\Omega n'+ik_zV_z'n_0+\operatorname{div}(n_0\mathbf{V}_\perp')=0, \tag{7.6}$$

and calculate $n' = n'(\psi)$. The resulting Poisson equation has the form

$$\Delta\psi+\sum_{e,\,i}\frac{4\pi e^2}{m}\left\{n_0\psi\left(\frac{k_z}{\Omega}\right)^2+\operatorname{div}\left(\frac{n_0\nabla_\perp\psi}{D}\right)-\frac{l\psi}{\Omega r}\cdot\frac{\partial}{\partial r}\left(n_0\frac{r\omega_B+2V_0}{rD}\right)\right\}=0, \tag{7.7}$$

where

$$D=\left(\omega_B+2\frac{V_0}{r}\right)\left[\omega_B+\frac{1}{r}\cdot\frac{\partial}{\partial r}(rV_0)\right]-\Omega^2.$$

3. Low-Frequency Centrifugal Instability. Let us consider perturbations with $k_z = 0$, $\omega \ll \omega_{B_i}$. We shall assume that the plasma is dense, $\omega_{p_i} \gg \omega_{B_i}$, and the electric field weak, $V_0 \ll r\omega_{B_i}$. It then follows from (7.7) that

$$\operatorname{div}(n_0 \nabla_\perp \psi) + \left\{ \left(\frac{lV_E}{\Omega_e}\right)^2 \frac{1}{r^3} \cdot \frac{\partial n_0}{\partial r} + \frac{l}{r\Omega_e} \cdot \frac{\partial}{\partial r} \left[\frac{n_0}{r} \cdot \frac{\partial}{\partial r} (rV_E) \right] \right\} \psi = 0. \tag{7.8}$$

Here,

$$V_E = -cE_0/B_0.$$

A. Approximation of Small-Scale Perturbations. Suppose $l \gg 1$. From (7.8) we then obtain a local dispersion equation analogous to Eq. (6.7) for the flute instability of a plasma in a gravitational field:

$$\Omega_e^2 - \frac{k_\varphi^2}{k^2} \cdot \frac{V_E^2}{r} \varkappa_n = 0, \qquad \varkappa_n \equiv \partial \ln n_0/\partial r, \qquad k_\varphi = l/r. \tag{7.9}$$

Instability occurs for a plasma with a density that decreases along the radius, $\varkappa_n < 0$. The growth rate is

$$\gamma = \left(\frac{|\varkappa_n|}{r}\right)^{1/2} V_E \left(\frac{k_\varphi}{k}\right)^{1/2}. \tag{7.10}$$

B. Exact Solution of Eq. (7.8) for a Plasma with a Gaussian Density Distribution and Homogeneous Angular Velocity. Let us consider the special case of a plasma with $n_0 = \bar{n}e^{-(r/a)^2}$ and $V_E/r \equiv \Omega_E = \text{const}$. (The quantity Ω_E can be interpreted as the angular velocity of rotation of the electrons.) Under these assumptions, Eq. (7.8) reduces to

$$\psi'' + \frac{\psi'}{r} - \frac{l^2}{r^2}\psi - \frac{2}{a^2}(r\psi' - \nu\psi) = 0, \tag{7.11}$$

where

$$\nu = \frac{\omega_E(\omega_E - 2\omega)}{(\omega - \omega_E)^2}, \qquad \omega_E \equiv k_\varphi V_E. \tag{7.12}$$

Equation (7.11) is identical with Eq. (6.42) investigated in §6.2. The eigenvalues ν are determined by Eq. (6.45). If ν has the form (7.12), the latter leads to the equation

$$\omega = \omega_E \left\{ 1 + \frac{1}{l + 2n} \left[-1 \pm \sqrt{1 - (l + 2n)} \right] \right\}, \tag{7.13}$$

where $n = 0, 1, 2, \ldots$, for the frequency of the perturbations.

For large l and n, Eq. (7.13) yields

$$\gamma = \frac{\omega_E}{\sqrt{l+2n}} \equiv \frac{l}{a} V_E \frac{1}{\sqrt{l+2n}}, \tag{7.14}$$

which is in qualitative agreement with (7.10), $\gamma \simeq V_E/a$.

Perturbations that correspond to the lowest level (n = 0) of the mode with l = 1 possess an interesting property – their frequency is, in accordance with (7.13)

$$\omega = 0 \quad \text{for} \quad n = 0, \quad l = 1. \tag{7.15}$$

The lowest levels of the other modes (n = 0, l > 1) have a complex frequency, ω = Re ω + iγ, where

$$\left.\begin{aligned} \operatorname{Re}\omega &= \Omega_E (l-1), \\ \gamma &= \pm \Omega_E \sqrt{l-1}. \end{aligned}\right\} \tag{7.16}$$

The influence of the thermal motion of the ions on the centrifugal instability will be considered in §7.3.

§7.2. Low-Frequency Instability of a Plasma with an Inhomogeneous Velocity Profile

Let us consider perturbations with $\omega \ll \omega_{B_i}$ in the approximation of planar symmetry, assuming that the electric field and the density gradient are directed along **x**. Under these conditions, the low-frequency centrifugal instability discussed in §7.1 is impossible. However, we shall show that the plasma may nevertheless be unstable if the electric field and the associated drift velocity of the particles, $V_E = -cE_0/B_0$, are inhomogeneous.

The equation for the potential of the perturbations can be obtained by making the substitutions $l/r \to k_y$ and $\partial/\partial r \to \partial/\partial x$ in Eq. (7.8) and omitting the terms $\sim 1/r$. The result can be reduced to the form

$$\frac{d}{dx}\left[n_0(\omega-\omega_E)^2\left(1+\frac{\omega_{Bi}^2}{\omega_{p_i}^2}\right)\frac{d\varphi}{dx}\right] - k_y^2 n_0(\omega-\omega_E)^2\left(1+\frac{\omega_{Bi}^2}{\omega_{p_i}^2}\right)\varphi = 0, \tag{7.17}$$

where

$$\omega_E = k_y V_E, \quad V_E = -\frac{cE_0}{B}, \quad \varphi = \frac{\psi}{\omega-\omega_E}. \tag{7.18}$$

1. Necessary Condition for Instability of a Plasma with $\nabla n_0 = 0$, $\omega_{p_i}^2 \gg \omega_{B_i}^2$ (Rayleigh's Theorem). If $\nabla n_0 = 0$ and $\omega_{p_i}^2 \gg \omega_{B_i}^2$, Eq. (7.17) reduces to an equation that is identical with the equation of oscillations of a plane-parallel flow of a perfect (inviscid) liquid:

$$\frac{d^2\psi}{dx^2} - k_y^2\psi - \psi \frac{\partial^2 V_0/\partial x^2}{V_0 - \frac{\omega}{k_y}} = 0. \tag{7.19}$$

Let us establish the conditions under which the oscillation frequency can be complex. We multiply both sides of Eq. (7.19) by ψ^* and integrate the result over x from x_1 to x_2, assuming $\psi(x_1) = \psi(x_2) = 0$. Then, if $\omega = \mathrm{Re}\,\omega + i\gamma$, the imaginary part of the integral is given by

$$\gamma \int_{x_1}^{x_2} \frac{\partial^2 V_0/\partial x^2}{\left(V_0 - \frac{\mathrm{Re}\,\omega}{k_y}\right)^2 + \left(\frac{\gamma}{k_y}\right)^2} |\psi|^2 dx = 0. \tag{7.20}$$

It can be seen that this equation cannot be satisfied for all profiles $V_0(x)$, but only if $\partial^2 V_0/\partial x^2$ vanishes somewhere between x_1 and x_2, i.e., if

$$\left(\frac{\partial^2 V_0}{\partial x^2}\right)_{x=x_0} = 0, \qquad x_1 < x_0 < x_2. \tag{7.21}$$

This is the necessary condition for instability and is also known as Rayleigh's theorem.

2. Long-Wavelength Perturbations. Sufficient Condition for Instability. Equation (7.19) can be readily solved for perturbations with finite $W \equiv \omega/k_y$ in the limit $k_y \to 0$. In this case the term with k_y^2 can be neglected and (7.19) can be reduced to [cf. (7.17)]:

$$\frac{d}{dx}\left[(V_0 - W)^2 \frac{d}{dx}\left(\frac{\psi}{V_0 - W}\right)\right] = 0. \tag{7.22}$$

Hence

$$\psi = (V_0 - W)\left[C_1 + C_2 \int \frac{dx}{(V_0 - W)^2}\right], \tag{7.23}$$

where C_1 and C_2 are constants.

Taking into account the boundary conditions $\psi(x_1) = \psi(x_2) = 0$, and using (7.23), we obtain the dispersion equation

$$\int_{x_1}^{x_2} \frac{dx}{(V_0 - W)^2} = 0. \quad (7.24)$$

If this equation has complex roots, the plasma is unstable against perturbations with $k_y \to 0$. This is a sufficient condition of instability.

3. Analogy with the Two-Stream Instability. Neccesary and Sufficient Condition for the Instability of a Plasma with a Monotonic Velocity Profile. Let $V_0(x)$ be a monotonic function of x, i.e., the derivative $dV_0(x)/dx$ vanishes nowhere (between x_1 and x_2). Then Eq. (7.24) can be represented in the form

$$\int_{V_1}^{V_2} \frac{(dx/dV_0)\,dV_0}{(V_0 - W)^2} = 0, \quad (7.25)$$

where $V_1 = V_0(x_1)$, $V_2 = V_0(x_2)$. Except for the notation, this equation is identical with the equation of long-wavelength ($k_y \to 0$) Langmuir (plasma) oscillations of a nonmagnetized plasma with a continuous velocity distribution function of the particles:

$$\int \frac{f_0(v)\,dv}{(v - w)^2} = 0, \qquad w = \frac{\omega}{k}. \quad (7.26)$$

The role of the distribution function $f_0(v)$ in the case of an inhomogeneous plasma in an electric field is played by the quantity $dx(V_0)/dV_0$:

$$\frac{dx(V_0)}{dV_0} \leftrightarrow f_0(v). \quad (7.27)$$

We have shown in §2.7 of Volume 1 that Eq. (7.26) can have complex solutions with $\mathrm{Im}\,\omega > 0$ if $f_0(v)$ has not less than two maxima, i.e., has at least one minimum. Using the analogy with (7.27), we conclude that a necessary condition for instability of a plasma in an electric field is the presence of a V_0 for which

$$\frac{d^2x}{dV_0^2} = 0 \quad \text{for} \quad \frac{d^3x(V_0)}{dV_0^3} > 0. \quad (7.28)$$

Expressing V_0 in this condition in terms of x, we find that the condition (7.28) is identical with the Rayleigh condition (7.21).

In §2.7 of Volume 1 we used the Nyquist method to obtain a necessary and sufficient condition for the instability of perturbations described by Eq. (7.26). It has the form

$$\int \frac{f_0(v)-f_0(v_0)}{(v-v_0)^2}\,dv<0, \tag{7.29}$$

where v_0 is the velocity corresponding to the minimum of $f_0(v)$. This result can be reformulated in the case of an electric field: for the instability of a plasma with a monotonic velocity profile it is necessary and sufficient that in the interval between x_1 and x_2 there should be a point x_0 at which

$$\mathcal{P}\int_{x_1}^{x_2} \frac{dx}{[V_0(x)-V_0(x_0)]^2}>0, \qquad \left(\frac{d^2V_0}{dx^2}\right)_{x=x_0}=0 \tag{7.30}$$

($\mathcal{P}$ is the principal value of the integral).

Integrating by parts, we can also represent this condition in the form

$$\left.\frac{1}{\frac{dV_0}{dx}[V_0-V_0(x_0)]}\right|_{x_1}^{x_2}+\int_{x_1}^{x_2}\frac{d^2V_0/dx^2}{\left(\frac{dV_0}{dx}\right)^3[V_0-V_0(x_0)]}\,dx<0. \tag{7.31}$$

4. Interval of Wave Numbers of Growing Perturbations in a Plasma with a Monotonic Velocity Profile. Suppose that the instability condition (7.31) is satisfied. One can show that there must then be growth of not only perturbations with $k_y \to 0$ but also of perturbations whose wave numbers lie in the interval $0 \le k_y^2 \le k_{max}^2$, where k_{max}^2 are the eigenvalues of the parameter $(-\lambda)$ determined from the equation

$$\frac{d^2\psi}{dx^2}+\lambda\psi-V(x)\psi=0, \tag{7.32}$$

and

$$U=\frac{d^2V_0/dx^2}{V_0-V_0(x_0)}\,. \tag{7.33}$$

It is assumed that $V_0(x)$ is positive and increases with x, so that the potential energy U is everywhere negative.

This assertion is proved in Lin's monograph.

§7.3. Influence of the Ion Temperature on the Low-Frequency Instabilities

Allowance for the temperature of the particles in perturbations with $k_z = 0$ means allowance for effects to order ω^*/ω, where $\omega^* = kV_L$ and V_{L_i} is the velocity of the ion Larmor currents [see §6.2]:

$$\mathbf{V}_{Li} = \frac{1}{m_i \omega_{B_i} n_0} [\mathbf{e}_z, \nabla p_i]. \tag{7.34}$$

In problems on the oscillations of a plasma in an electric field, the oscillation frequency ω is $\sim kV_E^{(0)}$. From the condition $V_{Li} \gtrsim V_E^{(0)}$ we obtain a condition for the ion temperature to be important:

$$T_i \gtrsim \frac{eE_0}{a}. \tag{7.35}$$

1. Basic Equations of Perturbations with $T_i \neq 0$. In investigating perturbations of a plasma with $T_i \neq 0$, $E_0 \neq 0$, and $\nabla T_{0i} = 0$ one can proceed from the system of macroscopic equations (4.51). In the limit $\frac{\partial}{\partial z} \to 0$ and when collisional viscosity can be neglected, this system becomes

$$\frac{\partial n_e}{\partial t} + (\mathbf{V}_E \nabla) n_e = 0,$$

$$\frac{\partial n_i}{\partial t} + (\mathbf{V}_E \nabla) n_i + \operatorname{div} \left\{ \left(\frac{\partial}{\partial t} + (\mathbf{V}_E \nabla) \right) \left[\frac{\mathbf{e}_z}{\omega_B}, \mathbf{V}_E + \mathbf{V}_L \right] \right\} = 0. \tag{7.36}$$

Although these equations have been obtained in the approximation of frequent collisions, they also remain in force as $\nu \to 0$. Linearizing this system of equations and using Poisson's equation, we obtain for the case of cylindrical symmetry

$$\frac{1}{r} \cdot \frac{\partial}{\partial r} \left[(\omega - \omega_E)^2 n_0 S r^3 \frac{\partial \varphi}{\partial r} \right] + \left[(1 - l^2)(\omega - \omega_E)^2 n_0 S + \omega^2 r \frac{\partial n_0}{\partial r} \right] \varphi = 0, \tag{7.37}$$

where

$$S = 1 + \left(\frac{\omega_{Bi}}{\omega_{p_i}} \right)^2 - \frac{\omega_i^*}{\omega - \omega_E}, \qquad \varphi = \frac{\psi}{r(\omega - \omega_E)},$$

$$\omega_E = \frac{l}{r} \cdot \frac{cE_0}{B_0}, \qquad \omega_i^* = \frac{l}{r} \cdot \frac{cT_i}{eB_0} \cdot \frac{\partial \ln n_0}{\partial r}. \tag{7.38}$$

Equation (7.37) can be understood as the appropriate generalization of Eqs. (7.8) and (7.17) to the case $\omega_i^* \neq 0$.

2. Centrifugal Instability and Stabilization Effects in the Case of a Dense Plasma with a Gaussian Density Distribution and Homogeneous Rotation. A finite ion temperature helps to stabilize the flute instability due to the centrifugal drift of ions. To see this, let us consider the example discussed in §7.1.3B, assuming $n_0 = \bar{n}e^{-(r/a)^2}$ and E_0 = const. We then obtain Eq. (7.11) but with a different value of ν:

$$\nu = -\frac{\omega_E^2 + 2\omega_E\Omega + \Omega\omega_i^*}{\Omega(\Omega - \omega_i^*)}, \qquad \Omega \equiv \omega - \omega_E. \tag{7.39}$$

The frequency of the perturbations is

$$\omega = -\omega_E + \frac{1}{\nu}\left\{\frac{\omega_i^*}{2}(\nu - 1) - \omega_E \pm \sqrt{\left[\frac{\omega_i^*}{2}(\nu - 1) - \omega_E\right]^2 - \nu\omega_E}\right\}. \tag{7.40}$$

It can be seen that the centrifugal instability cannot develop if

$$\left|\frac{\omega_i^*}{\omega_E}(\nu - 1) - 1\right| > \sqrt{\nu}, \quad \nu = 2, 3, \ldots \tag{7.41}$$

In contrast to the flute instability due to gravitational drift (§6.2), for which the mode $l = 1$ is unstable even for large ω_i^*, perturbations with all l are stabilized in this case.

3. Suppression of the Instability Due to Inhomogeneity of the Velocity Profile. At a sufficiently high ion temperature there is suppression of not only the centrifugal instability but also the instability due to inhomogeneity of the velocity profile (§7.2). To see this, let us consider perturbations with $k_y \to 0$. Using (7.37), we obtain an equation that is analogous to (7.22) but with finite V_L/V_0:

$$\frac{d}{dx}\left\{[(V_0 - W)^2 - V_L^2]\frac{d\psi}{dx}\right\} = 0. \tag{7.42}$$

This yields a dispersion equation of the type (7.24):

$$\int_{x_1}^{x_2} \frac{dx}{(V_0 - W)^2 - V_L^2} = 0. \tag{7.43}$$

It can be seen that for sufficiently large V_L the terms with V_0 can be omitted. In the absence of an electric field, the instability cannot develop, as follows from §6.2.

§7.4. Ion-Cyclotron and High-Frequency Instabilities

1. Excitation of Ion-Cyclotron Oscillations of a Plasma with $\omega_{p_i} \leqslant \omega_{B_i}$ by Centrifugal Drift. If the ratio $V_E/r\omega_{B_i}$ is not too small, there may be growth of not only the low-frequency perturbations but also perturbations with $\omega \geqslant \omega_{B_i}$ – cyclotron and high-frequency perturbations. In §7.4.1 we shall consider cyclotron perturbations; in §7.4.2, high-frequency perturbations.

We shall assume small-scale perturbations and go over from (7.7) to a local dispersion equation. If $\omega \simeq \omega_{B_i}$, it has the form

$$1-\frac{\omega_{p_i}^2}{(\omega-k_\varphi V_{0i})^2-\omega_{B_i}^2}-\left(\frac{k_z}{k}\right)^2\frac{\omega_{p_e}^2}{(\omega-k_\varphi V_{0e})^2}=0. \tag{7.44}$$

From this we find that oscillations of the type (7.44) grow if $k_\varphi(V_{0i}-V_{0e})\simeq\omega_{B_i}$, i.e, if

$$\frac{k_\varphi}{r}\cdot\frac{V_E^2}{\omega_{B_i}^2}\simeq 1. \tag{7.45}$$

If $\omega_{p_i} < \omega_{B_i}$, the growth rate is of order ω_{p_i}.

2. High-Frequency Centrifugal Instability of a Plasma with $\omega_{p_i} \gg \omega_{B_i}$. For small-scale perturbations with $\omega_{B_i} < \omega < \omega_{B_e}$, Eq. (7.7) yields the local dispersion equation

$$1+\left(\frac{\omega_{pe}}{\omega_{B_e}}\right)^2-\left(\frac{\omega_{pe}}{\Omega_e}\cos\theta\right)^2+\frac{l}{r}\cdot\frac{\varkappa_n\omega_{pe}^2}{\Omega_e k_\perp^2}-\frac{\omega_{pi}^2}{\left[\Omega_e-\frac{l}{r}(V_{0i}-V_{0e})\right]^2}=0,$$

$$\Omega_e\equiv\omega-\frac{l}{r}V_E. \tag{7.46}$$

If the substitutions $\Omega_e\to\omega$ and $V_{0i}-V_{0e}\to V_{0i}$ are made, it is similar to Eq. (2.60).

The instability conditions and the growth rates can be found by using the results of §13.1 of Volume 1 and §2.5.

§7.5. Instability of an Electron Cloud with a Nonmonotonic Density (Diocotron Instability)

For purely electron oscillations with $\omega \ll \omega_{B_e}$, $k_z = 0$, and $\nabla n_0 \parallel \mathbf{x}$, Eq. (7.7) yields

$$\Delta\psi - \frac{4\pi e^2 k_y}{m\omega_B} \cdot \frac{\frac{\partial n_0}{\partial x}}{\omega - k_y V_0}\psi = 0. \qquad (7.47)$$

In the absence of ions that compensate the space charge of the electrons, the equilibrium electron density is uniquely related to the field E_0:

$$\frac{\partial E_0}{\partial x} = 4\pi e n_0. \qquad (7.48)$$

Using this equation to express n_0 in terms of E_0 and then in terms of $V_0 \equiv -cE_0/B_0$ and substituting the result into (7.47), we obtain

$$\frac{\partial^2\psi}{\partial x^2} - k_y^2\psi + \frac{\partial^2 V_0/\partial x^2}{\frac{\omega}{k_y} - V_0}\psi = 0. \qquad (7.49)$$

This equation is exactly the same as (7.19), although it has been derived under different assumptions. In view of this identity and on the basis of the results of §7.2, we conclude that an uncompensated electron cloud is unstable against perturbations with sufficiently small k_y if the Rayleigh condition (7.21) is satisfied. Since $n_0 \sim \partial V_0/\partial x$, the instability condition means that the density distribution must have at least one maximum:

$$\left(\frac{\partial n_0}{\partial x}\right)_{x=x_0} = 0. \qquad (7.50)$$

The instability of a spatially inhomogeneous electron cloud is frequently called the diocotron instability.

Bibliography

1. C. C. Lin, The Theory of Hydrodynamic Stability, Cambridge University Press (1955). Contains an exposition of the stability theory of plane-parallel flows of an ordinary liquid. Rayleigh's theorem (§7.2) is also discussed.

2. A. A. Vedenov, E. P. Velikhov, and R. Z. Sagdeev, Usp. Fiz. Nauk, 73:701 (1961) [Sov. Phys. – Uspekhi, 4:332 (1962)]. The low-frequency centrifugal instability (§7.1) is considered.
3. M. N. Rosenbluth, N. A. Krall, and N. Rostoker, Nucl. Fusion, Suppl. 1:143 (1962). The influence of a finite ion Larmor radius on the low-frequency centrifugal instability is investigated.
4. M. N. Rosenbluth and A. Simon, Phys. Fluids, 7:557 (1964). A necessary and sufficient condition is obtained for the stability of the plane-parallel inviscid flow of a liquid with a monotonic velocity profile. The results of this paper are given in (§7.2).
5. M. N. Rosenbluth and A. Simon, Phys. Fluids, 8:1300 (1965). Derivation and investigation of the equations of low-frequency perturbations with $k_z = 0$ taking into account a finite ion Larmor radius and inhomogeneity of the electric field for the case of planar and cylindrical symmetry (§§7.1-7.3).
6. A. B. Mikhailovskii and V. S. Tsypin, ZhETF Pis. Red., 3:247 (1966) [JETP Letters, 3:158 (1966)]. Investigation of the high-frequency instabilities in a strongly inhomogeneous plasma with a strong electric field (§7.4).
7. I. S. Baikov, L. S. Bogdankevich, and A. A. Rukhadze, Nucl. Fusion, 5:318 (1965). Discussion of the importance of the centrifugal effect in oblique perturbations, $k_z \neq 0$, of a collisionless plasma. It is shown that this effect is manifested qualitatively in the same way as the effect of a gravitational force directed against the density gradient, $g > 0$ (cf. §6.5).
8. F. F. Chen, Phys. Fluids, 9:965 (1966). Investigation of the centrifugal instabilities of a collisional plasma in a radial electric field for both $k_z = 0$ and $k_z \neq 0$. Discussion of the physical meaning of the asymmetry of the stabilization condition (7.41). It is shown that this is due to a Coriolis force. Investigation of the effect of rotation of the plasma on the gradient-dissipative instabilities (the latter are discussed in the case $E_r = 0$ in §5.2).
9. I. S. Baikov, Zh. Tekh. Fiz., 36:1137 (1966) [Sov. Phys. – Tech. Phys., 11:837 (1966)]. Further development of [7]; perturbations of the first azimuthal modes are considered.
10. G. C. MacFarlane and H. G. Hay, Proc. Phys. Soc., 63B:409 (1953).
11. O. J. Buneman, Electronics, 3:1 (1957); J. Electron Control, 3:507 (1953).
12. R. W. Gould, J. Appl. Phys., 28:599 (1957).
13. R. H. Levy, Phys. Fluids, 8:1288 (1965).
14. W. Knauer, J. Appl. Phys., 37:602 (1966).
15. R. H. Levy and J. D. Callen, Phys. Fluids, 8:2298 (1965).
16. R. H. Levy, Phys. Fluids, 11:920 (1968). Papers [10-16] contain investigations of the diocotron instability (§7.5).
17. H. E. Wilhelm, Nucl. Fusion, 2:6 (1962). The centrifugal instability is considered in the framework of ideal hydrodynamics. The effect of a surrounding neutral gas is taken into account.
18. V. V. Arsenin, Nucl. Fusion, 5:152 (1965).
19. C. C. Damm et al., Phys. Fluids, 8:1472 (1965).
20. Yu. N. Dnestrovskii and D. P. Kostomarov, Dokl. Akad. Nauk SSSR, 167:1032 (1966) [Sov. Phys. – Doklady 11:326 (1966)].

21. Yu. N. Dnestrovskii, D. P. Kostomarov, and A. A. Chechina, Zh. Tekh. Fiz., 38:1205 (1968) [Sov. Phys. – Tech. Phys., 13:887 (1969)].
22. A. V. Timofeev, Nucl. Fusion, 6:93 (1966).
23. M. N. Rosenbluth and A. Simon, Phys. Fluids, 9:726 (1966).
24. A. V. Timofeev, Zh. Tekh. Fiz., 36:1787 (1966) [Sov. Phys. – Tech. Phys., 11:1331 (1967)].
25. A. V. Timofeev, Zh. Tekh. Fiz., 38:14 (1968) [Sov. Phys. – Tech. Phys., 13:9 (1968)].
26. A. V. Timofeev, Plasma Physics, 10:235 (1968). In [18-26] investigations are made of the flute instability of a low-density plasma, $\omega_{p_i} \ll \omega_{B_i}$, in an electric field.
27. T. E. Stringer and G. Schmidt, Plasma Physics, 9:53 (1967). Investigation of low-frequency instabilities of a plasma with arbitrary ω_{pi}/ω_{Bi}. Attention is drawn to the analogy between these instabilities and the Kelvin–Helmholtz instability.

Chapter 8

Plasmas in Fields with a Shear

§8.1. Definition of a Shear

Suppose that a current $\mathbf{j}_0$ with components

$$\mathbf{j}_0 = j_{0y}(x)\,\mathbf{e}_y + j_{0z}(x)\,\mathbf{e}_z \tag{8.1}$$

flows in a plasma. In accordance with Maxwell's equations, the equilibrium magnetic field $\mathbf{B}_0$ in this case must have components

$$\mathbf{B}_0 = B_{0z}(x)\,\mathbf{e}_z + B_{0y}(x)\,\mathbf{e}_y, \tag{8.2}$$

satisfying the relations

$$-\frac{\partial B_{0z}}{\partial x} = \frac{4\pi}{c} j_{0y}, \quad \frac{\partial B_{0y}}{\partial x} = \frac{4\pi}{c} j_{0z}. \tag{8.3}$$

The lines of force of the field (8.2) are straight. Because B_{0z} and B_{0y} depend on x, they are not all parallel to each other – the direction of the line of force is, in general, a function of the coordinate x. Thus, the field (8.2) is an example of a *field with straight but not parallel lines of force*. The study of plasma instabilities in a field of this kind is the subject of the present chapter.

Since the stationary parameters of both the field and the plasma depend only on x in the case (8.2), the coordinate dependence of the perturbations can be taken in the form

$$\psi(\mathbf{r}) = e^{ik_y y + ik_z z}\psi(x). \tag{8.4}$$

In contrast to the case $\mathbf{B}_0 \parallel \mathbf{z}$ discussed above, the wave numbers k_z and k_y do not now have the meaning of the longitudinal and transverse (in the plane perpendicular to $\mathbf{B}_0$ and ∇f_0) components of the wave vector **k**. The role of the latter is now played by quantities $k_\parallel$ and k_b satisfying the conditions

$$\left.\begin{aligned} ik_\parallel\psi &= \mathbf{e}_0\nabla\psi; \\ ik_b\psi &= \mathbf{e}_b\nabla\psi, \end{aligned}\right\} \tag{8.5}$$

where $\mathbf{e}_0$ and $\mathbf{e}_b$ are unit vectors along $\mathbf{B}_0$ and in the direction perpendicular to the plane $\mathbf{e}_0$, $\mathbf{e}_x$, i.e.,

$$\left.\begin{aligned} &\mathbf{e}_0 = h_z\mathbf{e}_z + h_y\mathbf{e}_y, \\ &\mathbf{e}_b \equiv [\mathbf{e}_0\mathbf{e}_x] = h_z\mathbf{e}_y - h_y\mathbf{e}_z, \\ &h_z = \frac{B_{0z}}{B_0}, \qquad h_y = \frac{B_{0y}}{B_0}. \end{aligned}\right\} \tag{8.6}$$

The subscript b is used to label the "binormal" – in the case of cylindrical geometry $\mathbf{e}_b$ is directed along the binormal to the line of force (see §11.6).

It follows from (8.5) and (8.6) that $(k_\parallel, k_b)$ are related to (k_y, k_z) by the equations

$$\left.\begin{aligned} k_\parallel &= k_zh_z + k_yh_y, \\ k_b &= k_yh_z - k_zh_y. \end{aligned}\right\} \tag{8.7}$$

Since the coefficients h_z and h_y are functions of x, it follows that $k_\parallel$ and k_b are also functions of x in the general case. In problems with perturbations satisfying $k_\parallel \ll k_b$ the coordinate dependence of $k_\parallel$ is the most important; for in the case $\mathbf{B}_0 \parallel \mathbf{z}$, small k_z/k_y correspond to all the types of gradient instability discussed above. Therefore, in accordance with the first equation of (8.7), the presence of even a small B_{0y} may lead to an appreciable difference between $k_\parallel$ and k_z (since $k_y \gg k_z$).

Let us now consider how the coordinate dependence of $k_\parallel(x)$ and, accordingly, the derivative $\partial k_\parallel/\partial x$ is determined. Suppose $k_\parallel(x) = 0$ at some point $x = x_0$, i.e.,

$$k_zh_z(x_0) + k_yh_y(x_0) = 0. \tag{8.8}$$

Then near x_0

$$k_\parallel(x) = \left(k_z\frac{\partial h_z}{\partial x} + k_y\frac{\partial h_y}{\partial x}\right)_{x=x_0}(x - x_0). \tag{8.9}$$

Eliminating the wave numbers k_z and k_y by means of (8.8) and the second equation of (8.7) (for $x = x_0$), we obtain

$$k_{\|}(x)=\left\{h_z^2 k_b \frac{\partial}{\partial x}\left(\frac{B_{0y}}{B_{0z}}\right)\right\}_{x=x_0}(x-x_0). \tag{8.10}$$

It can be seen that $k_{\|}$ depends on the coordinates if

$$\frac{d}{dx}\left(\frac{B_{0y}}{B_{0z}}\right)\neq 0. \tag{8.11}$$

A field to which there corresponds a coordinate-dependent $k_{\|}$ is known as a *magnetic field with skew lines of force* or a *magnetic field with shear.*

In the case considered (planar symmetry) a field with shear is the same as a field with nonparallel lines of force. (In more complicated cases this is not so; see §11.6.) The fact that the two lines of force that pass through the points x_0 and $x \neq x_0$ are not parallel means

$$[\mathbf{e}_0(x),\quad \mathbf{e}_0(x_0)]\neq 0. \tag{8.12}$$

Assuming that the difference $x-x_0$ is small and expanding $e_0(x)$ in a series in the neighborhood of x_0, we find that (8.12) reduces to (8.11).

Returning to Eqs. (8.3), we note that the condition for a nonvanishing shear can be expressed simply as the nonvanishing of the current in the direction of the magnetic field:

$$\mathbf{j}_0\mathbf{B}_0\neq 0. \tag{8.13}$$

Equation (8.10) can be transformed by means of (8.3) to

$$k_{\|}(x)=\frac{4\pi}{c}\frac{j_{0\|}}{B_0}k_b(x-x_0). \tag{8.14}$$

As a measure of the shear one can take a dimensionless parameter Θ, defined as the angle through which the lines of force turn as one traverses the inhomogeneity distance scale a of the plasma:

$$\Theta\simeq a_{\perp}h_z^2\partial(B_{0y}/B_{0z})/\partial x. \tag{8.15}$$

With allowance for (8.3), this relation can also be represented in the form

$$\Theta \simeq 4\pi a\,(\mathbf{j}_0\mathbf{B}_0)/cB_0^2. \tag{8.16}$$

We shall frequently refer simply to Θ as the *shear*.

Expressed in terms of Θ, Eq. (8.10) becomes

$$k_{\|} \simeq \Theta k_b\,(x - x_b)/a. \tag{8.17}$$

The relation (8.17) can be used conveniently to investigate perturbations for which $k_{\|} = 0$ when there is no shear. If this is not the case but nevertheless $k_{\|} \ll k_b$, the relation

$$k_{\|}(x) - k_{\|}(x_0) \simeq \Theta k_b\,(x - x_0)/a \tag{8.18}$$

has a more transparent meaning. This can be obtained by means of Eq. (8.8), which, however, must now be regarded as approximate.

§8.2. Shear Stabilization

If the presence of a shear results in $k_{\|}$ being a fairly steep function of x, it is possible for perturbations to be stabilized in a distinctive manner. This occurs as follows.

We recall that every wave packet has a certain finite extension in the x direction and that this is at least as much as $1/k_x$. On the other hand, the longitudinal wave number of a packet that grows in time is always bounded above by some $k_{\|\max}$. It follows that packets can only grow if the change of their longitudinal wave number over a distance $1/k_x$ does not exceed $k_{\|\max}$, i.e.,

$$\frac{\partial k_{\|}(x)}{\partial x} \cdot \frac{1}{k_x} < k_{\|\max} \tag{8.19}$$

The values of $k_{\|\max}$ and k_x can be determined from the problem of a plasma in a field with $\mathbf{B}_0 \| \mathbf{z}$ and are therefore independent of the shear. This makes it clear that the wave packet cannot grow if the derivative $\partial k_{\|}/\partial x$ is sufficiently large. In accordance with (8.18), this means

$$\Theta > (k_{\|\max}/k_b)\,k_x a. \tag{8.20}$$

The inequality (8.19) and the opposite inequality (8.20) refer to perturbations with $\gamma \geq \text{Re}\,\omega$. If $\gamma \ll \text{Re}\,\omega$, the longitudinal wave number must not change appreciably over a distance of order $\Delta x \simeq \gamma^{-1}\,\partial\,\text{Re}\,\omega/\partial k_x$ [see §5.1 in Volume 1] or over the inhomogeneity distance scale a of the plasma if $a < \gamma^{-1}\,\partial\,\text{Re}\,\omega/\partial k_x$. Therefore, the condition for the variation of $k_{\|}$ to be unimportant for such perturbations entails

$$\frac{\partial k_{\|}}{\partial x}\min\left(\frac{1}{\gamma}\cdot\frac{\partial\,\text{Re}\,\omega}{\partial k_x},\ a\right) < k_{\|\max} \tag{8.21}$$

Conversely, the presence of a shear renders the existence of perturbations of this kind impossible if

$$\Theta > (k_{\|\max}/k_b)\max\{1,\ \gamma a\,(\partial\,\text{Re}\,\omega/\partial k_x)^{-1}\}. \tag{8.22}$$

It must be borne in mind that the stabilizing effect of the shear may be vitiated by the destabilizing effect of the current in the plasma that produces the shear. The role of the current will be considered in §§8.6–8.8. We shall first consider the effect of a shear when the current is ignored (§§8.3–8.5).

§8.3. General Estimates of the Stabilizing Influence of a Shear

Plasma instabilities depend strongly on the relationship between the temperature and density gradients, the extent to which the plasma is collisional, and the transverse wave number. In this section we shall not specify the first two factors but assume that the plasma has parameters that are least favorable for stability (for example, fairly large $|\partial \ln T/\partial \ln n_0|$ with not too frequent collisions). We shall consider how a shear affects perturbations in such a plasma for different values of the transverse wave number.

Large-Scale Perturbations. Suppose $k_{\perp}a \simeq 1$, which corresponds to the largest scale perturbations. For these, either (8.20) or (8.22) yields an estimate for a shear to be significant:

$$\Theta > k_{\|\max}a. \tag{8.23}$$

The wave number $k_{\|\max}$ can attain values of the order of ω^*/v_{Ti}. For such perturbations, (8.23) entails

$$\Theta > \rho_i/a. \tag{8.24}$$

It can be seen that all large-scale instabilities can be suppressed by even a fairly small shear. However, the creation of even a small shear by the passage of a current through the plasma may entail the development of different kinds of current instabilities, both gradient and the ordinary nongradient instabilities. To estimate the possible destabilizing effect of the current, we shall use (8.16) and eliminate the parameter Θ from (8.20). The result can be represented in the form

$$\Pi_i > v_{T_i}/V_0, \tag{8.25}$$

where $\Pi_i \equiv (\omega_{p_i} a/c)^2$ is the dimensionless ion density number. To eliminate the possible development of ordinary current instabilities, we must have $v_{T_i} \gtrsim V_0$. This gives us a necessary condition for the suppression of large-scale instabilities:

$$\Pi_i > 1. \tag{8.26}$$

If this condition is not satisfied, then if $v_{T_i} \gtrsim V_0$ and $|\partial \ln T/\partial \ln n_0| \geq 1$, large-scale perturbations must grow, and estimates show that these will lead to turbulent transport coefficients of the order of the Bohm diffusion coefficient.

Small-Scale Perturbations. Let us now consider perturbations with $k_\perp a \gg 1$. If $\gamma \simeq \mathrm{Re}\,\omega$, the condition (8.20) and the relation $k_{\parallel \max} \simeq \omega^*/v_{Ti}$ yield an estimate for the shear that can stabilize such perturbations:

$$\Theta > \rho_i/\tilde{a}, \tag{8.27}$$

where $\tilde{a} \simeq 1/k_\perp$ is the minimal localization scale of the perturbation with the given $k_\perp$.

The inequality (8.27) can also be regarded as an upper bound on the localization region of the perturbations for a given shear:

$$\tilde{a} < \rho_i/\Theta. \tag{8.28}$$

A decrease in the localization region of the perturbations must also be regarded as a stabilizing effect, since the smaller the value $\tilde{a}$, the smaller are the turbulent fluxes at right angles to the magnetic field.

In accordance with (8.28), $\tilde{a}$ becomes of order ρ_i when

$$\Theta \simeq 1. \tag{8.29}$$

If the shear has this large value, the growth of perturbations with $k_\perp \rho_i \lesssim 1$ is impossible. On the other hand, one can show that if $\Theta \simeq 1$ it is also impossible for perturbations with $k_\perp \rho_i \gg 1$ to be excited. Thus, if $\Theta \simeq 1$, the plasma must be free of all instabilities.

The condition (8.29) can be expressed in terms of the dimensionless ion density number and the directed velocity of the electrons [cf. (8.25)]:

$$\Pi_i \simeq (a/\rho_i)(v_{T_i}/V_0). \tag{8.30}$$

It can be seen that if $V_0 \lesssim v_{T_i}$ this condition can be satisfied only for very large Π_i:

$$\Pi_i \gtrsim a/\rho_i. \tag{8.31}$$

A plasma with a lower Π_i can sustain either ordinary current instabilities if $V_0 > v_{T_i}$, or small-scale gradient instabilities if $V_0 < v_{T_i}$.

§8.4. Shear Stabilization of a Collisionless Plasma

In the case of a collisionless plasma, the estimates of §8.3 are to be interpreted as necessary conditions for stabilization if the temperature gradient is not small compared with the density gradient, $|\partial \ln T/\partial \ln n_0| \gtrsim 1$. Let us now consider a collisionless plasma with $\nabla T = 0$.

If a shear is absent, then in a plasma of sufficiently great length with $\nabla n_0 \neq 0$ and $\nabla T = 0$ instabilities – hydrodynamic and kinetic – can develop. Hydrodynamically unstable perturbations have very small $k_\parallel$ and are therefore very sensitive to both a shear and the current responsible for the shear. The effect of a current on such perturbations was discussed in §3.4. The joint effect of the current and the shear will be considered in §8.6.

The kinetic instability is sensitive to the current only for very small $k_\perp \rho_i$, $(k_\perp \rho_i)^2 < V/v_{T_e}$ [see (3.57)]. We shall assume the opposite, $(k_\perp \rho_i)^2 > V/v_{T_e}$, and consider the effect that a shear has on this instability if the current is ignored.

1. Condition for the Small-Scale Kinetic Instability to be Unaffected by a Shear. We have seen in §8.1 that if a shear is present an x-dependent longitudinal wave number $k_{\|}$ corresponds to a perturbation with given k_y and k_z. In the approximation $k_{\|} \ll k_b$, this dependence is determined by (8.18). Using this formula, we find that if the perturbation has the form of a wave packet localized in a region $\Delta x \simeq \tilde{a} \ll a$ the wave number $k_{\|}$ may nevertheless be regarded as a constant if

$$\tilde{a} \leqslant ak_{\|}(x_0)/k_b\Theta. \tag{8.32}$$

The growth in time of the amplitude of a packet with constant $k_{\|}$ can be characterized by a local dispersion equation written down for $|x-x_0| \leq \tilde{a}$ if the growth rate satisfies the condition

$$\gamma(x_0) > \tilde{a}^{-1}\,\partial\omega(x_0)/\partial k_x \tag{8.33}$$

($\tilde{a}$ is the minimal localization scale of the perturbation). This inequality can be most readily satisfied when $\tilde{a} \simeq \tilde{a}_{max}$, where $\tilde{a}_{max} \equiv ak_{\|}(x_0)/(k_b\theta)$ plays the role of the maximal dimension of the localization region of a packet with constant $k_{\|}$ [cf. (8.28)]. From (8.32) and (8.33) we then find that a shear does not significantly affect the behavior of a wave packet if

$$\Theta < ak_{\|}(x_0)\,\gamma(x_0)/(k_b\,\partial\omega/\partial k_x). \tag{8.34}$$

By means of this relation we obtain a condition for a shear to have no important effect on the kinetic instability of a plasma with an inhomogeneous density considered in §3.1.2. Suppose $z_i \ll 1$, i.e., we have long-wavelength perturbations, $k_b \simeq k_x$. Then

$$\left.\begin{aligned} k_{\|}/k_b &\simeq k_b\rho_i^2/a,\\ \gamma &\simeq (\omega^*)^2 z_i/(k_{\|}v_{Te}),\\ \partial\omega/\partial k_x &\simeq \omega^* z_i/k_b, \end{aligned}\right\} \tag{8.35}$$

and the condition (8.34) reduces to

$$\Theta < \left(\frac{m_e}{m_i}\right)^{1/2} k_b\rho_i. \tag{8.36}$$

It can be seen that perturbations with $k_b\rho_i \simeq 1$ are least sensitive to a shear: for them it is sufficient that $\Theta < (m_e/m_i)^{1/2}$.

2. Estimates for a Shear That Stabilizes Kinetically Unstable Perturbations. Let us now consider the case of a shear that is not too small:

$$\Theta > \left(\frac{m_e}{m_i}\right)^{1/2} k_b \rho_i. \tag{8.37}$$

If this condition is satisfied, one must consider the behavior of a packet in a region whose dimensions exceed the packet dimension $\tilde{a}$. The behavior of a packet in an inhomogeneous plasma was discussed in §5.4 of Volume 1. It was shown that the maximum amplitude of the packet grows in time in accordance with the law

$$|\psi| \sim \exp\left\{\int \gamma[x(t)]\,dt\right\}, \tag{8.38}$$

where x(t) is the coordinate of the packet maximum.

As the packet moves along the x direction, the sign of the growth rate may be reversed – growth may be replaced by decay. Therefore, the packet succeeds in growing appreciably in amplitude if

$$\int\limits_{t(x_1)}^{t(x_2)} \gamma[x(t)]\,dt > 1, \tag{8.39}$$

where x_2 and x_1 are the points at which γ vanishes. This inequality can also be expressed in terms of the rate of spatial growth [see §5.4]:

$$\left|\int\limits_{x_1}^{x_2} \operatorname{Im} k_x\,dx\right| > 1. \tag{8.40}$$

Let us consider what this condition means for the case of a kinetic instability with $z_i \ll 1$. Proceeding from a relation of the type (3.15), we obtain an equation for k_x:

$$\begin{gathered}\operatorname{Im} k_x = -\frac{\sqrt{\pi}\,(\omega - \omega^*)}{4\rho_i^2 v_{T_e} |k_\parallel| \operatorname{Re} k_x}, \\ \operatorname{Re} k_x \left[\left(\frac{2k_\parallel^2 T}{m_i \omega^2} - 1 + \frac{\omega^*}{\omega}\right)\frac{1}{2\rho_i^2} - k_y^2\right]^{1/2},\end{gathered} \tag{8.41}$$

where $k_\parallel = k'_\parallel x$, $\omega^* \equiv \omega_{n_e}$.

Substituting this result into (8.40), we arrive at an estimate for the shear at which this condition ceases to hold:

$$\Theta_{\max} \simeq \left(\frac{m_e}{m_i}\right)^{1/2} \int \frac{dX}{X} \frac{1-\omega/\omega^*}{\left[2X^2 \frac{\omega^*}{\omega} - 1 - 2(k_b\rho_i)^2 + \frac{\omega^*}{\omega}\right]^{1/2}}. \tag{8.42}$$

Here the integration with respect to X is carried out in the region in which the denominator is real:

$$X > \left[1 + 2(k_b\rho_i)^2 - \frac{\omega^*}{\omega}\right]^{1/2} \Big/ \sqrt{2}, \tag{8.43}$$

and the frequency ω is assumed to be slightly less than ω^*, $0 < 1-\omega^*/\omega \ll 1$.

If $\frac{\omega^*}{\omega} - (1 + 2k_b^2\rho_i^2) \gg (k_b\rho_i)^2$, we obtain from (8.42) and (8.43) an estimate for the shear that stabilizes such perturbations; this estimate is identical with (8.37).

There is less effective stabilization of perturbations with the frequency

$$\omega \simeq \frac{\omega^*}{1 + 2(k_b\rho_i)^2}. \tag{8.44}$$

In this case, (8.43) shows that X may become very small, so that the integral (8.42) diverges logarithmically. To eliminate this divergence we recall that X is bounded below by the condition $\mathrm{Re}\, k_x > \mathrm{Im}\, k_x$. This condition is violated when

$$k_{\parallel} \simeq \left(\frac{m_e}{m_i}\right)^{1/6} \frac{\omega^*}{v_{T_i}} (k_b\rho_i)^{2/3}, \tag{8.45}$$

and in this case

$$\mathrm{Im}\, k_x \simeq \mathrm{Re}\, k_x \simeq \frac{1}{\rho_i} \left(\frac{m_e}{m_i}\right)^{1/6} (k_b\rho_i)^{2/3}. \tag{8.46}$$

The condition (8.40) then gives an estimate for the shear:

$$\Theta_{\max} \simeq \left(\frac{m_e}{m_i}\right)^{1/3} (k_b\rho_i)^{4/3}. \tag{8.47}$$

It can be seen that even the most dangerous perturbations are stabilized if $\Theta > (m_e/m_i)^{1/3}$.

§8.5. Shear Stabilization of a Collisional Plasma

In a collisional plasma, perturbations with $k_\parallel \simeq \omega^*/v_{Ti}$ can be excited in two cases: (1) when the collisions are not too frequent, $\omega^* > \nu_{ei}$, and the temperature gradient is not too small, $|\partial \ln T/\partial \ln n_0| \geq 1$; (2) in a strongly collisional plasma, $\omega^* < \nu_{ei}$, when the temperature gradient is small, $|\partial \ln T/\partial \ln n_0| \leq 1$. Under these conditions the estimates for a stabilizing shear obtained in §8.3 have the nature of necessary conditions for stabilization. Otherwise, i.e., in the absence of growing perturbations with $k_\parallel \simeq \omega^*/v_{Ti}$, stabilization can occur at a much smaller shear. An example of this kind of a readily stabilized collisional plasma will be discussed in this section.

We shall consider here the case of a plasma with $\nabla T = 0$ and not too frequent collisions, $\omega^* > \nu_{ei}$. In a collisional plasma with $\nabla T = 0$ there is no transverse gradient of the current, so that the presence of a current in such a plasma leads only to a shear but not to current instabilities. Therefore, the same kind of instabilities arise in such a plasma as in a plasma without a current [see §5.2]. We shall consider how these instabilities are affected by a shear.

1. Inertial-Dissipative Instability with $\omega_s \simeq \omega^*$. A growth rate $\gamma \simeq \mathrm{Re}\,\omega \simeq \omega^*$ corresponds to perturbations with $\omega_s \simeq \omega^*$, i.e., these are the most rapidly growing perturbations. Using formula (8.20), we find that such perturbations are stabilized if

$$\Theta > z_i^{3/4} (\nu_{ie} a/v_{T_i})^{1/2}. \tag{8.48}$$

In particular, the smallest scale perturbations, $k_\perp a \simeq 1$, are stabilized with even a very small shear:

$$\Theta > (\nu_{ie} a/v_{T_i})^{1/2} (\rho_i/a)^{3/2}. \tag{8.49}$$

For the stabilization of perturbations with $k_\perp \rho \simeq 1$ one requires a shear of order

$$\Theta \simeq (\nu_{ie} a/v_{T_i})^{1/2}. \tag{8.50}$$

2. Inertial-Dissipative Instability when $\omega_s \simeq \omega^*$. In this case, it follows from §5.2 that $\gamma \ll \mathrm{Re}\,\omega$. An estimate for the shear can be obtained in the same way as in §8.4. To do this, we must substitute into the inequality (8.34) the characteristic values of $k_{\|}/k_b$, γ, and $\partial\omega/\partial k_x$, which are (see §5.2)

$$\left.\begin{array}{c} k_{\|}/k_b \simeq k_b\rho_i^2/a, \\ \gamma \simeq (\omega^*/k_{\|}v_{Te})^2 z_i\nu_{ei}, \qquad \partial\omega/\partial k_x \simeq \omega^* z_i/k_b. \end{array}\right\} \tag{8.51}$$

In this way we find that a shear affects the perturbations under consideration [the condition opposite to (8.34)] if

$$\Theta > \nu_{ie}/\omega^*. \tag{8.52}$$

In general, this is a more stringent condition than (8.48). At the limit of applicability of the collisional approximation, when $\nu_{ei} \simeq k_{\|}v_{Te}$, the condition (8.52) is identical with (8.37).

§8.6. Influence of a Shear on the Current Instabilities of a Collisionless Plasma

As we have already mentioned in §3.4, the passage of a current through a plasma may lead to the development of instabilities of two kinds – hydrodynamic and kinetic. The first kind has a large growth rate and small k_z; the second, a small growth rate and large k_z. We shall now consider how these are affected by a shear.

1. Hydrodynamic Instability. Using (3.84), we obtain an estimate for perturbations with $k_\perp a \simeq 1$:

$$k_{\|\max} a \simeq V_0/a\omega_{Be}. \tag{8.53}$$

Hence, and from (8.23), it follows that a shear affects large-scale perturbations if

$$\Theta > V_0/a\omega_{Be}. \tag{8.54}$$

Expressing Θ in terms of the current in accordance with (8.16), we find

$$\Pi_e > 1. \tag{8.55}$$

This means that a large-scale current instability can develop only in a plasma with a sufficiently low density, $\Pi_e < 1$.

Note that the hydrodynamic excitation of perturbations also occurs in the absence of a current [see §3.1]. For this instability, $ak_{\parallel max} \simeq (m_e/m_i)^{1/2} (\rho_i/a)^{1/2}$. From this we obtain an estimate for the shear that affects the instability without current: $\Theta > (m_e/m_i)^{1/2} (\rho_i/a)^2$. This is of the order of (8.54) exactly when the instability with a current begins to predominate over the instability without a current. Therefore the passage of a current in a plasma with $\Pi_e < 1$ does not lead to stabilization of the instability without a current but to its replacement by the instability with a current.

Let us now suppose that (8.55) is satisfied, $\Pi_e > 1$, and see whether perturbations that are strongly localized in the x direction, $k_x a \gg 1$, can grow under these conditions. We shall assume that the perturbations are electrostatic, taking $k_\perp > \omega_{p_e}/c$. We shall also assume $k_\perp \rho_i \ll 1$. Obviously the conditions $k_\perp \rho_i \ll 1$ and $k_\perp \gg \omega_{p_e}/c$ are compatible only if $\beta \ll m_e/m_i$.

A. Perturbations with $\omega \gg \omega^*$. It is precisely this condition which the perturbation frequency satisfies if the shear is neglected [see §3.4]. One can show that if $\omega \gg \omega^*$ the perturbations in a plasma with a shear are described by the equation

$$\frac{d^2\psi}{dx^2} - k_b^2\psi\,[1 + p\,(x - x_0)^2 + q\,(x - x_0)] = 0, \tag{8.56}$$

where

$$\left.\begin{aligned} p &= -\frac{m_i}{m_e}\left(\frac{\omega_{B_i}}{\omega}\right)^2 \left(\frac{k_{\parallel}'}{k_b}\right)^2, \\ q &= -\left(\frac{\omega_{B_i}}{\omega}\right)^2 \frac{k_{\parallel}'}{k_b}\cdot\frac{1}{e_e n_0}\cdot\frac{\partial j_{0\parallel}}{\partial x}, \\ k_{\parallel}' &\equiv \left(\frac{dk_{\parallel}}{dx}\right)_{x=x_0}. \end{aligned}\right\} \tag{8.57}$$

A linear substitution of the independent variable reduces Eq. (8.56) to the equation for a quantum-mechanical oscillator. The equation that follows for the eigenvalues is

$$\left(\frac{q^2}{4p} - 1\right)\frac{|k_b|}{\sqrt{p}} = 2n + 1, \quad n = 0, 1, 2, \ldots \tag{8.58}$$

The region of localization of the solution with n = 0 is of the order of

$$\Delta x \simeq \left(\frac{1}{k_b^2 p}\right)^{1/4}. \tag{8.59}$$

If p and q have the form (8.57), Eq. (8.58) means

$$\omega^2 + i\omega(2n+1)\left(\frac{m_i}{m_e}\right)^{1/2}\frac{|\omega_{B_i}k'_{\parallel}|}{k_b^2} + \frac{m_e}{4m_i}(\varkappa_j V_0)^2 = 0, \qquad (8.60)$$

where $\varkappa_j \equiv \partial \ln(n_0V)/\partial x$. The shear (the terms with $k'_{\parallel}$) plays a dual role – it restricts the region of localization in the x direction [see (8.59)] and decreases the growth rate of the current instability [Eq. (8.60)]. Let us consider the second of these effects. The growth rate of the lowest level, n = 0, is least sensitive to the shear. If the shear is not too large, the growth rate of the corresponding perturbations is

$$\gamma = \frac{1}{2}\left(\frac{m_e}{m_i}\right)^{1/2}\varkappa_j V_0, \qquad (8.61)$$

which is in qualitative agreement with (3.56).

Using (8.60) and (8.16) we find that in deriving (8.61) the shear can be ignored if

$$\Pi_e < k_b a. \qquad (8.62)$$

This agrees with the estimates of §8.3 for the case $k_b a \simeq 1$.

Now suppose

$$\Pi_e \gg k_b a. \qquad (8.63)$$

The growth rate of the lowest level is then approximately

$$\gamma \simeq \frac{1}{4}\left(\frac{m_e}{m_i}\right)^{1/2}\varkappa_j V_0 \frac{k_b a}{\Pi_e}. \qquad (8.64)$$

The region of localization of such perturbations is approximately

$$\Delta x \simeq \frac{a}{\Pi_e^{1/2}} \simeq \frac{c}{\omega_{pe}}. \qquad (8.65)$$

It must therefore be assumed that $k_b \Delta x \gg 1$.

B. Perturbations with $\omega \leq \omega^*$. As Π_e increases, the growth rate (8.64) decreases and may become of the order of ω^* if

$$\Pi_e \simeq \left(\frac{m_e}{m_i}\right)^{1/2}\frac{V_0}{v_{T_i}}\cdot\frac{a}{\rho_i}. \qquad (8.66)$$

(In accordance with (3.58) this Π_e exceeds unity.) At even greater Π_e, one must take into account the terms of order ω^*/ω in Eq. (8.56). Then p and q are determined by the expressions

$$\left.\begin{aligned} p &= -\frac{m_i}{m_e}\cdot\frac{\omega_{B_i}^2 (k_{\parallel}')^2 (1-\omega_{p_e}^*/\omega)}{\omega(\omega-\omega_{p_i}^*)k_b^2}, \\ q &= -\frac{\omega_{B_i}}{\omega(\omega-\omega_{p_i}^*)}\cdot\frac{k_{\parallel}'}{k_b}\cdot\frac{1}{e_e n_0}\cdot\frac{\partial j_{0\parallel}}{\partial x}. \end{aligned}\right\} \quad (8.67)$$

Instead of (8.60) we obtain

$$\omega(\omega-\omega_{p_i}^*)-(2n+1)\sqrt{-(\omega-\omega_{p_i}^*)(\omega-\omega_{p_e}^*)}\times$$
$$\times\left(\frac{m_i}{m_e}\right)^{1/2}\left|\frac{\omega_{B_i}k_{\parallel}'}{k_b^2}\right|+\frac{m_e}{4m_i}\frac{(\varkappa_j V_0)^2}{1-\dfrac{\omega_{pe}^*}{\omega}}=0. \quad (8.68)$$

In the limit of large $k_{\parallel}'\left[\Pi_e \gg \left(\frac{m_e}{m_i}\right)^{1/2}\frac{V_0 a}{\rho_i v_{T_i}}\right]$ this equation has a root with $\mathrm{Im}\,\omega > 0$ and a frequency near ω_{pe}^*:

$$\left.\begin{aligned} \mathrm{Re}\,\omega &\simeq \omega_{pe}^*, \\ \gamma &\simeq \omega_{pe}^*\left\{\left(\frac{m_e}{m_i}\right)^{1/2}\frac{V_0 a}{\rho_i v_{T_i}}\right\}^{2/3}\Pi_e^{-2/3}. \end{aligned}\right\} \quad (8.69)$$

It follows from the above analysis that the maximal growth rate of the current instability, $\gamma \simeq \left(\frac{m_e}{m_i}\right)^{1/2}\frac{V_0}{a}$, is attained only in a plasma with a low dimensionless electron density number, $\Pi_e \ll 1$. With increasing Π_e, the growth decreases first as $1/\Pi_e$ [Eq. (8.64)] and then (for $\gamma < \omega^*$) as $\Pi_e^{-2/3}$.

2. Kinetic Instability. We find an estimate for the effect that a shear has on this instability by means of formula (8.34) and the following characteristic values of $k_{\parallel}$, γ, and $\partial\omega/\partial k_x$ (see §3.4.2):

$$\left.\begin{aligned} k_{\parallel}/k_b &\simeq (\rho_i/a)(V_0/v_{T_i}), \\ \gamma \simeq \omega^* V_0/v_{Te}, &\quad \partial\omega/\partial k_x \simeq z_i\omega^* k_b. \end{aligned}\right\} \quad (8.70)$$

A shear is important if

$$\Theta \gtrsim (m_e/m_i)^{1/2}(V_0/v_{T_i})^2(k_b\rho_i)^{-1}. \quad (8.71)$$

In accordance with (3.62) this result applies to fairly long-wavelength perturbations, $k_b\rho_i < V_0/v_{T_i}$. This shows that the kinetic current instability is not suppressed so effectively by a shear as the instability without a current.

§8.7. Current-Convective Instability of a Collisional Plasma in a Field with a Shear

In a collisional plasma with $\nabla T \neq 0$ a current gradient can lead to the development of two forms of current-convective instability: the inertial and noninertial instabilities (§5.6). We shall now consider how these are affected by a shear.

1. Inertial Instability. The inertial current-convective instability has characteristic values of $k_{\parallel}$ determined by the condition (5.49). Using this and formula (8.23), we obtain an estimate for a shear that can affect this instability when $k_{\perp}a \approx 1$:

$$\Theta \gtrsim (\nu_{ie}^2 V_0/a)^{1/3}/\omega_{B_i}. \tag{8.72}$$

If we express Θ in this equation in terms of j_0 and then in terms of the dimensionless ion density number, we find

$$\Pi_i > (m_e/m_i)^{3/4}\,(\rho_i v_{T_i}/aV_0)^{3/2}\,S^{-3/2}. \tag{8.73}$$

On the other hand, considering (5.50), we find that the condition for the perturbations to be nonelectrostatic, $\omega/\nu_{ei} > \Pi_e$, reduces to an inequality of the same nature. Thus, the problem of establishing the effect of a shear on the inertial current-convective instability must, in general, be solved with allowance for the effects that arise when the perturbations are not electrostatic if one wishes to consider perturbations with $k_{\perp}a \approx 1$. However, we shall here assume $k_{\perp}a \gg 1$ and consider only electrostatic perturbations. We shall investigate large-scale nonelectrostatic perturbations in a plasma that satisfies the condition (8.73) in §8.8.

A. Perturbations with $\omega \gg \omega^*$. We shall first assume that, as in the case $\Theta = 0$, $\omega \gg \omega^*$. Then the original equation can be reduced to the form (8.56), in which

$$\left.\begin{aligned} p &= \mathrm{i}\,(m_i/m_e)\,(k_{\parallel}'/k_b)^2\,\omega_{B_i}^2/\nu_{ei}\omega, \\ q &= -\frac{3}{2}\,\omega_{B_i}k_{\parallel}'V_0\varkappa_T/k_b\omega^2. \end{aligned}\right\} \tag{8.74}$$

As in the case of (8.60), we obtain the dispersion equation

$$\omega^3-(2n+1)\,\omega^{5/2}\sqrt{i\frac{m_i}{m_e}\tau_e}\left|\frac{\omega_{B_i}k'_{\parallel}}{k_b^2}\right|+i\frac{9}{16}\cdot\frac{m_e}{m_i\tau_e}(\varkappa_T V_0)^2=0. \qquad (8.75)$$

If we neglect the term with $k'_{\parallel}$, Eq. (8.75) yields

$$\gamma\simeq \mathrm{Re}\,\omega\simeq(\varkappa_T^2V_0^2\nu_{ie})^{1/3}, \qquad (8.76)$$

which is approximately the same as (5.50) obtained without allowance for a shear.

The approximation of small $k'_{\parallel}$ reduces to the condition

$$\Pi_e<k_b a\left[\frac{\nu_{ei}a}{V_0}\left(\frac{m_i}{m_e}\right)^{1/2}\right]^{2/3}\equiv\Pi_e^{(0)}. \qquad (8.77)$$

At the limits of applicability of the collisional approximation ($\omega\simeq\nu_{ei}$), this result agrees with (8.62).

If Π_e is greater than is allowed by the condition (8.77), Eq. (8.75) does not yield (8.76) but

$$\gamma\simeq\mathrm{Re}\,\omega\simeq\left(\frac{\Pi_e^{(0)}}{\Pi_e}\right)^{2/5}(\varkappa_T^2V_0^2\nu_{ie})^{1/3}. \qquad (8.78)$$

As a function of ν_{ei}, $\gamma\propto\nu_{ei}^{3/5}$.

It can be seen that, as in the case of a collisionless plasma, the growth rate decreases with increasing Π_e, although in accordance with a different law [cf. (8.64)].

At the same values of V_0 and a, the collisional growth rate is greater than the collisionless growth rate. The localization region of the perturbations is also larger. In this sense, the instability of a collisional plasma is more interesting than that of a collisionless plasma.

B. Perturbations with $\omega\simeq\omega^*$. The approximation $\omega\gg\omega^*$ ceases to hold when

$$\Pi_e\gtrsim\left(\frac{\nu_{ei}}{\omega^*}\right)^{3/2}\frac{V_0}{v_{T_i}}\left(\frac{m_e}{m_i}\right)^{1/2}\frac{a}{\rho_i}\equiv\Pi_e^{(1)}. \qquad (8.79)$$

At such values of Π_e the dispersion equation (8.58) reduces to

$$p^{3/2}=\frac{|k_b|}{4}q^2, \qquad (8.80)$$

where

$$\left.\begin{aligned} p &= \mathrm{i}\,\frac{m_i}{m_e}\,\frac{\omega_{B_i}^2\tau_e(k_{\|}')^2}{\omega k_b^2}\,\frac{\left(1-\dfrac{\omega_{ne}}{\omega}-1.71\,\dfrac{\omega_{Te}}{\omega}\right)}{1-\dfrac{\omega_{p_i}^*}{\omega}}, \\ q &= -\frac{3}{2}\,\frac{\omega_{B_i}k_{\|}'\varkappa_T V_0}{k_b\omega^2}\cdot\frac{1}{1-\dfrac{\omega_{p_i}^*}{\omega}}. \end{aligned}\right\} \qquad (8.81)$$

In the zeroth approximation in q, Eq. (8.80) yields

$$\mathrm{Re}\,\omega = \omega_{ne} + 1.71\omega_{Te}. \qquad (8.82)$$

Taking into account the expression for q, we find that the growth rate is approxiamtely

$$\gamma \simeq \left(\frac{\Pi_e^{(1)}}{\Pi_e}\right)^{2/3}\mathrm{Re}\,\omega. \qquad (8.83)$$

In order to obtain a complete picture of the current instabilities of a plasma with $\Pi_e > 1$, the electrostatic instabilities discussed above must be augmented by the nonelectrostatic instabilities which are discussed in §8.8.

§8.8. Tearing-Mode Instability

Here we shall investigate the possibility of excitation by a current of nonelectrostatic perturbations in a collisional plasma. The assumption that the perturbations are nonelectrostatic and that collisions are important is justified if

$$(\omega_{pe}/ck_\perp)^2 < \omega/\nu_{ei} < 1. \qquad (8.84)$$

Under these assumptions the perturbations are described by the following system of equations:

$$\left.\begin{aligned} &\nabla_\perp\left(\frac{\omega_{p_i}}{\omega_{B_i}}\right)^2\nabla_\perp\psi + \frac{4\pi k_b}{B_0\omega}\cdot\frac{\partial j_{0\|}}{\partial x}A_{\|} - \\ &\qquad -\frac{\mathrm{i}\omega_{p_i}^2 k_{\|}^2}{\omega\nu_{ei}}\left(\psi - \frac{\omega A_{\|}}{ck_{\|}}\right) = 0, \\ &\Delta_\perp A_{\|} - \frac{\mathrm{i}k_{\|}\omega_{pe}^2}{c\nu_{ei}}\left(\psi - \frac{\omega A_{\|}}{ck_{\|}}\right) = 0. \end{aligned}\right\} \qquad (8.85)$$

In this system, $A_\parallel$ is the component of the vector potential **A** of the perturbation along $\mathbf{B}_0$.

We shall be interested in perturbations with $\omega \gg \omega^*$ and we have therefore omitted the terms of order ω^*/ω. The frequency of electron collisions is assumed to be fairly high: $\nu_{ci} > (k_\parallel v_{T_e})^2/\omega$.

In Eqs. (8.85), the longitudinal wave number $k_\parallel$ is a function of the coordinate x and vanishes somewhere at the point $x = x_0$. Far from x_0, where one can assume that

$$\left.\begin{aligned} &\omega\nu_{ei}/[(m_i/m_e)\,\omega_{B_i}^2 k_\parallel^2/k_\perp^2] \ll 1,\\ &\nu_{ei}k_b\,(\partial j_{0\parallel}/\partial x)/\omega k_\parallel e_e \omega_{B_e} n_0 \ll 1,\end{aligned}\right\} \tag{8.86}$$

both the first and the second of the equations (8.85) yield approximately

$$A_\parallel = (ck_\parallel/\omega)\,\psi. \tag{8.87}$$

This means that there is no longitudinal component of the perturbed electric field, $E_\parallel = 0$. Taking into account the small difference between ψ and $\omega A_\parallel/ck_\parallel$, we obtain an equation that describes the perturbations far from the point $x = x_0$:

$$\Delta_\perp A_\parallel - \frac{4\pi k_b}{k_\parallel c B_0}\cdot\frac{\partial j_{0\parallel}}{\partial x}A_\parallel = 0. \tag{8.88}$$

Here, k_b and B_0 are assumed to be weak functions of x. Therefore, using (8.3) and (8.7), we can make the following transformations in the last term on the left-hand side of (8.88):

$$\frac{4\pi k_b}{cB_0}\cdot\frac{\partial j_{0\parallel}}{\partial x} = \frac{\partial}{\partial x}\left(\frac{4\pi k_b j_{0\parallel}}{cB_0}\right) = \frac{\partial^2 k_\parallel}{\partial x^2}\,. \tag{8.89}$$

Equation (8.88) then reduces to

$$\frac{d^2 A_\parallel}{dx^2} - k_b^2 A_\parallel - \frac{1}{k_\parallel}\frac{d^2 k_\parallel}{dx^2}A_\parallel = 0. \tag{8.90}$$

Equation (8.90) ceases to hold at very small $k_\parallel(x)$ (i.e., for $x \approx x_0$) since the terms ignored above are then not small. Therefore, at $x \approx x_0$ the solution of the approximate equation (8.90) must be fitted to the solution of the more exact system of equations (8.85). If x_1

and x_2 are points near x_0 ($x_1 < x_0 < x_2$) at which Eq. (8.90) is still valid, the condition for the fitting of the solutions reduces to equating the differences

$$\Delta = [d \ln A_{\parallel}(x)/dx]\Big|_{x_1}^{x_2}, \tag{8.91}$$

calculated for the solution of the "external" and "internal" problems.

We find the solution in the external region by means of (8.90) and the boundary conditions

$$A_{\parallel}(x_+) = A_{\parallel}(x_-) = 0, \tag{8.92}$$

where x_+ and x_- are the coordinates of the plasma boundary (the plasma is assumed to be confined in a conducting vessel). For simplicity we assume $k_b\,|x_+ - x_-| \ll 1$. In (8.90) we can then neglect the term with k_b^2, and the solution can be found explicitly:

$$A_{\parallel}(x) = Ck_{\parallel}\begin{cases} \displaystyle\int_x^{x_+} dx/k_{\parallel}^2, & x > x_0, \\ \displaystyle -\int_{x_-}^{x} dx/k_{\parallel}^2, & x < x_0, \end{cases} \tag{8.93}$$

where C is an arbitrary constant. In deriving (8.93) we have assumed that at $x \simeq x_0$ the values of $A_{\parallel}$ to the right and the left of the point x_0 are the same. Substituting (8.93) into (8.91), we find Δ for the external problem:

$$\Delta = -[k_{\parallel}'(x_0)]^2\left[\frac{1}{k_{\parallel}(x_+)\,k_{\parallel}'(x_+)} - \frac{1}{k_{\parallel}(x_-)\,k_{\parallel}'(x_-)} + \int_{x_-}^{x_+}\frac{k_{\parallel}''\,dx}{k_{\parallel}(k_{\parallel}')^2}\right]. \tag{8.94}$$

In the internal region ($x \approx x_0$), the system (8.85) can be written thus:

$$\left.\begin{aligned} \frac{d^2\psi}{dx^2} - \frac{m_i}{m_e}\cdot\frac{\omega_{B_i}^2}{\nu_{ei}\gamma}(k_{\parallel}')^2(x-x_0)^2\psi &= \\ = \frac{i}{c}\cdot\frac{m_i}{m_e}\cdot\frac{\omega_{B_i}^2}{\nu_{ei}}\,k_{\parallel}'(x-x_0)\,A_{\parallel}&, \\ \frac{d^2A_{\parallel}}{dx^2} = \left(\frac{\omega_{pe}}{c}\right)^2\frac{\gamma}{\nu_{ei}}\left(A_{\parallel} - \frac{ck_{\parallel}'(x-x_0)}{i\gamma}\psi\right)&, \end{aligned}\right\} \tag{8.95}$$

where $\gamma \equiv -i\omega$. Here we have again assumed that the second inequality in (8.86) holds.

From the second equation of (8.95)

$$\Delta = \left(\frac{\omega_{pe}}{c}\right)^2 \frac{\gamma}{\nu_{ei}} \int_{x_1}^{x_2} \left(1 - \frac{ck'_{\parallel}(x - x_0)}{i\gamma} \cdot \frac{\psi}{A_{\parallel}}\right) dx. \qquad (8.96)$$

We have used the fact that $A_{\parallel} \approx$ const near x_0. As $|x-x_0|$ increases, the integrand in (8.96) tends to zero [by virtue of (8.87)], so that the limits of integration x_1 and x_2 can be replaced by infinite limits.

It remains to find the ratio $\psi/A_{\parallel}$ from the first equation (8.95), calculate the integral (8.96), and equate the result to the right-hand side of (8.94). The change of variables

$$\left.\begin{aligned} x - x_0 &= \xi\delta, \\ \psi/A_{\parallel} &= (-i\gamma/ck'_{\parallel}\delta)\,\Phi, \\ \delta &\equiv (m_e/m_i)^{1/4} (\gamma\nu_{ei}/\omega_{B_i}^2 k'^2_{\parallel})^{1/4}, \end{aligned}\right\} \qquad (8.97)$$

reduces (8.96) and the first equation of (8.95) to the canonical form

$$\Delta = QI, \qquad (8.98)$$

$$\frac{d^2\Phi}{d\xi^2} - \xi^2\Phi = \xi, \qquad (8.99)$$

where

$$I = \int_{-\infty}^{\infty} (1 + \Phi\xi)\, d\xi, \qquad (8.100)$$

$$Q = \left(\frac{\omega_{pe}}{c}\right)^2 \frac{\gamma\delta}{\nu_{ei}}. \qquad (8.101)$$

The integral I has been calculated by Furth et al., and is equal to

$$I = 4\pi\Gamma\left(\frac{3}{4}\right) \Big/ \Gamma\left(\frac{1}{4}\right). \qquad (8.102)$$

Since Q and I are positive (for $\gamma > 0$), Eq. (8.98) can be satisfied provided $\Delta > 0$, i.e., in accordance with (8.94), if

$$\int_{x_-}^{x_+} \frac{k''_{\parallel}\, dx}{k_{\parallel}(k'_{\parallel})^2} + \frac{1}{k_{\parallel}(x_+)\, k'_{\parallel}(x_+)} - \frac{1}{k_{\parallel}(x_-)\, k'_{\parallel}(x_-)} < 0. \qquad (8.103)$$

This is an instability condition. It can be expressed solely in terms of the steady-state parameters of the plasma and is ultimately determined by the distribution of the current over the cross section of the plasma.

Suppose the inequality (8.103) is satisfied, so that there is instability (frequently called the tearing-mode instability). Using (8.98), (8.101), and the last equation of (8.97), we can estimate the growth rate of the perturbations and the width of the internal region:

$$\gamma \simeq \gamma_0^{2/5} \nu_{ei}^{3/5} \Pi_e^{-2/5}, \tag{8.104}$$

$$\delta \simeq a\Pi_e^{-3/5} (\nu_{ei}/\gamma_0)^{2/5}, \tag{8.105}$$

where $\gamma_0 \simeq (m_e/m_i)^{1/2} V_0/a$ is the growth rate of the current instability of a collisionless plasma in the absence of a shear [see Eq. (3.56)].

The growth rate (8.104) satisfies the original assumptions (8.84) if

$$\Pi_e^{1/6} < \gamma_0/\nu_{ei} < \Pi_e. \tag{8.106}$$

Here we have used $k_\perp \simeq 1/\delta$. The assumption $\omega > \omega^*$ is justified if

$$\beta < (\gamma_0/\nu_{ei})^{2/5} \Pi_e^{3/5} (\nu_{ei}/\omega_{Be}). \tag{8.107}$$

In accordance with (8.106), $\max \gamma_0/\nu_{ei} \le \Pi_e$, so that the β_{max} at which our solution is valid must satisfy the condition

$$\beta_{max} < (\nu_{ei}/\omega_{Be}) \Pi_e. \tag{8.108}$$

The width of the internal region (8.105) must be greater than the ion Larmor radius, which is possible provided

$$\beta < (m_e/m_i)(\nu_{ei}/\gamma_0)^{4/5} \Pi_e^{-1/5}. \tag{8.109}$$

This yields an upper bound for β [cf. (8.108)]:

$$\beta_{max} < (m_e/m_i) \Pi_e^{-1/3}. \tag{8.110}$$

The largest of the values of β satisfying the conditions (8.108) and (8.109) is attained for $\Pi_e \simeq (\omega_{Bi}/\nu_{ei})^{3/4}$. It does not exceed

$$\beta_{max} < (m_e/m_i)(\nu_{ei}/\omega_{Bi})^{1/4}. \tag{8.111}$$

At larger values of β this instability is impossible.

Bibliography

Collisionless Plasma

1. L. V. Mikhailovskaya and A. B. Mikhailovskii, Nucl. Fusion, 3:28 (1963).
2. L. V. Mikhailovskaya and A. B. Mikhailovskii, Nucl. Fusion, 3:113 (1963).
3. A. A. Galeev, Dokl. Akad. Nauk SSSR, 150:503 (1963) [Sov. Phys. – Doklady, 8:444 (1963).
4. A. A. Galeev, Zh. Eksp. Teor. Fiz., 44:1920 (1963) [Sov. Phys. – JETP, 17:1292 (1963)]. In [1-4] a study is made of the effect of a shear on the instabilities due to a density gradient. (In [3, 4] it is also assumed that $\nabla T \neq 0$, but the most "dangerous" instabilities with $\omega \simeq k_{\|} v_{Ti}$) are not considered.) A necessary condition for the stabilization of large-scale perturbations, $\Theta > \rho_i/a$, (§8.3) is obtained. Both electrostatic and nonelectrostatic perturbations, which are important for $\beta > m_e/m_i$, are studied. For the latter case a number of stabilization conditions that depend on β are obtained. In [2] the stabilization condition $\Theta > (m_e/m_i)^{1/2}$ (§8.4) is obtained for wave-packet type perturbations.
5. A. B. Mikhailovskii, in: Reviews of Plasma Physics, Vol. 3, Consultants Bureau, New York (1967), p. 159.
6. A. A. Galeev, S. S. Moiseev, and R. Z. Sagdeev, At. Energ., 15:451 (1963). The reviews [5, 6] include a discussion of the effect of a shear on plasma instability.
7. B. Coppi, Phys. Lett., 11:226 (1964).
8. A. B. Mikhailovskii, Zh. Tekh. Fiz., 35:1933 (1965) [Sov. Phys. – Tech. Phys., 10:1490 (1966)]. In [7, 8] a study is made of the effect of a shear on the hydrodynamic current-convective instability (§8.6.1).
9. J. D. Jukes, Phys. Fluids, 7:1468 (1964). A condition of the type (8.37) (§8.4) is obtained from the solution of an eigenvalue problem.
10. N. A. Krall and M. N. Rosenbluth, Phys. Fluids, 8:1488 (1965). In the same manner as in [3, 4] the condition $\Theta > \rho_i/a$ is obtained but with allowance for the numerical coefficient (§8.3).
11. B. B. Kadomtsev and O. P. Pogutse, Plasma Physics and Controlled Nuclear Fusion Research, IAEA, Vienna (1966), Vol. 1, p. 365.
12. B. B. Kadomtsev and O. P. Pogutse, in: Reviews of Plasma Physics, Vol. 5, Consultants Bureau, New York (1970), p. 249. In [11, 12] it is shown that perturbations with $k_{\|} \simeq \omega^*/v_{Ti}$ play an important role in a plasma with an inhomogeneous temperature (§8.3); such perturbations are the hardest of all to stabilize by a shear. Estimates are given for the effect of a shear on the main types of plasma instability. The role of turbulent processes in an unstable plasma in a field with a shear is also discussed.
13. B. Coppi et al., Nucl. Fusion, 6:261 (1966). Discussion of the behavior of wave packets in an inhomogeneous plasma in a field with shear. For such perturbations a stabilization condition of the type $\Theta > (m_e/m_i)^{1/2} k_b \rho_i$ is obtained with a numerical coefficient (§8.4).
14. O. P. Pogutse, Zh. Eksp. Teor. Fiz., 52:759 (1967) [Sov. Phys. – JETP, 25:498 (1967)].
15. B. Coppi, M. N. Rosenbluth, and R. Z. Sagdeev, Phys. Fluids, 10:582 (1967). In [14, 15] there is a discussion of the effect of a shear on an instability with $k_{\|} \simeq \omega^*/v_{Ti}$ in a plasma with $\nabla T \neq 0$ (§8.3).

16. A. B. Mikhailovskii, Zh. Tekh. Fiz., 37:1365 (1967) [Sov. Phys. – Tech. Phys., 12:993 (1968)]. It is shown that the instability of the electron-acoustic branch in a plasma with $\nabla T \neq 0$ is, like the ion-acoustic branch, one of the most difficult instabilities to stabilize by means of a shear.

17. E. Ya. Kogan and S. S. Moiseev, Zh. Tekh. Fiz., 37:805 (1967) [Sov. Phys. – Tech. Phys., 12:579 (1967)]. It is shown that very short-wavelength perturbations, $k_\perp \rho_i \simeq (m_i/m_e)^{1/2}$, of a plasma with $\nabla n_0 \neq 0, \nabla T = 0$ are stabilized only when $\Theta \simeq 1$.

18. P. H. Rutherford and E. Frieman, Phys. Fluids, 10:1007 (1967). Investigation of the behavior of wave packets. A condition is obtained for the stabilization of the most dangerous perturbations of a plasma with $\nabla n_0 \neq 0$, $\nabla T = 0$; it has the form $\Theta > (m_e/m_i)^{1/3} k_b \rho_i$ (§8.4).

19. B. Coppi, Phys. Fluids, 8:2273 (1965). Discussion of an instability of the tearing-mode type for $\omega > \nu_{ei}$. An instability of this kind can only arise for very small β.

20. E. Frieman, K. Weimer, and P. Rutherford, Plasma Physics and Controlled Nuclear Fusion Research, Vol. 1, IAEA, Vienna (1966) p. 595.

21. K. Kitao, Plasma Physics, 9:523 (1967). In [20, 21] the effect of a shear on nonelectrostatic instabilities of a plasma with $\beta > m_e/m_i$ is discussed.

22. J. N. Davidson and T. Kammash, Nucl. Fusion, 8:203 (1968). A review article in which the method of eigenfunctions is compared with the method of wave packets. Detailed exposition of the mathematical apparatus employed in the investigation of the instabilities of a plasma in a magnetic field with shear.

23. A. A. Rukhadze and V. P. Silin, Usp. Fiz. Nauk, 82:499 (1964) [Sov. Phys. Uspekhi, 7:209 (1964)]. Review article containing an estimate of the effect of a shear on plasma instability.

Collisional Plasma

24. H. P. Furth, J. Killen, and M. N. Rosenbluth, Phys. Fluids, 6:459 (1963). Investigation of dissipative current instabilities in a magnetic field with shear. The possibility of the development of an instability of the tearing-mode type is demonstrated (§8.8). Study of the effect of a shear on the inertial current-convective instability due to a current gradient (§8.7). Classification of the dissipative instabilities.

25. J. L. Johnson, J. M. Greene, and B. Coppi, Phys. Fluids, 6:1169 (1963). Discussion of the same questions as in [24].

26. F. C. Hoh, Phys. Fluids, 7:956 (1964).

27. B. Coppi, Phys. Fluids, 8:2273 (1965).

28. A. A. Galeev, Zh. Tekh. Fiz., 36:1740 (1966) [Sov. Phys. – Tech. Phys., 11:1297 (1967). In [26-28] an investigation is made of the effect of a finite value of an ion Larmor radius on the inertial current instabilities. In [27] references to earlier papers on this subject can be found.

29. B. B. Kadomtsev and O. P. Pogutse, See [11]. The noninertial current-convective instability in a field with shear is discussed.

30. T. E. Stringer, Plasma Physics and Controlled Nuclear Fusion Research, Vol. 1, IAEA, Vienna (1966), p. 571. Study of perturbations of the quasi-mode type (wave packets in k_z) in a plasma with current.

31. J. D. Jukes, Phys. Fluids, 10:1107 (1967). Attention is drawn to the fact that a shear has little effect on the instability of a plasma with $\nabla T \neq 0$. The same questions are discussed in the papers of Moiseev, Baikov, and Mikhailovskii cited in Chapter 5.

32. E. Frieman et al., see [20]. Study of nonelectrostatic perturbations of a plasma with $\nabla n_0 \neq 0$ with allowance for viscosity.

Chapter 9

Influence of a Shear on the Instability of a Plasma in a Gravitational Field

§9.1. Shear Suppression of the Flute Instability. Suydam's Condition

If there is a gravitational force directed against the density gradient, a flute instability can develop in a plasma (see §6.1). A shear hinders the development of the flute instability since the very existence of flute-type perturbations with $k_{\|} = 0$ is impossible if $\Theta \neq 0$. At the same time, it follows from (8.20) that the characteristic value of $k_{\|}$ cannot be appreciably less than

$$k_{\| \mathrm{opt}} \simeq \Theta/a. \tag{9.1}$$

On the other hand, perturbations cannot grow if $k_{\|} \gtrsim \sqrt{|g\varkappa|}/c_A$. (This applies to a plasma with a not too low density; see §6.4.) Hence, we obtain an estimate for the shear that has an appreciable effect on the flute instability:

$$\Theta \gtrsim \sqrt{|g|a}/c_A. \tag{9.2}$$

Let us now consider the stabilizing role of a shear quantitatively, assuming that the condition (9.2) is satisfied.

If a shear and a gravitational force are present, oscillations with small $k_{\|}$ are described by the differential equation

$$\frac{1}{k_{\|}} \Delta_{\perp} (k_{\|}\psi) - \frac{\omega^2}{c_A^2 k_{\|}^2} \Delta\psi - \frac{g\varkappa k_b^2}{c_A^2 k_{\|}^2} \psi = 0. \tag{9.3}$$

If $k_{\|} = \text{const} \neq 0$, this equation reduces to Eq. (6.57) (for the case $\omega \gg \omega_{ni}$); if $k_{\|} = 0$, to Eq. (6.4).

We shall now assume $dk_{\|}/dx \neq 0$ and, in accordance with (8.10), we take the coordinate dependence of $k_{\|}$ to be linear:

$$k_{\|} = k'_{\|}(x - x_0). \tag{9.4}$$

We replace ω by $\gamma = -i\omega$ and $x - x_0$ by ξ:

$$\xi = \frac{(x - x_0)\, c_A k'_{\|}}{\gamma}. \tag{9.5}$$

Then (9.3) can be written in the form

$$(\xi^2 + 1)\frac{d^2\psi}{d\xi^2} + 2\xi\frac{d\psi}{d\xi} + \left(\frac{k_b}{c_A k'_{\|}}\right)^2 [\gamma_0^2 - \gamma^2(1 + \xi^2)]\,\psi = 0,$$
$$\gamma_0 \equiv \sqrt{-g\varkappa}. \tag{9.6}$$

We shall be interested in a solution of (9.6) localized near $x - x_0$ over a distance that is small compared with the inhomogeneity scale a, i.e., in a solution which decreases for large ξ. For large ξ ($\xi \gg 1$), Eq. (9.6) reduces to

$$\xi^2\frac{d^2\psi}{d\xi^2} + 2\xi\frac{d\psi}{d\xi} + \frac{k_b^2}{(c_A k'_{\|})^2}(\gamma_0^2 - \gamma^2\xi^2)\,\psi = 0. \tag{9.7}$$

The solution of this equation that is bounded at infinity is a Bessel function of the second kind of imaginary argument multiplied by $\xi^{-1/2}$:

$$\psi = \frac{C}{\sqrt{\xi}} K_{i\alpha}\left(\frac{k_b\gamma}{c_A k'_{\|}}\xi\right), \tag{9.8}$$

where

$$\alpha = \sqrt{\left(\frac{\gamma_0 k_b}{c_A k'_{\|}}\right)^2 - \frac{1}{4}}.$$

We now make an assumption (justified a posteriori) that if a shear of the order (9.2) is present the growth rate is appreciably decreased, $\gamma \ll \gamma_0$, or that the instability disappears entirely. At the same time we can assume $k_b\gamma \ll k'_{\|}c_A$, since the argument of the function K is small if $1 \ll \xi \ll c_A k'_{\|}/k_b\gamma$. If the argument is small, K remains finite at certain points at least if α is real. Otherwise the external solution (9.6) cannot be fitted to the internal solution, so that there are no eigenmodes. The inequality $\alpha^2 \leq 0$ is therefore the stabilization condition. Written out in full, this condition (Suydam's condition) is

$$\frac{|g|\varkappa k_b^2}{c_A^2 (k'_{||})^2} < \frac{1}{4}. \tag{9.9}$$

Approximately, this means the same as (9.2).

However suppose $\alpha^2 > 0$. We shall find the growth rate of the perturbations, assuming that α is small compared with unity, i.e., when

$$\left(\frac{\gamma_0 k_b}{c_A k'_{||}}\right)^2 - \frac{1}{4} \ll 1. \tag{9.10}$$

For this it is necessary to know the solution of (9.6) for all ξ and not only for $\xi \gg 1$.

Assuming $\xi \ll \frac{\gamma_0}{\gamma} \simeq \frac{c_A k'_{||}}{k_b \gamma}$ in (9.6), we reduce this equation to the form

$$(\xi^2 + 1)\frac{d^2\psi}{d\xi^2} + 2\xi\frac{d\psi}{d\xi} - \nu(\nu+1)\psi = 0, \tag{9.11}$$

where $\nu = -{}^1\!/_2 + i\alpha$. The Legendre functions of index ν and argument $i\xi$ give a solution of (9.11):

$$\psi = AP_\nu(i\xi) + BQ_\nu(i\xi). \tag{9.12}$$

If ψ is even, the constants A and B must be chosen such that $A/B = \frac{\pi}{2}\left(i - \cos\frac{\pi\nu}{2}\right)$.

For the interval $1 \ll \xi \ll \frac{c_A k'_{||}}{k_b \gamma}$ we obtain from (9.12), using the asymptotic representation for Q and P, the expression

$$\psi = \frac{A}{\sqrt{\xi}}\cos\left[\alpha \ln 8\xi + 2\,\mathrm{amp}\,\Gamma(1 + i\alpha) - \mathrm{amp}\,\Gamma(1 + 2i\alpha) - \tan^{-1}\exp(-\pi\alpha) - \frac{\pi}{4}\right]. \tag{9.13}$$

For such values of ξ we can, making an expansion in a series, obtain from (9.8)

$$\psi = \mathrm{const}\,\frac{1}{\sqrt{\xi}}\cos\left[\alpha \ln\left(\frac{k_b\gamma}{c_A k'_{||}}\xi\right) - \mathrm{amp}\,\Gamma(1 + i\alpha) + \frac{\pi}{2}\right]. \tag{9.14}$$

These two expressions must be identical. This leads to a dispersion equation corresponding to the lowest level:

$$\gamma = 8\left| c_A \frac{k'_{\|}}{k_b} \right| \exp\left[-\frac{\pi+2}{\alpha} \right]. \tag{9.15}$$

It can be seen that if $\alpha \ll 1$ the growth rate γ is exponentially small.

To be able to apply our results to the problem of the stability of a plasma in a curved magnetic field we must make the substitution $g \rightarrow 2(T_e + T_i)/m_i R$, where R is the radius of curvature. The condition (9.9) then shows that for a given shear the plasma pressure must not exceed a critical value,

$$\beta \equiv 8\pi (p_e + p_i)/B_0^2 < \beta_{crit},$$

where

$$\beta_{crit} = \Theta^2 R/4a. \tag{9.16}$$

A plasma in a field with a shear and real curvature will also be considered in Chapters 11, 13, 17, and 18.

§9.2. Suydam's Instability for a Finite Ion Larmor Radius

In Eq. (9.3) we now include terms of order ω_{n_i}/ω. If ω_i^*/ω is finite, the perturbations are described by an equation that differs from (9.3) only by the substitution

$$\omega^2 \rightarrow \omega(\omega - \omega_{n_i}). \tag{9.17}$$

This equation can be reduced to the form (9.6), where the role of γ^2 is now played by

$$\Gamma^2 \equiv -\omega(\omega - \omega_{n_i}). \tag{9.18}$$

We can therefore apply our above analysis of Eq. (9.6), replacing γ^2 by Γ^2.

We then conclude that localized perturbations exist only if the condition opposite to (9.9) holds. At the same time, $\Gamma^2 > 0$. However, in contrast to §9.1, this does not yet mean there is instability. In accordance with (9.18), the oscillation frequency is real when $\Gamma^2 > 0$ if

$$|\omega_{n_i}| > 2\Gamma. \qquad (9.19)$$

Far from the instability boundary, $\Gamma \simeq \gamma_0$, and the condition (9.19) has qualitatively the same meaning as in the absence of a shear – it describes the effect of stabilization of the flute instability by a finite ion Larmor radius (cf. §6.2).

Near the instability boundary, i.e., when the condition (9.10) holds, the expression for Γ is given by the right-hand side of Eq. (9.15). In this case Γ is exponentially small so that the effect of the finite ion Larmor radius is manifested for even very small ρ_i/a:

$$\rho_i/a \sim \exp(-\pi/\alpha). \qquad (9.20)$$

Kulsrud has solved Eq. (9.6) numerically in the region in which the condition (9.10) is not satisfied. According to Kulsrud, the effect of the finite Larmor radius for the parameters of a plasma contained in a C stellarator leads to a twofold increase in the value of the critical β compared with the value that follows from Suydam's condition (9.16).

§9.3. Influence of a Shear on the Flute Instability of a Low-Density Plasma

Suydam's condition (9.9) ceases to hold if the plasma density is too low. This is due to the inapplicability at small Π_e of Eq. (9.3) itself, from which the condition is derived. This is because in Eq. (9.3) we have neglected terms of order $(k_\perp c/\omega_{p_e})^2$. If $k_\perp \simeq 1/a$, this neglect is valid if $\Pi_e \equiv (\omega_{p_e} a/c)^2 > 1$.

Now suppose that the dimensionless electron density number is low, $\Pi_e < 1$. The perturbations can then be assumed to be electrostatic. If a shear and a gravitational force are present, such perturbations are described by the equation

$$\left(1 - \frac{\omega_{n_i}}{\omega}\right)\Delta\psi - k_b^2\psi\left(\frac{\gamma_0}{\omega}\right)^2 + k_\parallel^2 \frac{m_i}{m_e}\left(\frac{\omega_{B_i}}{\omega}\right)^2\left(1 - \frac{\omega_{n_e}}{\omega}\right)\psi = 0. \qquad (9.21)$$

Equation (9.21), like (8.56), reduces to the equation for a quantum-mechanical oscillator. It is then simple to obtain an equation for the eigenfrequencies of the oscillations. If $\omega \gg \omega^*$, this equation is

$$\omega^2 + \gamma_0^2 [1 + x(1 - \sqrt{1 + 2/x})] = 0, \tag{9.22}$$

where $x \equiv [(2n+1)\,\omega_{B_i} k_{\|}' / \gamma_0 k_B^2]^2 (m_i/2m_e)$. It can be seen that the effect of a shear on the flute instability can be characterized by the dimensionless parameter x – the effect is appreciable if x > 1, i.e , if

$$\Theta \gtrsim k_b \rho_i (a/R)^{1/2} (m_e/m_i)^{1/2}. \tag{9.23}$$

At the limit of its applicability, this inequality is qualitatively the same as Suydam's condition (9.16). However, in contrast to (9.16), it is not a stabilization condition. It follows from (9.22) that even when $x \gg 1$ a plasma is unstable, the following relations holding:

$$\gamma \simeq \gamma_0 / x^{1/2} \simeq (m_e/m_i)^{1/2} (a/R\Theta)\, \omega^*. \tag{9.24}$$

The growth rate (9.24) becomes comparable with ω^* when $\Theta \simeq (m_e/m_i)^{1/2} (a/R)$. If $T_i > T_e$, the effect of the finite ion Larmor radius, considered for a high-density plasma in §9.2, also becomes effective. Equation (9.21) shows that this effect stabilizes the perturbations if

$$\Theta > 4 (m_e/m_i)^{1/2} a/R. \tag{9.25}$$

§9.4. Influence of a Shear on the Gravitational-Kinetic Instability

If $g\varkappa < 0$ in a collisionless plasma, a gravitational force can excite perturbations with both $\omega > k_z v_{T_e}$ and $\omega < k_z v_{T_e}$ (see §6.5). The effect of a shear on perturbations with $\omega > k_z v_{T_e}$ was discussed in §9.3. We shall now consider perturbations with $\omega < k_z v_{T_e}$, which correspond to the gravitational-kinetic instability.

An estimate of the shear that stabilizes these perturbations can be obtained from (8.34). In accordance with the results of §6.5, we must substitute into this formula the same $\partial\omega/\partial k_x$ as in (8.35) and the following values of $k_{\|}$ and γ:

$$\left.\begin{aligned} k_{\|}/k_b &\simeq (g\varkappa)^{1/2}/\omega_{B_i}, \\ \gamma &\simeq (m_e/m_i)^{1/2} (g\varkappa)^{1/2} k_b \rho_i. \end{aligned}\right\} \tag{9.26}$$

We then find that a shear is effective if

$$\Theta > (m_e/m_i)^{1/2} (g/\varkappa v_{T_i}^2) (k_b\rho_i)^{-1}. \qquad (9.27)$$

Expressing g in terms of the radius of curvature, we can also write this condition in the form

$$\Theta > (m_e/m_i)^{1/2} (a/R) (k_b\rho_i)^{-1}. \qquad (9.28)$$

Since the gravitational force is important if $g > \varkappa v_{T_i}^2 z_i$, the condition (9.27) is more stringent than (8.37).

§9.5. Influence of a Shear on the Gravitational-Dissipative Instability

If $\nu_{ei} > \omega$, a plasma without shear can sustain both the flute instability and the gravitational-dissipative instability discussed in §6.6. Let us consider the effect that a shear has on the two different forms – inertial and noninertial – of this instability.

1. Inertial Instability (Small $k_{\|}$). We shall assume that the perturbation frequency is fairly low so that $\omega < \nu_{ei} (ck/\omega_{p_e})^2$. We can then assume that the perturbations are electrostatic. The basic equation that describes such perturbations has the form

$$\Delta\psi - k_b^2 (\gamma_0/\omega)^2 \psi - ik_{\|} (m_i/m_e) (\omega_{p_i}^2/\omega\nu_{ei}) \psi = 0. \qquad (9.29)$$

Here we have omitted terms of order ω^*/ω. This equation can be solved in the same manner as (8.56) and (9.21). As a result we obtain the dispersion equation

$$k_b^2 (\gamma_0^2/\gamma^2 - 1) = (2n+1) \; [(k_{\|}')^2 (m_i/m_e) \omega_{B_i}^2/\gamma\nu_{ei}]^{1/2}, \qquad (9.30)$$
$$n = 0, 1, 2, \ldots$$

For sufficiently small $k_{\|}'$, this equation describes a flute instability with $\gamma = \gamma_0$. A shear affects this instability appreciably if

$$\Theta \gtrsim (\gamma_0\nu_{ie}/\omega_{B_i}^2)^{1/2} k_b a. \qquad (9.31)$$

The growth rate depends strongly on the collision frequency:

$$\gamma \simeq \gamma_0 (\gamma_0 k_b^4 \nu_{ie}/(k_{\|}')^2 \omega_{B_i}^2)^{1/3}. \qquad (9.32)$$

Recalling that $\Theta \simeq k_{\parallel}/k_{\perp}$ when $k_{\perp}a \simeq 1$, we see that there is qualitative agreement between (9.31) and the inequality (6.71). In the case of a field without shear, the latter characterizes the interval of $k_{\parallel}/k_{\perp}$ at which the flute instability is replaced by the gravitational-dissipative instability.

In accordance with (9.32), a shear has a stabilizing effect, decreasing the growth rate, $\gamma \propto \Theta^{-2/3}$. This kind of instability is completely stabilized only if $\gamma \le \omega^*$, in particular, due to an increase in the importance of ion–ion collisions.

2. Noninertial Instability (Large $k_{\parallel}$). In accordance with §6.6, ion inertia ceases to play a role for sufficiently large $k_{\parallel}$. In this case eigenmode-type solutions of the kind considered in §9.5.1 are absent, so that one can only speak of wave packets. We shall estimate the effect of a shear on the growth of a packet in the same way as in §9.4, using the results of §8.4 [see (8.34)]. We take the values of $k_{\parallel}$ and γ needed for this estimate from §6.6:

$$k_{\parallel}/k_b \simeq (g\varkappa)^{1/2}/\omega_{B_i}, \qquad \gamma \simeq \nu_{ie}, \tag{9.33}$$

and $\partial\omega/\partial k_x$ is determined by (8.35). We then find that a shear is important if

$$\Theta > \frac{1}{z_i}\,\frac{\nu_{ie}a^2(g\varkappa)^{1/2}}{v_{T_i}^2} \simeq \frac{1}{z_i}\left(\frac{a}{R}\right)^{1/2}\left(\frac{m_e}{m_i}\right)^{1/2}\frac{a}{\lambda}, \tag{9.34}$$

where $\lambda \simeq v_{T_i}/\nu_{ii}$ is the ion mean free path.

If $\nu_{ei} \simeq k_{\parallel}v_{T_e}$, the relation (9.34) gives results of the same order as (9.28).

§9.6. Nonelectrostatic Gravitational-Dissipative Instability

In contrast to §9.5, we shall now consider nonelectrostatic perturbations. These were investigated under conditions when dissipation can be ignored in §9.1. In this approximation we showed that the perturbations grow if Suydam's condition (9.9) is not satisfied.

We shall now assume that Suydam's condition is satisfied and show that as a result of dissipation (friction of electrons against ions) the Suydam perturbations may nevertheless grow if a force of gravity is directed against the density gradient, $(-g\varkappa) > 0$.

The system of basic equations that describes this type of perturbation is

$$\left.\begin{aligned} &\Delta_\perp \psi + \frac{g\varkappa}{\omega^2} k_b^2 \psi - i \frac{m_i}{m_e} \frac{\omega_{B_i}^2}{\nu_{ei}\omega} k_\parallel^2 \left(\psi - \frac{\omega}{ck_\parallel} A_\parallel\right) = 0, \\ &\Delta_\perp A_\parallel - \frac{ik_\parallel \omega_{p_e}^2}{c\nu_{ei}} \left(\psi - \frac{\omega}{ck_\parallel} A_\parallel\right) = 0. \end{aligned}\right\} \tag{9.35}$$

As in §8.8, we distinguish two characteristic regions of x values – a region exterior to the dissipative layer and an interior region. Assuming that the perturbation frequency is sufficiently low, we neglect ion inertia in the external region, and we then obtain from (9.35)

$$\frac{d^2 A_\parallel}{ax^2} - k_b^2 \left(1 - \frac{g\varkappa}{c_A^2 k_\parallel^2}\right) A_\parallel = 0. \tag{9.36}$$

This is the same kind of equation as (9.7) but written down for $A_\parallel \sim x\psi$ and not ψ. The solution of (9.36) has a form similar to (9.8):

$$A_\parallel = C (k_b x)^{1/2} K_s (k_b x), \tag{9.37}$$

where $s = \left[\frac{1}{4} + g\varkappa (k_b / c_A k_\parallel')^2\right]^{1/2}$. If $k_b x \ll 1$, we find $A_\parallel \sim x^{(1-s)}$. Therefore, at the boundary between the perfectly conducting and the dissipative regions the logarithmic derivative of the vector potential $A_\parallel$ is approximately

$$d \ln A_\parallel / dx \simeq (1 - s)/\delta, \tag{9.38}$$

where δ is the width of the dissipative region [approximately equal to (8.97)]. On the other hand, it follows from the solution of the internal problem [see §8.8] that the logarithmic derivative is approximately

$$d \ln A_\parallel / dx \simeq \Delta \simeq (\omega_{p_e}/c)^2 (\gamma/\nu_{ei})\, \delta. \tag{9.39}$$

Equating (9.38) and (9.39) and taking into account (8.97), we find that the growth rate is approximately

$$\gamma \simeq \gamma_0 \left(\gamma_0 k_b^4 \nu_{ei} / k_\parallel'^2 \omega_{B_i}^2\right)^{1/3}, \tag{9.40}$$

where $\gamma_0 = \sqrt{-g\varkappa}$ is the growth rate of the flute instability. A more rigorous calculation of the growth rate can be found in the papers cited at the end of this chapter.

A comparison of (9.40) and (9.39) shows that the growth rate of the large-scale nonelectrostatic instability is of the same order as the growth rate of the small-scale electrostatic instability.

Bibliography

1. B. R. Suydam, Proceedings of the Second United Nations International Conference on the Peaceful Uses of Atomic Energy, Geneva (1958), Vol. 31, Publ.: U.N., Geneva (1958), p. 157.
2. L. V. Mikhailovskaya and A. B. Mikhailovskii, Dokl. Akad. Nauk SSSR, 150:531 (1963) [Sov. Phys. – Doklady, 8:491 (1963)].
3. R. M. Kulsrud, Phys. Fluids, 6:904 (1963). In [2,3] a study is made of the effect of a finite ion Larmor radius on a Suydam-type instability (§9.2). In [3], Eq. (9.6) is solved and an expression is obtained for the frequency of Suydam-type perturbations near the instability boundary [Eq. (9.15)].
4. H. P. Furth, J. Killeen, and M. N. Rosenbluth, Phys. Fluids, 6:459 (1963). Investigation of the characteristic oscillations of a collisional plasma. It is shown that a flute gravitational-dissipative instability (g modes) can arise under conditions when Suydam's stability condition is satisfied (§9.6). There is also a brief discussion of the localized gravitational-dissipative instability (§9.5.1).
5. J. L. Johnson, J. M. Greene, and B. Coppi, Phys. Fluids, 6:1169 (1963). Discussion of the nonlocal and local g-mode instability (§9.6 and §9.5.1).
6. L. V. Mikhailovskaya and A. B. Mikhailovskii, Nucl. Fusion, 5:234 (1965). Study of the effect of a shear on the flute instability of a low density collisionless plasma (§9.3).
7. J. D. Jukes, Phys. Fluids, 7:1468 (1964). Study of the effect of a shear on the gravitational-kinetic instability (§9.4).
8. J. D. Jukes, Phys. Fluids, 7:52 (1964).
9. B. Coppi, Phys. Rev. Lett., 12:417 (1964).
10. B. Coppi, Phys. Fluids, 7:1096 (1964).
11. B. Coppi, Phys. Fluids, 7:1501 (1964). In [8-11] a study is made of the effect of a finite ion Larmor radius on the gravitational-dissipative instability (§§9.5 and 9.6).
12. K. V. Roberts and J. B. Taylor, Phys. Fluids, 8:315 (1965). A study of perturbations of the type of wave packets in k_z (quasimodes) corresponding to the gravitational-dissipative instability.
13. T. E. Stringer, Plasma Physics and Controlled Nuclear Fusion Research, Vol. 1, IAEA, Vienna (1966), p. 571. The quasimode method is used to study the gravitational-dissipative instability. The effect of a shear on the noninertial branch of this instability is considered (§9.5.2).
14. B. Coppi, Phys. Rev. Lett., 12:6 (1964).
15. B. Coppi, Ann. Phys. (New York), 30:178 (1964).

16. B. Coppi, J. M. Greene, and J. L. Johnson, Nucl. Fusion, 6:101 (1966). In [14-16] the correspondence between the different instability modes of a collisional plasma is discussed in the approximation of a vanishing ion Larmor radius.
17. B. Coppi et al., Nucl. Fusion, 6:261 (1966). Study of the effect of a shear on the gravitational-dissipative instability when $|\omega + i\nu_{ei}| > k_{\parallel}v_{Te}$ (§§9.4.1 and 9.5.1).
18. B. Coppi and M. N. Rosenbluth, Plasma Physics and Controlled Nuclear Fusion Research, Vol. 1, IAEA, Vienna (1966), p. 617.
19. B. Coppi and B. W. Roos, Plasma Physics, 9:585 (1967).
20. T. E. Stringer, Phys. Fluids, 10:418 (1967). In [18-20] a study is made of the effect of ion–ion collisions on the inertial gravitational-dissipative instability. The noninertial instability is also considered in [20].
21. F. F. Chen, Phys. Fluids, 9:905 (1966). Study of the effect of a shear on the centrifugal instability of a plasma in a radial electric field.
22. A. A. Rukhadze and V. P. Silin, Usp. Fiz. Nauk, 96:87 (1968) [Sov. Phys. – Uspekhi, 11:659 (1969)]. Review article containing an estimate of the effect of a shear on the instability of a plasma in a gravitational field.

Chapter 10

Weakly Ionized Plasmas

§10.1. Low-Frequency Instability due to a Density Gradient

If a plasma is weakly ionized, the electron temperature generally exceeds the ion temperature, $T_e \gg T_i$. In this case the inhomogeneity of the plasma can be characterized by the ratio ρ_0/a, where $\rho_0 = (T_e/m_i\omega_{B_i}^2)^{1/2}$ is the effective ion Larmor radius and a is the characteristic inhomogeneity scale of the plasma. In this chapter we shall assume that ρ_0/a is fairly small (weakly inhomogeneous plasma). It is then possible to restrict the investigation to low-frequency perturbations, $\omega \ll \omega_{B_i}$ (perturbations with $\omega > \omega_{B_i}$ in a plasma with finite ρ_0/a will be considered in §10.2).

We shall assume that kinetic effects are not important. This is justified if the frequency of collisions between the electrons and neutrals is sufficiently high ($\nu_e > k_z v_{T_e}$). It is then possible to base the treatment on a system of hydrodynamic equations. For the case $B_0 = 0$, equations of this type were derived in §6.4 in Volume 1. The magnetic field is taken into account by adding the magnetic term of the Lorentz force to the right-hand side of the equations of motion. The linearized equations of hydrodynamics for $\mathbf{E}_0 = 0$ (in the absence of a static electric field) have the form

$$\left.\begin{aligned}
&-i\omega n_e' + ik_z V_{ze}' n_0 + \operatorname{div}(n_e \mathbf{V}_{\perp e})' = 0,\\
&-i\omega n_i' + ik_z V_{zi}' n_0 + \operatorname{div}(n_i \mathbf{V}_{\perp i})' = 0,\\
&0 = \frac{e_e}{m_e}\mathbf{E}' - \frac{T_{0e}}{m_e}(\nabla \ln n_e)' + [\mathbf{V}_e' \boldsymbol{\omega}_{B_e}] - \nu_e \mathbf{V}_e',\\
&-i\omega \mathbf{V}_i' = \frac{e_i}{m_i}\mathbf{E}' + [\mathbf{V}_i', \omega_{B_i}] - \nu_i \mathbf{V}_i'.
\end{aligned}\right\} \qquad (10.1)$$

It is assumed that $\nabla T_{0e} = 0,\ T_{0i} = 0,\ T'_e = 0$. In addition, we have ignored electron inertia, which is justified if $\omega < \nu_e$.

The magnetic field B_0 is important if $\nu_e < \omega_{Be}$. If this condition is satisfied, the electron equation of motion yields

$$\mathbf{V}'_{\perp e} = \mathbf{V}'_E - \frac{T_e}{m_e \omega_{B_e}} [(\nabla \ln n_e)', \mathbf{e}_z] + \frac{\nu_e (e_e \mathbf{E}'_\perp - T_e (\nabla \ln n_e)')}{m_e \omega^2_{B_e}}$$

$$V'_{ze} = -\frac{\mathrm{i} k_z}{\nu_e m_e} \left(\frac{n'_e}{n_0} + \frac{e_e \psi'}{T_e} \right), \tag{10.2}$$

where the prime denotes, as above, the perturbed part; the subscript zero has been omitted from T_{0e} and $\mathbf{V}'_E = \frac{c\,[\mathbf{E}', \mathbf{e}_z]}{B_0}$.

Assuming $\omega \ll \omega_{Bi}$, we obtain similar equations for the ions:

$$\left.\begin{aligned} \mathbf{V}'_{\perp i} &= \mathbf{V}'_E - \frac{\mathrm{i}\,(\omega + \mathrm{i}\nu_i)\, e_i \mathbf{E}'_\perp}{m_i \omega^2_{B_i}}, \\ V'_{zi} &= \frac{e_i k_z \psi'}{m_i (\omega + \mathrm{i}\nu_i)}. \end{aligned}\right\} \tag{10.3}$$

After the substitution of (10.2) and (10.3) into the corresponding continuity equations (10.1) and the use of the conditions of electrical quasineutrality ($n'_e = n'_i$), we obtain the following dispersion equation for perturbations with $k_\perp a \gg 1$:

$$(k_\perp \rho_0)^2 \left(1 + \frac{\mathrm{i}\nu_i}{\omega}\right) - \frac{k_z^2 T_e}{m_i \omega (\omega + \mathrm{i}\nu_i)} + \frac{\mathrm{i}\Delta_e (\omega - \omega_{ne})}{\omega (\omega + \mathrm{i}\Delta_e)} = 0. \tag{10.4}$$

Here, as in the earlier chapters, $\omega_{ne} = k_y \varkappa_n c T_e / e_e B_0$, and Δ_e stands for

$$\left.\begin{aligned} &\Delta_e = \Delta_{\perp e} + \Delta_{\parallel e},\ \Delta_{\perp e} \equiv (k_\perp \rho_e)^2 \nu_e, \\ &\Delta_{\parallel e} \equiv k_z^2 T_e / m_e \nu_e, \end{aligned}\right\} \tag{10.5}$$

where ρ_e is the electron Larmor radius.

If $\omega > \nu_i$ and $\Delta_{\parallel e} < \Delta_{\perp e}$, Eq. (10.5) reduces to the dispersion equation (4.79) (investigated in §5.2) for a fully ionized plasma (with the substitution $\nu_e \to \nu_{ei}$). Under these conditions, oscillations characterized by the equations of §5.2.1 grow.

One of the differences between (10.5) and (4.79) is manifested at fairly small k_z when $\Delta_{\perp e} > \Delta_{\parallel e}$, i.e., when

$$\cos\theta \equiv \frac{k_z}{k} \lessdot \frac{\nu_e}{\omega_{Be}}. \tag{10.6}$$

This difference is important if the condition (10.6) is satisfied for $\cos\theta = (\cos\theta)_{opt}$, where $(\cos\theta)_{opt}$ is defined by (5.12). It follows that the instabilities of a weakly ionized plasma must be reinvestigated if

$$\frac{\omega_{ne}}{\nu_e} < \frac{m_e}{m_i}. \tag{10.7}$$

If $k_\perp \simeq 1/a$, this inequality means

$$\left(\frac{\rho_0}{a}\right)^2 < \frac{\nu_e}{\omega_{B_e}}. \tag{10.8}$$

Assuming (10.7) [or (10.8)] is satisfied and neglecting the terms with $\Delta_{\parallel e}$ in (10.5) and also the contribution of the longitudinal ion motion, we obtain

$$(\omega + i\nu_i)(\omega + i\nu_e k_\perp^2 \rho_e^2) + i\frac{m_e}{m_i}\nu_e(\omega - \omega_{ne}) = 0. \tag{10.9}$$

By virtue of (10.7) the quantity $m_e/m_i\nu_e$ in this equation must be regarded as a large parameter. An approximate solution of (10.9) therefore has the form

$$\left.\begin{aligned} &\operatorname{Re}\omega = \omega_{ne},\\ &\gamma = \frac{m_i}{m_e\nu_e}(\omega_{ne}^2 - \nu_i\nu_e k_\perp^2\rho_e^2). \end{aligned}\right\} \tag{10.10}$$

Instability occurs if

$$\frac{\nu_i}{\nu_e} < \left(\frac{\lambda_e}{a}\right)^2, \tag{10.11}$$

where $\lambda_e = v_{T_e}/\nu_e$ is the electron mean free path. The growth rate of the instability is small compared with the gradient frequency ω_{ne}.

Thus, under conditions when the properties peculiar to a weakly ionized plasma are manifested, the instability growth rate is smaller than in the case of a fully ionized plasma.

§10.2. High-Frequency Instability due to a Density Gradient

In this section we shall, in contrast to §10.1, consider a plasma with $\rho_0/a > 1$. Then, as in the case of a collisionless plasma (see §2.2), instabilities with $\omega > \omega_{Bi}$ are the most important.

The expressions (10.2) for V_e' also remain in force when $\omega > \omega_{Bi}$, whereas (10.3) no longer follows from (10.1) for the ion velocity but

$$\mathbf{V}_i' = \frac{e_i \mathbf{k}\psi'}{m_i(\omega + i\nu_i)} . \tag{10.12}$$

The dispersion equation for perturbations with $\omega > \omega_{B_i}$ is

$$\frac{\omega_{ne} + i\Delta_e}{\omega + i\Delta_e} - \frac{k^2 T_e}{m_i \omega (\omega + i\nu_i)} = 0. \tag{10.13}$$

We shall analyze this equation approximately. Remembering that the instability can occur only when $\nu_i < \omega \leq \omega_{ne}$, and neglecting the terms of order ω/ω_{ne} and ν_i/ω in (10.13), we obtain

$$\omega^2 = \frac{k^2 T_e}{m_i} \cdot \frac{i\Delta_e}{\omega_{ne} + i\Delta_e} . \tag{10.14}$$

From this we conclude that for fixed $k_\perp \geq k_z$ the growth rate attains a maximum that is approximately equal to

$$\gamma \simeq \operatorname{Re} \omega \simeq k \left(\frac{T_e}{m_i}\right)^{1/2} \tag{10.15}$$

for $\Delta_e \simeq \omega_{ne}$, which corresponds to the wave number

$$k_z \simeq (k_y \varkappa)^{1/2} (\nu_e/\omega_{Be})^{1/2}. \tag{10.16}$$

The solution (10.15) satisfies the condition $\omega < \omega_{ne}$ if $\rho_0/a > 1$, which agrees with the instability condition for a collisionless plasma (cf. §2.2). The assumption that the ion damping is small, $\nu_i/\omega < 1$, leads in conjunction with $\omega > \omega_{B_i}$ to the sufficient condition $\nu_i/\omega_{B_i} < 1$.

The applicability of the solution (10.15)-(10.16) is also restricted by the condition for electron collisions to be frequent, $\nu_e > k_z v_{Te}$, which in the present case entails

$$\omega_{ne} < \nu_e. \tag{10.17}$$

In the opposite limiting case, when $\omega_{ne} > \nu_e$, the instability is essentially kinetic and is described by the equations of §2.2. The transition region, $\omega_{ne} \simeq \nu_e$, has been considered by Timofeev.

§10.3. Current-Convective Instability

Let us consider the effect that a longitudinal electric field $E_0 \parallel B$ has on the stability of a weakly ionized plasma ($\rho_0 < a$). This field gives rise to a longitudinal electron drift with velocity

$$V_{0e} = \frac{e_e E_0}{m_e \nu_e}. \tag{10.18}$$

In this case, the perturbations are described by a system of equations of the type (10.1) but with the substitution $\omega \to (\omega - k_z V_{0e})$ in the electron continuity equation. The dispersion equation (10.4) is replaced by (we ignore the longitudinal ion motion)

$$(k\rho_0)^2 \left(1 + \frac{i\nu_i}{\omega}\right) + \frac{k_z V_0 \omega_{ne} + i\Delta_e (\omega - \omega_{ne})}{\omega(\omega + i\Delta_e - k_z V_0)} = 0. \tag{10.19}$$

If $\nu_e < (m_i/m_e)\omega_{ne}$ [the opposite condition to (10.7)], the longitudinal current has an important effect if a condition of the type (5.47) with the substitutions $v_{Ti} \to (T_e/m_i)^{1/2}$, $\rho_i \to \rho_0$, $\nu_{ei} \to \nu_e$ is satisfied and we recall that the expression for ω^* is now determined by the density gradient. The corresponding growth rate is large compared with ω_{ne}. At not too small k_z, when the first term on the left-hand side of (10.19) can be ignored, the growth rate is

$$\gamma = \frac{k_y}{k_z} \cdot \frac{\nu_e}{\omega_{Be}} \varkappa_n V_0. \tag{10.20}$$

This is the Kadomtsev–Nedospasov current-convective instability.

As k_z decreases the growth rate behaves in the same way as in the case of a fully ionized plasma (see §5.6).

Let us now consider a plasma for which the conditions (10.7) and (10.8) are satisfied. The expression (10.20) is replaced by

$$\gamma = \frac{k_y}{k_z} \cdot \frac{\nu_e}{\omega_{Be}} \cdot \frac{\varkappa_n V_0}{1 + \left(\frac{k_\perp}{k_z} \cdot \frac{\nu_e}{\omega_{Be}}\right)^2}. \tag{10.21}$$

In this case, the growth rate attains a maximum that is less than (5.50):

$$\gamma_{\max} \simeq \varkappa_n V_0. \tag{10.22}$$

At the same time, $k_z/k_\perp \simeq \nu_e/\omega_{Be}$. If k_z is further decreased, the growth rate decreases and then goes over into (10.10).

§10.4. Gravitational-Dissipative Instability

As in Chapter 6, we shall assume that the ions are subjected to a force of gravity $g \parallel x$, which simulates a curvature of the lines of force. In the ion equations (10.1), we must then make the substitution $\omega \to \omega + gk_y/\omega_B$. If $\omega \gg gk_y/\omega_B$, the dispersion equation reduces to

$$(\omega + i\nu_i)(\omega + i\Delta_e) - g\varkappa\left(\frac{k_y}{k_\perp}\right)^2 + i\frac{m_e}{m_i}\nu_e(\omega - \omega_{ne}) = 0. \quad (10.23)$$

This equation describes the flute instability considered in §6.1 if

$$|g\varkappa|^{1/2} > \left(\nu_i, \frac{m_e}{m_i}\nu_e, (k_\perp\rho_e)^2\nu_e\right). \quad (10.24)$$

If $\nu_i < \sqrt{|\varkappa g|} < (m_e/m_i)\nu_e$, there is a gravitational-dissipative instability similar to that considered in §6.6. Its growth rate is

$$\gamma = \frac{|g\varkappa|}{\Delta_e + \frac{m_e}{m_i}\nu_e}. \quad (10.25)$$

In the case of a strongly collisional plasma $[\omega < (\nu_i, \Delta_e)]$, Eq. (10.23) yields

$$\omega = \omega_{ne} - i\frac{m_i}{m_e\nu_e}(\nu_i\nu_e k_\perp^2\rho_e^2 + \varkappa g). \quad (10.26)$$

From this we find that for $\varkappa g < 0$ the plasma is unstable if

$$|g\varkappa| > \nu_i\nu_e(k_\perp\rho_e)^2. \quad (10.27)$$

For $g \simeq T_e/m_iR$ and $k_\perp \simeq a^{-1}$, this entails approximately

$$\frac{a}{R} > \frac{\nu_e}{\omega_{Be}} \cdot \frac{\nu_i}{\omega_{Bi}}. \quad (10.28)$$

The growth rate of the instability satisfies

$$\gamma \simeq \frac{|\varkappa g|m_i}{m_e\nu_e} \simeq \frac{T}{m_e\nu_e aR}. \quad (10.29)$$

It is small compared with the growth rate of the flute instability (6.8) as $\sqrt{|g\varkappa|}\, m_i/m_e\nu_e$.

§10.5. Instability of a Weakly Ionized Plasma in Crossed Electric and Magnetic Fields

A transverse electric field, $\mathbf{E}_0 \perp \mathbf{B}_0$, may be the cause of an instability of an inhomogeneous, weakly ionized plasma. As a result of collisions with neutrals, the drift velocity of a charged particle moving in crossed fields $\mathbf{E}_0$ and $\mathbf{B}_0$ differs from $\mathbf{V}_E \equiv c\,\frac{[\mathbf{E}_0, \mathbf{B}_0]}{B_0^2}$ by the factor $[1 + (\nu/\omega_B)^2]^{-1}$, where ν and ω_B refer to the ions or the electrons. If $\mathbf{E}_0 \parallel \mathbf{x}$ and $\mathbf{B}_0 \parallel \mathbf{z}$, the relative velocity of the ions and electrons due to this effect is

$$V_{0iy} - V_{0ey} = \frac{cE_{0x}}{B_0}\,\xi, \tag{10.30}$$

where

$$\xi = \left[\left(\frac{\nu}{\omega_B}\right)_i^2 - \left(\frac{\nu}{\omega_B}\right)_e^2\right] \Big/ \left[1 + \left(\frac{\nu}{\omega_B}\right)_i^2\right]\left[1 + \left(\frac{\nu}{\omega_B}\right)_e^2\right]. \tag{10.31}$$

In accordance with (6.1) the same velocity difference would arise if collisions were neglected but the plasma were subjected to a gravitational force g given by

$$g = g_{\text{eff}} \equiv -\frac{e_i E_{0x}}{m_i}\,\xi. \tag{10.32}$$

If $g_{\text{eff}}\,\varkappa < 0$ $\left(\varkappa = \frac{\partial \ln n_0}{\partial x}\right)$, the field g_{eff} renders the plasma more unstable. Therefore, in view of (10.32) and the fact that $\xi > 0$, an electric field has a destabilizing effect if

$$E_{0x}\frac{\partial \ln n_0}{\partial x} > 0. \tag{10.33}$$

We find the dispersion equation for small-scale perturbations when $\mathbf{E}_0 \neq 0$ as follows. We first find the total transverse electron velocity $\mathbf{V}_e \equiv \mathbf{V}_{0e} + \mathbf{V}_e'$ by using the equation of motion in the form (10.1) with the prime omitted. If terms of order $(\nu_e/\omega_{B_e})^2$ are neglected, the expression for $\mathbf{V}_e$ is determined by the first equation of (10.2) without the prime. This result in conjunction with the expression (10.2) for V_{ze}' is then substituted into the electron continuity equation (10.1), and we then find the perturbed electron density:

$$n_e' = -\frac{e_e \psi n_0}{T_{0e}} \cdot \frac{\omega_{ne} + i\Delta_e}{\omega - \omega_E + i\Delta_e}, \tag{10.34}$$

where $\omega_E = k_y V_{yE} \equiv -k_y c E_{0x}/B_0$, and Δ_e was introduced in §10.1.

To calculate the transverse ion velocity, we use the ion equation of motion analogous to (10.1) but with the prime omitted and $(-i\omega V_i')$ on the left-hand side replaced by dV_i/dt. We solve this equation by assuming that the left-hand side is small and the ratio ν_i/ω_{Bi} finite.

Then

$$\mathbf{V}_{\perp i}=\left(\mathbf{V}_E+\frac{\nu_i}{\omega_{B_i}^2}\cdot\frac{e_i}{m_i}\mathbf{E}_\perp\right)\left[1+\left(\frac{\nu_i}{\omega_{B_i}}\right)^2\right]^{-1}+\delta\mathbf{V}_{\perp i}. \tag{10.35}$$

Here, the term $\delta\mathbf{V}_{\perp i}$ corresponds to ion inertia:

$$\delta\mathbf{V}_{\perp i}=\left\{\frac{e_i}{m_i}\cdot\frac{d\mathbf{E}}{dt}\left(1-\frac{\nu_i^2}{\omega_{B_i}^2}\right)-2\nu_i\frac{d\mathbf{V}_E}{dt}\right\}\Big/\omega_{B_i}^2\left[1+\left(\frac{\nu_i}{\omega_{B_i}}\right)^2\right]^2. \tag{10.36}$$

The total derivative d/dt denotes $\partial/\partial t+(\mathbf{V}_{\perp i}^{(0)}\nabla)$, where $\mathbf{V}_{\perp i}^{(0)}$ is the first term of the right-hand side of (10.35).

We substitute the expressions (10.35) and (10.36) into the ion continuity equation and then linearize. The perturbed ion density is then found to be

$$n_i'=\frac{cn_0}{B_0}\psi\left\{\frac{\varkappa k_y+i(\nu_i/\omega_{B_i})k_\perp^2}{\omega-\omega_{E_i}}\Bigg|+\frac{k_\perp^2(1-\nu_i^2/\omega_{B_i}^2)}{\omega_{B_i}(1+\nu_i^2/\omega_{B_i}^2)}\right\}\left(1+\frac{\nu_i^2}{\omega_{B_i}^2}\right)^{-1}, \tag{10.37}$$

where $\omega_{E_i}=\omega_E(1+\nu_i^2/\omega_{B_i}^2)^{-1}$.

Equating the right-hand sides of (10.34) and (10.37), we obtain the dispersion equation

$$\frac{\omega_{ne}+i\Delta_e}{\omega-\omega_E+i\Delta_e}-\frac{\omega_{ne}-i\nu_ik_\perp^2\rho_0^2}{(\omega-\omega_{E_i})(1+\nu_i^2/\omega_{B_i}^2)}\Bigg|+\frac{k_\perp^2\rho_0^2(1-\nu_i^2/\omega_{B_i}^2)}{(1+\nu_i^2/\omega_{B_i}^2)^2}=0. \tag{10.38}$$

In the approximation $(\nu_i/\omega_{B_i})<1$, Eq. (10.38) yields

$$\frac{i\Delta_e(\omega-\omega_E-\omega_{ne})}{(\omega-\omega_E)(\omega-\omega_E+i\Delta_e)}+\left(\frac{\nu_i}{\omega_{B_i}}\right)^2\frac{\omega\omega_{ne}}{(\omega-\omega_E)^2}+k_\perp^2\rho_0^2\left(1+\frac{i\nu_i}{\omega-\omega_E}\right)=0. \tag{10.39}$$

Let us consider perturbations corresponding to the flute instability of a collisionless plasma in a field of gravity. We shall assume that the terms with Δ_e and $\nu_i/(\omega-\omega_E)$ are small. Then (10.39) yields

$$(\omega-\omega_E)^2-\frac{\varkappa k_y}{k_\perp^2}\cdot\frac{\omega\nu_i^2}{\omega_{B_i}}=0. \tag{10.40}$$

The solutions of this equation correspond to instability if the condition (10.33) holds and if E_{0x} is not too small:

$$eE_{0x} > \frac{1}{4} m_i \varkappa \left(\frac{\nu_i}{k_\perp}\right)^2. \tag{10.41}$$

The growth rate is

$$\gamma \simeq \frac{\nu_i}{\omega_{B_i}} \left| \frac{e}{m_i} E_{0x} \varkappa \right|^{1/2}. \tag{10.42}$$

This result also follows from the expression for the growth rate $\gamma \simeq |g\varkappa|^{1/2}$ of the flute instability if g in this expression is replaced by g_{eff} defined by (10.32). The instability condition (10.41) for perturbations with $k_\perp \simeq 1/a$ entails approximately

$$\frac{e\psi_0}{T_{0e}} > \frac{T_{0i}}{T_{0e}} \cdot \frac{a^2}{\lambda_i^2}, \tag{10.43}$$

where ψ_0 is the potential difference of the static field across the plasma and $\lambda_i = v_{T_i}/\nu_i$ is the ion mean free path.

Bibliography

Weakly Ionized Plasma without Longitudinal Current

1. A. V. Timofeev, Zh. Tekh. Fiz., 33:900 (1963) [Sov. Phys. – Tech. Phys., 8:682 (1964)]. It is shown that a weakly ionized plasma with an inhomogeneous density is unstable. The excitation of oscillations with $k_z = 0$ is considered.
2. G. M. Zaslavskii, S. S. Moiseev, and V. N. Oraevskii, Zh. Prikl. Mekh. Tekh. Fiz., No. 6, p. 29 (1963). Study of growth of excitations with $k_z \neq 0$; see also [6] below.
3. J. C. Woo and D. J. Rose, Phys. Rev. Lett., 19:104 (1967). The authors develop ideas similar to those put forward in [1] concerning the possible development of an instability in a current-free plasma. An instability condition of a fairly general form is obtained.
4. I. S. Baikov and A. A. Rukhadze, Zh. Tekh. Fiz., 38:1619 (1963) [Sov. Phys. – Tech. Phys., 13:1315 (1969)]. Allowance is made for a temperature perturbation and temperature gradient.
5. M. Popovic and H. Melchior, Plasma Phys., 10:495 (1968). Numerical analysis of "drift" waves in a weakly ionized plasma with allowance for ion viscosity.

Strongly Inhomogeneous Plasma without Longitudinal Current

6. A. V. Timofeev, Dokl. Akad. Nauk SSSR, 152:84 (1963) [Sov. Phys. – Doklady, 8:890 (1964)]. Discussion of the high-frequency instability of a weakly ionized plasma. Instability conditions are given.

7. A. M. Fridman, Dokl. Akad. Nauk SSSR, 154:567 (1963) [Sov. Phys. – Doklady, 9:75 (1964)]. Study of the high-frequency instability; an expression is given for its growth rate.

Current-Convective Instability

8. B. B. Kadomtsev and A. V. Nedospasov, J. Nucl. Energy, C1:230 (1960). It is shown that perturbations of the helical type can be excited by a current. The instability boundary is calculated.
9. B. B. Kadomtsev, Zh. Tekh. Fiz., 31:1273 (1961).
10. F. C. Hoh and B. Lehnert, Phys. Rev. Lett., 7:75 (1961).
11. B. B. Kadomtsev, in: Reviews of Plasma Physics, Vol. 2, Consultants Bureau, New York (1966), p. 153. In [9-11] the physical interpretation of the current-convective instability is one of the subjects discussed.
12. F. C. Hoh, Phys. Fluids, 5:22 (1962).
13. G. Guest and A. Simon, Phys. Fluids, 5:503 (1962).
14. R. R. Johnson and D. A. Jerde, Phys. Fluids, 5:988 (1962).
15. E. V. Ivash, Phys. Fluids, 8:699 (1965).
16. F. Holter and R. R. Johnson, Phys. Fluids, 9:622 (1960).
17. D. A. Huchital and E. H. Holt, Phys. Rev. Lett., 16:677 (1966). In [12-17] the Kadomtsev–Nedospasov theory is developed further and the limits of applicability of this theory are discussed.
18. L. E. Belovsova, Zh. Tekh. Fiz., 36:892 (1966) [Sov. Phys. – Tech. Phys., 11:658 (1966)]. A study of the instability of a hollow cylindrical column.
19. J. F. Reynolds and E. F. Holt, Phys. Rev., 175:205 (1968). A study of the effect of a shear on the current-convective instability.

Gravitational-Dissipative Instability

20. A. V. Timofeev, Zh. Tekh. Fiz., 33:776 (1963) [Sov. Phys. – Tech. Phys., 8:586 (1964)]. It is shown that an instability of this kind is possible and its boundary is established.

Plasmas in Crossed Electric and Magnetic Fields

21. A. Simon, Phys. Fluids, 6:382 (1963).
22. F. C. Hoh, Phys. Fluids, 6:1184 (1963). It is shown in [21, 22] that an instability can arise as a result of the difference between the ion and the electron drifts in crossed fields. Discussion (in [21]) of the instability mechanism.
23. D. L. Morse, Phys. Fluids, 8:1339 (1965). Numerical analysis of the instability of a plasma in crossed fields.

Other Aspects of the Stability of an Inhomogeneous Weakly Ionized Plasma

24. A. B. Timofeev, Zh. Tekh. Fiz., 32:1297 (1962) [Sov. Phys. – Tech. Phys., 7:959 (1963)]. Discussion of the instability of a plasma with an inhomogeneous temperature.
25. V. V. Vladimirov, Zh. Eksp. Teor. Fiz., 48:175 (1965) [Sov. Phys. – JETP, 21:119 (1965)]. Discussion of the high-frequency instability of a plasma in a

radial electric field with allowance for the processes of ionization and acceleration of the ions.

26. A. B. Mikahilovskii and O. P. Pogutse, Zh. Tekh. Fiz., 36:205 (1966) [Sov. Phys. – Tech. Phys., 11:153 (1966)]. Investigation of the instabilities of a weakly ionized plasma by means of the Bhatnagar–Gross-Krook model.

Part II

PLASMAS IN CURVED MAGNETIC FIELDS

Chapter 11

Role of Magnetic Drift of the Particles in Problems of Plasma Stability in a Curved Magnetic Field

§11.1. Magnetic Drift of Particles and the Flute Instability It Induces in an Inhomogeneous Plasma

When there is motion in a curved magnetic field, the particles drift at right angles to the lines of force. We shall give the name magnetic drift to this effect. The magnetic drift velocity is

$$\mathbf{V}_{\mathrm{dr}} = \frac{1}{\omega_B}[\mathbf{e}_0, \nabla \ln B_0]\left(\frac{v_\perp^2}{2} + v_\parallel^2\right), \tag{11.1}$$

where $\mathbf{e}_0$ is a unit vector along $\mathbf{B}_0$. Here we have assumed that the magnetic field in the plasma is irrotational (zero-pressure plasma without longitudinal current), so that the gradient of the magnetic field is uniquely related to the radius of curvature R:

$$\nabla_\perp \ln B_0 = \frac{\mathbf{n}}{R}, \tag{11.2}$$

where $\mathbf{n}$ is the principal normal to the line of force.

Equation (11.1) shows that the magnetic drift of charges of opposite sign – ions and electrons – is in opposite directions, i.e., there is a relative motion of the ions and the electrons at right angles to the magnetic field. The same situation arises in a plasma in crossed gravitational and magnetic fields when the particles drift at right angles to the lines of force with velocity

$$\mathbf{V}_{\mathrm{dr}} = -\frac{1}{\omega_B}[\mathbf{e}_0, \mathbf{g}]. \tag{11.3}$$

As was shown in §6.1, gravitational drift of particles leads to a flute instability of an inhomogeneous plasma when $\mathbf{g}\nabla n_0 < 0$. It is natural to expect the same type of instability to arise as a result of magnetic drift. That this is so can be seen by means of the following simple example. Suppose that the magnetic field is cylindrically symmetric and has only an azimuthal component:

$$\mathbf{B}_0 = (0,\, B_0,\, 0). \tag{11.4}$$

In this case, the vector ∇B_0 is directed against the radius and the particles drift along the **z** axis, $\mathbf{V}_{dr} \parallel \mathbf{z}$. Sausage-type perturbations, $\partial\psi/\partial\varphi = 0$, then correspond to the flute perturbations. We obtain the dispersion equation of such perturbations in the same way as in §6.1, taking into account the transverse ion inertia and the convection of the ions and electrons in fields of effective frequency $\omega' = \omega - \mathbf{k}\mathbf{V}_{dr}$. We represent this equation in a form similar to (6.3):

$$\frac{\omega_{p_i}^2}{\omega_{B_i}^2} + \frac{1}{k_\perp^2\omega_{B_i}}\,[\nabla\omega_{p_i}^2,\, \mathbf{k}]_\varphi \left\langle \frac{1}{\omega - \mathbf{k}\mathbf{V}_{dr}^{(i)}} - \frac{1}{\omega - \mathbf{k}\mathbf{V}_{dr}^{(e)}} \right\rangle = 0. \tag{11.5}$$

Here $\langle\cdots\rangle$ is the mean value with respect to the distribution function. Hence

$$\omega^2 = -\frac{2\nabla p\nabla \ln B_0}{m_i n_0}\cdot\frac{k_z^2}{k_\perp^2}. \tag{11.6}$$

It can be seen that if

$$\nabla p\nabla \ln B_0 > 0 \tag{11.7}$$

there is an instability with growth rate

$$\gamma \simeq \frac{v_{T_i}}{\sqrt{aR}}. \tag{11.8}$$

This example shows that magnetic drift of particles leads to the same effects as gravitational drift for

$$\frac{v_{T_i}^2}{R} \simeq g. \tag{11.9}$$

This analogy has already been noted in Chapters 6, 9, and 10. We shall also use it in what follows.

§11.2. Stability Condition for a Plasma in a Field of Complicated Geometry. Mean Magnetic Drift

As we have shown in §6.1, gravitational drift gives rise to a flute instability if $\mathbf{g}\nabla n_0 < 0$. In the opposite case, $\mathbf{g}\nabla n_0 > 0$, it makes the plasma more stable – it stabilizes various perturbations that grow if g = 0 (see §§6.5 and 6.6). Taking this into account and exploiting the analogy between curvature and gravitational force effects, let us attempt to formulate a condition that enables us to distinguish magnetic-field configurations that are favorable or unfavorable from the point of view of stability.

Comparing the expressions (11.1) and (11.3), we note that the role of the gravitational force $m_i\mathbf{g}$ in the case of a curved magnetic field is played by

$$m_i g_{\text{eff}} = -\nabla \ln B_0 \sum_{i,e} m\left(\frac{v_\perp^2}{2} + v_\parallel^2\right). \tag{11.10}$$

The stability condition of a plasma in a curved field must differ from that for a plasma in a gravitational field, $\mathbf{g}\nabla n_0 > 0$, above all by the replacement of $\mathbf{g}$ by $\mathbf{g}_{\text{eff}}$. However, g depends neither on the particle velocities nor on the coordinate along the line of force, whereas g_{eff} depends in general on both. In addition, in the case of a rectilinear field, ∇n_0 is not a function of the longitudinal coordinate, whereas in the case of a curved field the spatial distribution of the particles may depend on this coordinate. Therefore, apart from the replacement of g by g_{eff}, the stability condition (6.22) must also be averaged over the velocities and the coordinates in an appropriate manner.

As a first step in this direction, we must represent $\mathbf{g}\nabla n_0$ as an integral with respect to the velocities:

$$\mathbf{g}\nabla n_0 \rightarrow \int \mathbf{g}\nabla f_0 \, d\mathbf{v}. \tag{11.11}$$

We then average this result along a line of force by integrating over the volume of a force tube surrounding this line of force. We then find that a field configuration favorable for stability arises when

$$\int d\mathbf{r} \int m_i g_{\text{eff}} \nabla f_0 \, d\mathbf{v} > 0, \tag{11.12}$$

i.e.,

$$\int d\mathbf{r} \sum_{i,e} m \int d\mathbf{v} \left(\frac{v_\perp^2}{2} + v_\parallel^2 \right) \nabla f_0 \nabla \ln B_0 < 0. \tag{11.13}$$

In the case of a plasma with a scalar pressure this means

$$\int d\mathbf{r} \nabla p \nabla \ln B_0 < 0. \tag{11.14}$$

The conditions (11.13) and (11.14) cannot pretend to rigor, since they have only been derived by general arguments. We shall subsequently make them more precise. It will then be found that the rigorous conditions differ significantly from those given above under conditions when the relative gradients of the magnetic field and the pressure are comparable. At this juncture, we shall therefore make the opposite assumption – the relative gradient of the pressure is large compared with the magnetic-field gradient, $a \ll R$.

We represent the volume element in the form

$$d\mathbf{r} = \frac{dl}{B_0} d\Phi, \tag{11.15}$$

where dl is the element of length along the force tube and $d\Phi$ is the magnetic flux that passes through the cross section dS of the tube:

$$d\Phi = B_0 \, dS. \tag{11.16}$$

Using the constancy of the magnetic flux along the tube and the fact that p is constant along the tube, we represent (11.14) in the form

$$\int d\Phi \nabla_\perp p \int \frac{dl}{B_0} \nabla_\perp \ln B_0 < 0. \tag{11.17}$$

We now take into account one of the properties of an irrotational magnetic field of weak curvature – the quantity $B_0 dl$ is almost independent of the transverse coordinates, $\nabla_\perp (B_0 dl) = 0$. [In the special case of the field (11.5), $B_0 \sim 1/r$ and $dl = r d\varphi$, so that this equation is then exact.] Using this property and making the transformation

$$\frac{dl}{B_0} \nabla_\perp \ln B_0 = -\frac{1}{2} \nabla_\perp \frac{dl}{B_0} \tag{11.18}$$

in the left-hand side of the inequality (11.17), we reduce the latter to

$$\int d\Phi \nabla_\perp p \nabla_\perp U > 0, \tag{11.19}$$

where

$$U = \int \frac{dl}{B_0}. \qquad (11.20)$$

It follows from (11.19) that the curvature of the magnetic field must improve the stability if

$$\nabla p \nabla U > 0. \qquad (11.21)$$

Since U increases with decreasing magnetic field, the condition (11.21) means that a favorable situation arises when the modulus of the magnetic field has a minimum in the region of the plasma and increases to the plasma boundary. A field with this configuration is called a magnetic well or a field with a minimum-B configuration.

Bearing in mind what we have said above, we arrive at the following procedure for introducing a g_{eff} in the case of a field of complicated geometry: it must be assumed that g_{eff} is related to the integral U by the equation

$$g_{eff} = \frac{2T}{m_i} \nabla \ln U. \qquad (11.22)$$

In such an approach the cause of the flute instability must be regarded as the drift of the particles in the crossed fields g_{eff} and B_0.

§11.3. Flute Instability of a Plasma in an Axisymmetric Adiabatic Trap

When there is motion in a magnetic field whose strength varies along the line of force, the particles may be reflected from magnetic stoppers (mirrors), i.e., regions with a higher magnetic field. One speaks of an adiabatic trap if the plasma is confined in a restricted region because all of its particles execute oscillatory motion in the space between the two stoppers.

We shall show that in the case of an axisymmetric adiabatic trap the condition (11.21) is not satisfied, i.e., a plasma in a trap of this kind is unstable (that is, if we make the above assumptions that the particles have a Maxwellian velocity distribution and the inhomogeneity of the magnetic field is small compared with the plasma inhomogeneity). We shall omit the subscript zero of the stationary magnetic field **B**.

We shall assume that the plasma pressure and the current associated with the pressure gradient in the plasma are negligibly small. The magnetic field can then be described by the equations rot **B** = 0 and div **B** = 0. We replace **B** by a potential Ψ, $\mathbf{B} = \nabla\Psi$. In the axisymmetric case the equation $\Delta\Psi = 0$ for the potential reduces to

$$\frac{1}{r}\cdot\frac{\partial}{\partial r}\left(r\frac{\partial\Psi}{\partial r}\right)+\frac{\partial^2\Psi}{\partial z^2}=0. \tag{11.23}$$

This equation can be readily solved for small r, to which we restrict ourselves. If the magnetic field at r = 0 has the value $B_0(z)$, then it follows from (11.23) that for $r \neq 0$

$$\Psi=\int^{z} B_0(z)\,dz-\frac{r^2}{4}\cdot\frac{\partial B_0(z)}{\partial z}. \tag{11.24}$$

Hence

$$\left.\begin{aligned} B_z(z, r) &= B_0(z)-\frac{r^2}{4}\cdot\frac{\partial B_0(z)}{\partial z^2},\\ B_r(z, r) &= -\frac{r}{2}\cdot\frac{\partial B_0(z)}{\partial z}. \end{aligned}\right\} \tag{11.25}$$

In this approximation, the integral U reduces to

$$U=\int\frac{dl}{B}=\int\frac{dz}{B_z}=\int\frac{dz}{B_0}\left(1+\frac{r^2}{4}\cdot\frac{1}{B_0}\cdot\frac{\partial^2 B_0}{\partial z^2}\right). \tag{11.26}$$

Here r is the radial coordinate of the line of force, depending on z. It can be found from the equation

$$\frac{dr}{dz}=\frac{B_r}{B_z}\simeq-\frac{r}{2}\cdot\frac{\partial \ln B_0}{\partial z}, \tag{11.27}$$

i.e., approximately

$$r^2=\frac{r^2(0)\,B_0(0)}{B_0(z)}, \tag{11.28}$$

where r(0) is the radial distance to the line of force at z = 0.

The integration in (11.26) is extended over the region occupied by the plasma. It is assumed that the particles are confined to a region of finite z because of reflection from magnetic stoppers at large z.

It can be seen that U is an increasing function of r(0), as a result of which the condition (11.21) cannot be satisfied ($\partial p/\partial r < 0$). The average magnetic drift is therefore unfavorable, and the flute instability must arise in a plasma confined in a trap.

In accordance with (11.26), the rate at which the function U increases with the radius is determined by the integral of the square of the derivative $\partial B_0/\partial z$. We see that any further increase in the field fluting worsens the plasma stability.

§11.4. Instability of a Plasma in an Adiabatic Trap with a Minimum-B Configuration

In the foregoing section we have proved that plasmas in adiabatic traps with axial symmetry are unstable. However, it does not follow that a plasma in an axially asymmetric trap is unstable. To see that a plasma may be more stable in an axially asymmetric field, let us calculate $\int dl/B$ for some examples of such fields and show that, in contrast to the case of axial symmetry, this quantity decreases and not increases as the periphery of the trap is approached.

Let us suppose that the ordinary field of the mirror configuration is augmented by a multipole field in the plasma, i.e.,

$$\mathbf{B} = \mathbf{B}_{\text{mirr}} + \mathbf{B}_{\text{mult}} \tag{11.29}$$

where $\mathbf{B}_{\text{mirr}}$ is the field (11.25) and $\mathbf{B}_{\text{mult}}$ has the components

$$\left.\begin{aligned} B_{r.\,\text{mult}} &= -B_1 n \left(\frac{r}{a_0}\right)^{n-1} \cos n\varphi, \\ B_{\varphi.\text{mult}} &= B_1 n \left(\frac{r}{a_0}\right)^{n-1} \sin n\varphi. \end{aligned}\right\} \tag{11.30}$$

The field $\mathbf{B}_{\text{mult}}$ can be created by passing currents along 2n rectilinear rods parallel to the trap axis (with the currents passing in opposite directions in neighboring rods). If the rods are sufficiently long and thin, $B_1 = 2I/ca_0$, where I is the current in the rods and a_0 is the distance from the trap axis to the rod.

The approximate formula (11.26) for U remains in force, but r(z) must now be calculated with allowance for the fact that B_r is now derived from (11.25) and (11.30). Equation (11.27) is replaced by

$$\frac{dr}{dz} = -\frac{r}{2} \cdot \frac{\partial \ln B_0}{\partial z} - \frac{B_1}{B_0} n \left(\frac{r}{a_0}\right)^{n-1} \cos n\varphi. \tag{11.31}$$

The function $\cos n\varphi$ must be taken at $\varphi = \varphi(z)$, where $\varphi(z)$ is determined by the equation

$$\frac{d\varphi}{dz} = \frac{1}{r} \cdot \frac{B_\varphi}{B_z} \simeq \frac{B_1}{B_0} \cdot \frac{n}{r} \left(\frac{r}{a_0}\right)^{n-1} \sin n\varphi. \tag{11.32}$$

Let us consider the simplest case n = 2. It then follows from (11.31) that

$$r^2 = r_0^2 \frac{B_0(0)}{B_0(z)} \exp\left[-4 \int_0^z \frac{B_1}{a_0 B_0} \cos 2\varphi \, dz'\right]. \tag{11.33}$$

Substituting this result into (11.26) and integrating by parts, we obtain

$$U = \int \frac{dz}{B_0} + \frac{r_0^2}{4} B_0(0) \int dz \left\{\frac{(B_0')^2}{B_0^4} + \right.$$

$$\left. + 4 \frac{B_1 B_0'}{B_0^4 a_0} \cos 2\varphi \exp\left[-4 \int_0^z \frac{B_1}{a_0 B_0} \cos 2\varphi \, dz'\right]\right\}. \tag{11.34}$$

If B_1/B_0 is small, we can restrict ourselves in this equation to terms that are not higher than quadratic in B_1/B_0. We also take

$$B_0(z) = B_0(0)(1 + z^2/l^2), \tag{11.35}$$

assuming $z \ll 1$. After integrating in (11.32), substituting $\varphi(z)$ into (11.34), and taking into account the various assumptions, we reduce the expression for U to the form

$$U = \int \frac{dz}{B_0} + \frac{r_0^2}{l^4 B_0(0)} \left[1 - 8\left(\frac{l B_1}{a_0 B_0(0)}\right)^2\right] \int z^2 \, dz. \tag{11.36}$$

It can be seen that the function U decreases as r_0 increases if

$$\frac{B_1}{B_0} > \frac{1}{2\sqrt{2}} \cdot \frac{a_0}{l}. \tag{11.37}$$

If this condition is satisfied, stable confinement of a plasma is possible in a trap of this kind.

We shall now show that the nature of the radial dependence of the specific volume U is related qualitatively to the spatial dependence of the magnetic field modulus. Suppose $B_0(z)$ has the parabolic form (11.35) and B_{mult} is determined by Eqs. (11.30) with n = 2, so that

$$\left.\begin{aligned} B_z(r, z) &= B_0\left(1+\frac{z^2}{l^2}-\frac{r^2}{2l^2}\right), \\ B_r &= -\frac{rz}{l^2}B_0 - 2B_1\frac{r}{a_0}\cos 2\varphi, \\ B_\varphi &= 2B_1\frac{r}{a_0}\sin 2\varphi. \end{aligned}\right\} \tag{11.38}$$

Restricting ourselves to small r and z and using (11.38), we find

$$|B(r, z, \varphi)|^2 = B_0^2 + B_0^2\left\{\frac{2z^2}{l^2}+\left(\frac{2B_1}{B_0}\right)^2\left(\frac{r}{a_0}\right)^2\left[1-\left(\frac{B_0 a_0}{2B_1 l}\right)^2\right]\right\}. \tag{11.39}$$

The surfaces of equal |B| have the form of ellipsoids contained within one another if

$$\frac{B_1}{B_0} > \frac{1}{2}\cdot\frac{a_0}{l}. \tag{11.40}$$

It can also be seen from (11.39) that on any interior ellipsoid |B| is less than on an exterior ellipsoid, i.e., B increases outwards. A field of this kind is therefore called at minimum-B field. It follows from a comparison of (11.40) and (11.37) that in a minimum-B field the specific volume decreases toward the periphery, and this corresponds to stability.

§11.5. End Stabilization. Balloon-Mode Instability

1. End Stabilization. We have shown in §11.2 that a flute instability can develop in a plasma if the mean curvature of the field has an unfavorable direction (on the average the field is convex outward), U'/U > 0. However, this is not a sufficient condition for instability – it is also necessary that the flute-type perturbations belong to the class of characteristic plasma oscillations, and this is not always the case. In particular, flute-type pertur-

bations are absent if the plasma is bounded by conducting ends. For a plasma in a gravitational field this was proved in §6.4.

If there is contact with ends, one must consider instabilities associated with non-flute perturbations, $k_{\parallel} \neq 0$. In accordance with §6.4, we find that such perturbations grow if they do not differ too much from flute-type perturbations, i.e., if $k_{\parallel}$ is sufficiently small, $(k_{\parallel} c_A)^2 \leqslant |g\varkappa|$. Since $k_{\parallel} \simeq \pi/L$, this means that in the case of contact of the plasma with ends instability can arise if the system is sufficiently long. The instability condition obtained in §6.4 has the form

$$L \geqslant \pi \left(\frac{Ra}{\sqrt{\beta}} \right)^{1/2}. \tag{11.41}$$

If the radius of curvature R depends strongly on the coordinate along the line of force (and is possibly of alternating sign), R must be replaced by $(U'/U)^{-1}$. We then obtain the approximate instability condition

$$L \geqslant \pi \left(\frac{a}{\sqrt{\beta}\,(\ln U)'} \right)^{1/2}, \tag{11.42}$$

where the prime denotes the derivative along the radius.

The inequality (11.42) can be understood as an upper bound on the pressure p at which the plasma can still be stably confined in a plasma with a minimum of B_0,

$$\beta_{\max} \simeq \pi^2 \frac{a}{L^2 (\ln U)'}. \tag{11.43}$$

The quantity $\beta_{\max}$ is sometimes called the *depth of the magnetic well*.

It must be remembered that the condition (11.41) refers to a plasma whose dimensionless density number is not too low, $\Pi_e \equiv \left(\frac{\omega_{pe} a}{c}\right)^2 \gg 1$, i.e., $\beta > \frac{m_e}{m_i} \left(\frac{\rho_i}{a}\right)^2$. At lower values of Π_e it must be replaced by the condition (6.55).

2. *Balloon-Mode Instability.* Suppose that on the average $U'/U < 0$. This means that the plasma is stable against the flute instability. However, it may happen that there are sec-

tions of a line of force on which the radius of curvature is directed outward. If the length L_1 and the radius of curvature R_1 of these sections are such that a condition of the type (11.41) is satisfied,

$$L_1 \gg \pi \left(\frac{R_1 a}{\sqrt{\beta}} \right)^{1/2}, \tag{11.44}$$

then, in accordance with §11.5.1, there must be excitation of non-flute perturbations localized over the length L_1. An instability of this type is known as a balloon-mode instability.

§11.6. Shear in the Case of a Field of Cylindrical Symmetry. Suydam's Condition for a Plasma Cylinder

We have shown in Chapters 8 and 9 that one of the factors that has an important influence on plasma stability is a shear. In these chapters we introduced the concept of a shear for the case of a field with rectilinear lines of force. We shall now generalize this concept for the simplest case of a curved field – a cylindrically symmetric field with helical lines of force:

$$\mathbf{B}_0 = (0, B_{0\varphi}, B_{0z}) \tag{11.45}$$

and we shall obtain a condition for stabilization of the flute instability of a plasma cylinder by a shear.

1. Shear. The unit vectors $\mathbf{e}_0 \equiv \mathbf{B}_0/B_0$ and $\mathbf{e}_B = [\mathbf{e}_0, \mathbf{e}_r]$ corresponding to the field (11.45) are related to $\mathbf{e}_z$ and $\mathbf{e}_\varphi$ by the equations

$$\left.\begin{aligned} \mathbf{e}_0 &= h_z \mathbf{e}_z + h_\varphi \mathbf{e}_\varphi, \\ \mathbf{e}_b &= h_z \mathbf{e}_\varphi - h_\varphi \mathbf{e}_z, \end{aligned}\right\} \tag{11.46}$$

where

$$h_z = B_{0z}/B_0, \quad h_\varphi = B_{0\varphi}/B_0.$$

Since the steady-state parameters of the plasma and the magnetic field are independent of φ and z, we can assume [cf. (8.4)]

$$\psi(r) = e^{ik_z z + il\varphi} \psi(r). \tag{11.47}$$

The wave numbers $k_{\parallel}$ and k_b can be expressed in terms of k_z and l as follows:

$$k_{\parallel}=k_z h_z+\frac{l}{r}h_{\varphi},$$
$$k_b=\frac{l}{r}h_z-k_z h_{\varphi}. \tag{11.48}$$

Assuming $k_{\parallel}(r_0)=0$, we obtain for $k_{\parallel}(r)$ an expression analogous to (8.10):

$$k_{\parallel}(r)=\left(h_z^2 r k_b\frac{\partial\mu}{\partial r}\right)_{r=r_0}(r-r_0), \tag{11.49}$$

where

$$\mu=\frac{B_{0\varphi}}{rB_{0z}}. \tag{11.50}$$

The longitudinal wave number $k_{\parallel}$ depends on the coordinates if

$$\frac{d\mu}{dr}\neq 0. \tag{11.51}$$

The quantity $s_B \equiv 2\pi/\mu$ is the pitch of the line of force. Therefore, in the case of cylindrical symmetry, there is a shear if the field pitch depends on the radius. A measure of the shear [see §8.1] is now

$$\Theta\simeq arh_z^2\frac{d}{dr}\left(\frac{B_{0\varphi}}{rB_{0z}}\right). \tag{11.52}$$

The derivative $d\mu/dr$ can be related to the current flowing along the lines of force, $j_{0\parallel}$. To do this, multiply the Maxwell equation $\text{rot}\,\mathbf{B}_0 = (4\pi/c)\,\mathbf{j}_0$ by the unit vector $\mathbf{e}_0$. Then

$$\frac{d\mu}{dr}=\frac{4\pi}{c}\cdot\frac{j_{0\parallel}}{rh_z B_0}-\frac{2B_{0\varphi}}{r^2 B_{0z}}. \tag{11.53}$$

It can be seen that the radial dependence of the pitch depends strongly on $j_{0\parallel}$ if

$$j_{0\parallel}>\frac{c}{2\pi}\cdot\frac{B_{0\varphi}B_{0z}}{rB_0}. \tag{11.54}$$

In particular, this is the case if the field $B_{0\varphi}$ is produced by a current sheath situated far from the axis of the cylinder. At the same time

$$B_{0\varphi} \simeq \frac{4\pi}{c} a j_{0\parallel}, \tag{11.55}$$

where a is the width of the sheath. It can be seen that if $a \ll r$ the right-hand side of (11.54) is small compared with the left-hand side. In this limiting case, Eqs. (11.49) and (11.53) can be given a meaning that is not related to the concept of the pitch of the lines of force. Since $B_{0\varphi}$ changes more rapidly than r,

$$\frac{d\mu}{dr} \simeq \frac{1}{r} \cdot \frac{d}{dr}\left(\frac{B_{0z}}{B_{0\varphi}}\right), \tag{11.56}$$

where, in accordance with (11.53),

$$\frac{d}{dr}\left(\frac{B_{0\varphi}}{B_{0z}}\right) = \frac{4\pi}{c} j_{0\parallel} \frac{B_0}{B_{0z}^2}. \tag{11.57}$$

The radius r then disappears from Eq. (11.49) and the latter goes over into (8.10). The radial dependence of $k_\parallel$ in this approximation is entirely determined by the derivative of the tangent of the angle between the line of force and the z axis, and this derivative is, in its turn, uniquely related to the current $j_{0\parallel}$. A transition of this kind from a field of cylindrical symmetry to the case of planar symmetry was considered in §8.1.

It follows from (11.53) that the pitch of the lines of force also depends on the radius if the current $j_{0\parallel}$ in the element of volume under consideration is absent or is very small:

$$j_{0\parallel} \ll \frac{c}{2\pi} \cdot \frac{B_{0\varphi} B_{0z}}{r B_0}. \tag{11.58}$$

In this case, the field $B_{0\varphi}$ is due to the currents that flow near the cylinder axis in the plasma or in external conductors. At the same time, μ depends on the radius according to the law

$$\mu \sim 1/r^2. \tag{11.59}$$

An unusual situation arises in the case of a spatially homogeneous current distribution when there is a strong magnetic field B_{0z},

$j_{0\|} \simeq j_{0z} = \text{const}$, $B_{0z} \approx B_0$. In this case $B_{0\varphi} = (2\pi/c)\, r j_{0z}$, and the pitch of the lines of force is constant, $d\mu/dr = 0$, so that the wave number $k_{\|}$ is independent of the radius.

2. Flute Instability and Suydam's Stabilization Condition. The lines of force of the field (11.45) have a curvature directed toward the center of the cylinder. Therefore, in a plasma with a pressure that decreases radially outward, a flute instability may arise because of unfavorable magnetic drift of the particles. On the other hand, the magnetic field (11.45) has a shear and this, in accordance with Chapter 9, plays a stabilizing role. This means that the flute instability may be suppressed after all.

The effect of a shear on the flute instability in the case when the latter is due to gravitational drift was considered in §9.1. To transfer the results of §9.1 to the case of a plasma cylinder it is necessary to replace g in these results by g_{eff} and to substitute for $k_{\|}'/k_b$ the result that follows from (11.49). We find the expression for g_{eff} by means of the correspondence rule (11.9), taking into account

$$\frac{1}{R} = \frac{1}{r}\left(\frac{B_{0\varphi}}{B_0}\right)^2. \tag{11.60}$$

Then

$$\begin{aligned}(g\varkappa)_{\text{eff}} &= \frac{2}{n_0 m_i}\cdot\frac{\partial p_0}{\partial r}\left(\frac{B_{0\varphi}}{B_0}\right)^2,\\ k_{\|}'/k_b &= r\,\frac{d\mu}{dr}\left(\frac{B_{0z}}{B_0}\right)^2.\end{aligned} \tag{11.61}$$

By analogy with §9.1, we conclude that if the shear is fairly small, $d\mu/dr \to 0$, a flute instability must arise in the plasma cylinder with a growth rate $\sqrt{-(g\varkappa)_{\text{eff}}}$, i.e.,

$$\gamma = \sqrt{-\frac{2\partial p}{\partial r}\cdot\frac{1}{m_i n_0 R}} \simeq \frac{v_{T_i}}{\sqrt{aR}}. \tag{11.62}$$

These perturbations are localized near the surface $r = r_0$ at which $k_{\|}(r_0) = 0$, i.e. [see (11.48)], near r_0 satisfying the relation

$$k_z B_{0z}(r_0) + \frac{l}{r} B_{0\varphi}(r_0) = 0. \tag{11.63}$$

If the shear is not too small, the flute instability must be suppressed. Taking into account (9.9) and using the correspondence rules (11.61), we find that for this to happen it is necessary that

$$\frac{1}{4}\left(\frac{d\ln\mu}{dr}\right)^2 > -\frac{8\pi}{rB_z^2}\cdot\frac{\partial p}{\partial r}\,. \tag{11.64}$$

This is Suydam's stabilization condition for a cylindrical plasma column.

§11.7. Stabilization by a Shear due to Currents in External Conductors

1. Introduction. Above we have considered the stabilization of a plasma by a shear for various simple examples: in §9.1, for a field of planar symmetry; in §11.6, for a field of cylindrical symmetry. In the first case the shear at each point of the plasma is uniquely related to the current flowing at this point. The shear at the point r of a cylindrical plasma is determined by the currents that flow within the cylinder of radius r. In the latter case, the current need not flow in the plasma itself; for example, it may flow along a solid core in a plasma cylinder. However, it is important that in the case of cylindrical symmetry the shear does not depend on the currents in the conductors situated outside the plasma cylinder.

We shall now show that cylindrically asymmetric external currents can lead to a shear. The magnetic field of such currents can be represented in the form*

$$\left.\begin{aligned} B_r &= \sum_n b_n \frac{1}{\alpha}\cdot\frac{\partial}{\partial r} I_n(n\alpha r)\sin n(\varphi-\alpha z),\\ B_\varphi &= \sum_n n b_n \frac{1}{\alpha r} I_n(n\alpha r)\cos n(\varphi-\alpha z),\\ B_z &= B_0 - \sum_n n b_n I_n(n\alpha r)\cos n(\varphi-\alpha z). \end{aligned}\right\} \tag{11.65}$$

We shall assume that, in addition to the ordinary solenoidal coil that induces the field B_0, there is a system of helical conductors with pitch $L = 2\pi/\alpha$ which induce a helical field with amplitudes B_r, B_φ, and $B_z - B_0$. In particular, if there are n_0 pairs of sufficiently thin helical conductors with an alternating direction of the cur-

*The subscript zero is omitted from the static field B in this section.

rent, only the term with $n = n_0$ in the sums of (11.65) is nonvanishing [this question is discussed in more detail in the paper of Morozov and Solov'ev].

We shall assume that the helical field is small compared with B_0. For this case we solve the equations of the lines of force and show that, on the average, they have the form of spirals. Knowing the radial dependence of the spiral pitch, we can calculate the shear. An average shear of this kind can stabilize the flute instability that arises if the mean curvature of the lines of force is unfavorable. Below, we shall obtain a condition for shear stabilization of the flute instability in the case when the curvature is entirely due to the field (11.65) producing the shear.

We shall calculate the shear and find a stabilization condition on the basis of geometrical arguments and the analogy with the problem of a plasma in a gravitational field.

2. Mean Rotation of the Lines of Force and the Mean Shear. The equations of the lines of force have the form

$$\left.\begin{aligned} \frac{dr}{dz} &= \frac{B_r}{B_z} \simeq \frac{\widetilde{B}_r}{B_0} - \frac{\widetilde{B}_r \widetilde{B}_z}{B_0^2}, \\ \frac{d\varphi}{dz} &= \frac{1}{r} \cdot \frac{B_\varphi}{B_z} \simeq \frac{1}{r} \cdot \frac{\widetilde{B}_\varphi}{B_0} - \frac{\widetilde{B}_\varphi \widetilde{B}_z}{r B_0^2}. \end{aligned}\right\} \tag{11.66}$$

Here $\widetilde{B}_r = B_r$, $\widetilde{B}_\varphi = B_\varphi$, and $\widetilde{B}_z = B_z - B_0$ are oscillating functions of φ and z. In (11.66) we have carried out an expansion with respect to the small parameter $\widetilde{B}_z/B_0$ and retained only the small terms of second order.

We split r and φ as functions of z into parts averaged over the period L and parts that oscillate with this period:

$$r = \bar{r} + \tilde{r}, \quad \varphi = \bar{\varphi} + \tilde{\varphi}. \tag{11.67}$$

The oscillations of $\tilde{r}$ and $\tilde{\varphi}$ are due to the fields $\widetilde{B}$ and are therefore small compared with the mean fields. This means that on the right-hand sides of (11.66) we can expand all the functions of r and φ in series in r and $\tilde{\varphi}$. We then obtain the following two systems of equations for the oscillating and the mean parts of r and φ:

$$\frac{d\tilde{r}}{dz} = \frac{\widetilde{B}_r}{B_0}, \qquad \frac{d\tilde{\varphi}}{dz} = \frac{\widetilde{B}_\varphi}{\bar{r} B_0}, \tag{11.68}$$

$$\left.\begin{aligned}\frac{d\bar{r}}{dz} &= -\frac{\langle\tilde{B}_r\tilde{B}_z\rangle}{B_0^2} - \frac{1}{B_0^2}\left\langle\frac{\partial\tilde{B}_r}{\partial\bar{r}}\tilde{r} + \frac{\partial\tilde{B}_r}{\partial\bar{\varphi}}\tilde{\varphi}\right\rangle,\\ \frac{d\bar{\varphi}}{dz} &= -\frac{\langle\tilde{B}_\varphi\tilde{B}_z\rangle}{\bar{r}B_0^2} - \frac{1}{B_0}\left\langle\frac{\partial}{\partial\bar{r}}\left(\frac{\tilde{B}_\varphi}{r}\right)\tilde{r} + \frac{\partial}{\partial\bar{\varphi}}\left(\frac{\tilde{B}_\varphi}{\bar{r}}\right)\tilde{\varphi}\right\rangle.\end{aligned}\right\} \tag{11.69}$$

Here $\langle\ \rangle$ has the same meaning as a bar. On the right-hand side of these equations all the functions of r and φ are taken at $r = \bar{r}$, $\varphi = \bar{\varphi}$. Using (11.68) and (11.65), we find

$$\begin{gathered}\tilde{r} = \frac{1}{B_0}\sum_n \frac{b_n}{n\alpha}\cdot\frac{\partial}{\partial\rho} I_n(n\rho)\cos n(\bar{\varphi} - \alpha z),\\ \tilde{\varphi} = -\frac{1}{B_0}\sum_n \frac{b_n}{\rho^2} I_n(n\rho)\sin n(\bar{\varphi} - \alpha z),\\ \rho \equiv \alpha\bar{r}.\end{gathered} \tag{11.70}$$

Substituting this result into (11.69) and averaging, we obtain

$$\begin{gathered}\frac{d\bar{r}}{dz} = 0,\\ \frac{d\bar{\varphi}}{dz} = \frac{\alpha}{4B_0^2}\sum_n b_n^2\left(\frac{1}{\rho}\cdot\frac{\partial}{\partial\rho}\right)^2 I_n^2(n\rho).\end{gathered} \tag{11.71}$$

It can be seen that along the radius the line of force is not displaced on the average, $\bar{r}$ = const, whereas its mean rotation with respect to the azimuth is nonvanishing, $d\bar{\varphi}/dz \neq 0$. The rate of rotation depends in general on the radius, and this means that an average shear is present.

Turning to the analogy with the case of a cylindrical field (see §11.6), we find that $d\bar{\varphi}/dz$ plays the role of the parameter μ:

$$\frac{d\bar{\varphi}}{dz} \to \mu \equiv \frac{B_\varphi}{rB_z}. \tag{11.72}$$

Using this analogy and taking into account the second equation in (11.61), we find an expression for the $k'_\parallel/rk_b$ that characterizes the average shear:

$$\frac{k'_\parallel}{k_b} = \bar{r}\frac{d}{dr}\cdot\frac{d\bar{\varphi}}{dz}. \tag{11.73}$$

3. Mean Magnetic Drift. In accordance with §11.2, the average velocity of magnetic drift in the case of an almost homogeneous magnetic field [the field (11.65) belongs to this class]

can be expressed in terms of the derivative along the radius of the integral $U = \oint \frac{dl}{B}$. This integral is approximately equal to

$$U = \oint \frac{dl}{B} \equiv \oint \frac{dz}{B_z} \approx \oint \frac{dz}{B_0} - \oint \left[\frac{\widetilde{B}_z}{B_0^2} - \frac{\widetilde{B}_z^2}{B_0^3} \right] dz. \tag{11.74}$$

Expanding B_z in a series in the oscillations about $\bar{r}$ and $\bar{\varphi}$,

$$\widetilde{B}_z \approx \widetilde{B}_z(\bar{r}, \bar{\varphi}, z) + \frac{\partial \widetilde{B}_z}{\partial \bar{r}} \tilde{r} + \frac{\partial \widetilde{B}_z}{\partial \bar{\varphi}} \tilde{\varphi} \tag{11.75}$$

and averaging over the oscillations in (11.47), we find

$$U = \oint \frac{dz}{B_0} \left\{ 1 + \frac{1}{2} \sum_n \left(\frac{b_n}{B_0} \right)^2 \left[\left(\frac{\partial}{\partial \rho} I_n \right)^2 + \frac{n^2 I_n^2}{\rho^2} + n^2 I_n^2 \right] \right\}. \tag{11.76}$$

In the sums over n we have functions of positive powers of $\rho \equiv \alpha \bar{r}$. The integral U therefore increases with the radius, which for $\partial p / \partial \bar{r} < 0$ corresponds to an unfavorable direction of the mean magnetic drift. This drift is the same as that due to a gravitational force [see Eq. (11.22)]:

$$g_{\text{eff}} = \frac{2T}{m_i} \cdot \frac{\partial \ln U}{\partial \bar{r}} = \frac{T}{m_i} \sum_n \left(\frac{b_n}{B_0} \right)^2 \left[\left(\frac{\partial}{\partial \rho} I_n \right)^2 + \left(\frac{n I_n}{\rho} \right)^2 + (n I_n)^2 \right]. \tag{11.77}$$

4. Stabilization Condition. The magnetic drift in the effective gravitational field (11.77) can lead to a flute instability. However, this instability cannot develop if the shear (11.73) is sufficiently large. We find the stabilization condition from the correspondence rules by means of Eqs. (11.61) and (11.64):

$$\frac{\alpha^2}{4} \left[\frac{d}{d \ln r} \sum_n \left(\frac{b_n}{2B_0} \right)^2 \left(\frac{1}{\rho} \cdot \frac{\partial}{\partial \rho} \right)^2 I_n^2(n\rho) \right]^2 >$$

$$> - \frac{4\pi}{B_0^2} \cdot \frac{\partial p}{\partial r} \sum_n \left(\frac{b_n}{B_0} \right)^2 \left[\left(\frac{\partial I_n}{\partial \rho} \right)^2 + \left(\frac{n I_n}{\rho} \right)^2 + (n I_n)^2 \right]. \tag{11.78}$$

We shall return to the discussion for this result in §17.2.

§11.8. Mean Magnetic Drift and Shear for a Plasma Confined in Toroidal Trap

1. Toroidal Traps with Magnetic Surfaces. A plasma may remain in a confined volume if the particles (some

or all of them) are not reflected from mirrors, i.e., if they move in one particular direction along the lines of force. In this case the lines of force themselves must remain in a bounded region. Magnetic systems of this kind are known as closed traps.

From the point of view of applications, a central position is occupied in the large class of closed traps by those in which the lines of force lie on the surfaces of toroids that nest within one another. These are toroidal traps with magnetic surfaces.

2. Condition of Favorable Magnetic Drift for Traps with Magnetic Surfaces. The magnetic field of closed traps is curved, so that the particles in such a trap must execute magnetic drift. This can lead to a flute instability if the integral $U = \int dl/B$ increases as the plasma periphery is approached (see §11.2). Conversely, magnetic drift must play a stabilizing role if U decreases toward the periphery, i.e., in the case of a field with a mean minimum of B.

The form in which the condition for favorable drift, $\nabla p \nabla U > 0$, is written [see (11.20)] is not suited for the study of the problem of closed traps, since U refers to a line of force that, in general, passes around the toroid many times with respect to the major radius. However, this condition can be expressed differently – in terms of surface functions. We note that on every rational magnetic surface, which is such that on it the lines of force close after N revolutions, the integral U is uniquely related to the derivative of the volume V of the corresponding toroid with respect to the longitudinal magnetic flux Φ in this toroid:

$$\int \frac{dl}{B} = N \frac{dV}{d\Phi}. \tag{11.79}$$

Therefore, the condition for favorable magnetic drift (11.21) can be written as

$$\frac{dp}{dV} \cdot \frac{\partial}{\partial V} \cdot \frac{dV}{d\Phi} > 0. \tag{11.80}$$

Noting that $d\Phi/dV > 0$ and $dp/dV < 0$, we then obtain

$$\frac{d^2V}{d\Phi^2} < 0. \tag{11.81}$$

Closed traps for which this condition is satisfied are known as $V'' < 0$ systems.

3. Shear in the Case of the Field of a Toroidal Trap. Let us now consider what is a shear in the case of the field of a toroidal trap. We shall arrive at this concept in the same way as in §§8.1 and 11.6 by considering the coordinate dependence of the longitudinal wave number. We shall assume that the perturbed potential has the form

$$\psi(\mathbf{r}) = \psi(V) \exp[2\pi i (l\theta + n\zeta)]. \qquad (11.82)$$

Here we have used a system of coordinates in which the lines of force are rectilinear and the volume V of the toroid is taken as the radial coordinate. (The coordinates θ and ζ are the analogs of the azimuth and the z coordinate in the case of cylindrical symmetry.)

Taking the definition of $k_{\|}$ [see Eq. (8.5)], using the rules for differentiation in curvilinear coordinates, and repeating calculations similar to those made in §§8.1 and 11.6, we arrive at the desired result:

$$k_{\|} = \frac{V - V_0}{\sqrt{g_{11}}} k_b \frac{(d\Phi/dV)^2}{B^2 \sqrt{g}} \cdot \frac{d}{dV} \left(\frac{d\chi}{d\Phi} \right). \qquad (11.83)$$

Here, k_b is determined by the equation

$$ik_b\psi = \mathbf{e}_b \nabla\psi = \left(\left[\frac{\mathbf{B}}{B}, \frac{\nabla V}{|\nabla V|} \right] \nabla\psi \right); \qquad (11.84)$$

g_{11} is the component of the metric tensor g_{ik} in the coordinates (V, θ, ζ); $g = |g_{ik}|$; χ and Φ are the transverse and longitudinal magnetic fluxes. (For more details on the curvilinear coordinates and the integral characteristics of the magnetic field in a toroidal trap the reader is referred to the paper of Solov'ev and Shafranov.)

Thus, there is a "radial" dependence of $k_{\|}$ if the angle between the lines of force and the coordinate axes θ and ζ depends on the radius: $d\chi/d\Phi \neq \text{const}$.

§11.9. Trapped-Particle Instability in a Collisionless Plasma

Suppose that a plasma confined in a closed trap is stable against flute-type perturbations. If the effect of a shear is ignored, this means that the magnetic drift of the particles averaged over the distribution function and over the length of the line of force has a favorable direction – the same as the drift of the particles under the influence of a force of gravity directed into the plasma. In such a plasma, different kinds of flute instability can, however, arise. Some of them have the same nature as the instabilities of a plasma in a homogeneous field; for example, the instabilities due to a density or temperature gradient considered in Chapter 3. The quantitative relationships for such instabilities in closed systems will be given later in §16.4. Here we shall consider a new class of flute instabilities which arise for the same reason as the ordinary flute instability – magnetic drift of the particles and inhomogeneity of the plasma.

The magnetic field of closed traps is inhomogeneous along a line of force. Particles which move in such a trap with longitudinal velocity that is not very high are reflected from regions with higher magnetic field and are for this reason localized in the regions of weaker magnetic field. We shall say that such particles are *trapped*, in constrast to *untrapped* particles which move along the line of force without reflection.

The magnetic drift of the trapped particles is due to the field curvature in only that region in which the particles are localized. This drift must therefore be averaged over only the region of relatively weak field and not over the whole length of the line of force. In this region, the field curvature is, in general, unfavorable as, for example, in the case of a probkotron (mirror machine) without stabilizing rods. If the amplitude of the perturbations also has a maximum in such regions, the trapped particles must have a destabilizing effect, in contrast to the case of flute-type perturbations, for which the presence of trapped particles does not by itself lead to instability.

This effect of unfavorable drift of trapped particles can be deduced from the following elementary consideration. Taking into account only regions with unfavorable curvature and assuming that in these regions the perturbation amplitude is constant, we find the contribution of the trapped particles to the permittivity (cf. §6.1):

$$\varepsilon_0^{\text{trap}} \simeq \frac{\omega_{p_i\,\text{trap}}^2}{\omega_{B_i}^2} \frac{g_{\text{eff}}\,\varkappa}{\omega^2}. \tag{11.85}$$

The untrapped particles find themselves in a perturbed electric field that varies along the line of force, and they are therefore distributed in accordance with Boltzmann's law and have a permittivity given approximately by

$$\varepsilon_0^{\text{untr}} \simeq 4\pi e^2 n_{\text{untr}} / k^2 T. \tag{11.86}$$

Adding (11.85) and (11.86) and equating the result to zero, we obtain the growth rate

$$\gamma \simeq \gamma_0 k_\perp \rho_i \left(n_{\text{trap}} / n_{\text{untr}}\right)^{1/2}, \tag{11.87}$$

where $\gamma_0 \simeq v_{T_i}/(aR)^{1/2}$ is the growth rate of the flute instability.

The instability we have considered here is frequently known as the trapped-particle instability. We shall return to a quantitative analysis of this instability in Chapter 16.

Bibliography

1. M. N. Rosenbluth and C. Longmire, Ann. Phys., 1:120 (1957). It is suggested that the physical effects of gravitational and magnetic drift are idential (§11.1). The stability condition (11.21) is obtained. It is shown that a plasma in an axisymmetric field configuration is unstable (§11.3).
2. B. R. Suydam, in: Hot Plasma Physics and Thermonuclear Reactions [Russian Collection of Translations], Atomizdat (1959), p. 89. It is shown that a shear has a stabilizing effect in a plasma with cylindrical symmetry (§11.6).
3. J. L. Johnson et al., Phys. Fluids, 1:281 (1958). Study of the effect of stabilization by a shear produced by external conductors (§11.7). Energy considerations yield the stability condition V" < 0 (§11.8).
4. B. B. Kadomtsev, in: Plasma Physics and the Problem of Controlled Thermonuclear Reactions, Vol. 4, Pergamon Press, Oxford (1960), p. 17. Contains a proof of the equation $[\nabla p, \nabla U] = 0$. Curvilinear coordinates are used. The stability condition (11.21) is obtained.
5. I. B. Bernstein et al., Proc. Roy. Soc., A244:17 (1958). It is shown that balloon-type perturbations can develop (§11.5). The condition (11.44) is derived.
6. D. V. Sivukhin, in: Reviews of Plasma Physics, Vol. 1, Consultants Bureau, New York (1966), p. 1.
7. A. I. Morozov and L. S. Solov'ev, in: Reviews of Plasma Physics, Vol. 2, Consultants Bureau, New York (1966), p. 201.
8. M. B. Kruskal, in: Plasma Physics, IAEA (1965), p. 67. In the reviews [6-8] the reader can find a detailed exposition of the theory of the motion of charged particles in a curved magnetic field.
9. A. I. Morozov and L. S. Solov'ev, in: Reviews of Plasma Physics, Vol. 2, Consultants Bureau, New York (1966), p. 1. In §11.3 we have followed this paper

in proving the instability of the axisymetric trap and in §11.7 in calculating the shear produced by external conductors.

10. J. B. Taylor, Phys. Fluids, 6:1529 (1963).
11. B. A. Trubnikov, in: Plasma Physics and Controlled Nuclear Fusion Research, Vol. 1, IAEA, Vienna (1966), p. 83.
12. H. P. Furth and M. N. Rosenbluth, Phys. Fluids, 7:764 (1964). The results of [10-12] were used in §11.4 to calculate the integral $\int dl/B$ and $|B|$ in the case of a trap with stabilizing rods.
13. M. Kruskal and R. Kulsrud, Phys. Fluids, 1:253 (1958).
14. L. S. Solov'ev and V. D. Shafranov, Reviews of Plasma Physics, Vol. 5, Consultants Bureau, New York (1970), p. 1. Some results of [13, 14] were used in §11.8 in the discussion of the properties of a plasma confined in closed traps with magnetic surfaces.
15. B. B. Kadomtsev, ZhETF Pis Red., 4:15 (1966) [JETP Letters 4:10 (1966)].
16. B B. Kadomtsev and O. P. Pogutse, Zh. Eksp. Teor. Fiz., 51:1734 (1966) [Sov. Phys. – JETP, 24:1172 (1966)].
17. B. B. Kadomtsev and O. P. Pogutse, in: Reviews of Plasma Physics, Vol. 5, Consultants Bureau, New York (1970), p. 249. In considering the trapped-particle instability in §11.9 we have followed the ideas developed in [15-17].

Chapter 12

General Statement of the Problem of the Instabilities of a Plasma in a Curved Magnetic Field

§12.1. Classification of Plasma Instabilities in a Curved Magnetic Field

Above we have analyzed two classes of low-frequency, $\omega \ll \omega_{Bi}$, gradient instabilities. Some of these instabilities are insensitive to curvature effects and can even arise if $R = \infty$. These are the instabilities (discussed in Chapters 3-5) due to density and temperature gradients. Their frequency and growth rate are such that

$$\frac{|\omega|}{\omega_{B_i}} \simeq \left(\frac{\rho_i}{a}\right)^2. \tag{12.1}$$

An unfavorable field curvature is responsible for another class of instabilities, in particular, the flute instability, for which (§11.1)

$$\frac{|\omega|}{\omega_{B_i}} \simeq \frac{\rho_i}{(aR)^{1/2}}. \tag{12.2}$$

If there is a strong curvature, $R \simeq a$, this means

$$\frac{|\omega|}{\omega_{B_i}} \simeq \frac{\rho_i}{a}. \tag{12.3}$$

It can be seen that the frequency (12.1) is low compared with (12.3). The instabilities of the type (12.1) may therefore be called *slow gradient instabilities*; those of the type (12.3), *fast*.

If a plasma is in a rectilinear magnetic field, only the slow instabilities are possible (we assume g = 0 and the absence of directed streams). The stability investigation in this case therefore reduces to an analysis of only the slow perturbations, which we have done in Chapters 3-5. If the magnetic field is curved, it is above all necessary to investigate the possibility of growth of fast perturbations since these, at the same wave numbers, have growth rates that are larger by a factor of a/ρ_i than those of the slow instabilities. It is only necessary to investigate the slow instabilities if the fast instabilities cannot arise.

Our above division of instabilities into fast and slow is based on the assumption that $R \simeq a$. In a number of cases the field curvature may be very weak, $a \ll R$. The characteristic frequency of perturbations of the type (12.2) is then small compared with (12.3). It is comparable with the frequency of slow perturbations of the type (12.1) if

$$\frac{a}{R} \leqslant \left(\frac{\rho_i}{a}\right)^2. \tag{12.4}$$

As in §6.2.1, we shall call a plasma whose parameters satisfy this condition a plasma with finite ion Larmor radius. We shall see that, as in the case of a plasma in a field of gravity, the condition (12.4) is the condition for the stabilization of the flute instability. Thus, the division of perturbations into fast and slow becomes impossible just when the effect of a finite ion Larmor radius becomes important.

If the situation opposite to (12.4) obtains, i.e.,

$$\frac{a}{R} \geqslant \left(\frac{\rho_i}{a}\right)^2, \tag{12.5}$$

fast perturbations can be distinguished from the slow.

§12.2. Magnetohydrodynamic Description of Fast Gradient Perturbations

We have already drawn attention to a division into fast and slow perturbations for a plasma in a curved magnetic field. For the former, $\omega/\omega_{Bi} \simeq \rho_i/a$, so that the problem contains a small parameter $(\rho_i/a)^2\omega_{Bi}/\omega \simeq \omega^*/\omega$. As a result, the equations that describe these perturbations can be greatly simplified. We now turn to the derivation of simplified equations of this kind.

We begin by considering the order of the terms of the linearized equation of motion (4.52):

$$mn_0\left\{\frac{\partial \mathbf{V}'}{\partial t}+(\mathbf{V}^{(0)}\nabla)\,\mathbf{V}'+(\mathbf{V}'\nabla)\,\mathbf{V}^{(0)}\right\}+$$
$$+mn'(\mathbf{V}^{(0)}\nabla)\,\mathbf{V}^{(0)}=en_0\left(\mathbf{E}'+\frac{1}{c}[\mathbf{V}',\mathbf{B}_0]+\right.$$
$$\left.+\frac{1}{c}[\mathbf{V}^{(0)},\mathbf{B}']\right)+en'\left(\mathbf{E}^{(0)}+\frac{1}{c}[\mathbf{V}^{(0)},\mathbf{B}_0]\right)-\nabla p'-\operatorname{div}\overleftrightarrow{\pi}'+\mathbf{R}'. \quad (12.6)$$

Here the primed quantities are the perturbations and those with index zero the equilibrium parameters. In accordance with (4.52), the latter satisfy the equation

$$(\mathbf{V}^{(0)}\nabla)\,\mathbf{V}^{(0)}=en_0\left(\mathbf{E}^{(0)}+\frac{1}{c}[\mathbf{V}^{(0)},\mathbf{B}_0]\right)-\nabla p^{(0)}-\operatorname{div}\overleftrightarrow{\pi}^{(0)}+\mathbf{R}^{(0)}. \quad (12.7)$$

Equations (12.6) and (12.7) describe both a collisional and a collisionless plasma; in the latter case $\mathbf{R} = 0$.

1. <u>Components of Equation (12.6) at Right Angles to B_0.</u> We shall assume that the equilibrium electric field is not greater in order of magnitude than the remaining terms of Eq. (12.7). Then

$$V_\perp^{(0)}\simeq\frac{\rho_i v_{T_i}}{a},\quad E_\perp^{(0)}\simeq\frac{T}{ea}. \quad (12.8)$$

Using the continuity equations and the heat balance [see Eq. (4.52)], we estimate the order of magnitude of the density and pressure perturbations:

$$n'\simeq\frac{n_0}{\omega a}V'_\perp,\quad p'\simeq\frac{p_0}{\omega a}V'_\perp. \quad (12.9)$$

We obtain an estimate for $\operatorname{div}\overleftrightarrow{\pi}'$ by means of (12.9), assuming $|\operatorname{div}\pi'|\leqslant|\nabla p'|$. Using the Maxwell equation $\operatorname{rot}\mathbf{E}'=-\frac{1}{c}\cdot\partial\mathbf{B}'/\partial t$ and the first relation of (12.8), we obtain an estimate for B':

$$\left|\frac{1}{c}[\mathbf{V}^{(0)},\mathbf{B}']\right|\simeq\left(\frac{\rho_i}{a}\right)^2\frac{\omega_{B_i}}{\omega}E'_\perp\simeq\frac{\omega}{\omega_{B_i}}E'_\perp. \quad (12.10)$$

In a strong magnetic field, $\nu \ll \omega_B$, the transverse part of the force of friction $\mathbf{R}'$ is small as $(\nu/\omega_B)^2$.

Zeroth Approximation in ω/ω_{B_i}. Using the above estimates, we find that when terms of order ω/ω_{B_i} can be neglected the components of Eq. (12.6) at right angles to $\mathbf{B}_0$ reduce to

$$\mathbf{E}'_{\perp} + \frac{[\mathbf{V}', \mathbf{B}_0]}{c} = 0. \tag{12.11}$$

In this approximation the perturbed motion of the electron and ion components at right angles to the magnetic field occurs at the same velocity, equal to the electric drift velocity of the particles:

$$\mathbf{V}' = c\,[\mathbf{E}'_{\perp}, \mathbf{B}_0]/B_0^2. \tag{12.12}$$

We recall that, in contrast to (12.12), the perturbed velocity of waves with $\omega/\omega_{Bi} \simeq (\rho_i/a)^2$ is made up of the velocities of electric drift and the perturbed velocity of the Larmor currents [compare (12.12) with (4.53)], which are of the same order.

First Approximation in ω/ω_{Bi}. In the zeroth approximation in ω/ω_{Bi}, the electron and ion equations (12.6) become degenerate – they reduce to the single equation (12.11). To obtain a second independent equation we must include in (12.6) the terms of order ω/ω_{Bi}.

The left-hand side of the electron equation (12.6) is small compared with the right-hand side as ω/ω_{B_e} and can therefore be neglected in the approximation of finite ω/ω_{Bi}. On the left-hand side of the ion equation (12.6) the terms with the spatial derivatives of the velocity are small compared with $\partial \mathbf{V}'/\partial t$ as ω/ω_{Bi}; this can be seen by taking into account (12.8) and (12.9). Omitting these small terms and adding the electron to the ion equation, we obtain an equation of motion of the order ω/ω_{Bi}:

$$m_i n_0 \frac{\partial \mathbf{V}'_{\perp}}{\partial t} = -\nabla_{\perp}(p'_i + p'_e) - \operatorname{div}_{\perp}(\overleftrightarrow{\pi}'_i + \overleftrightarrow{\pi}'_e) + \frac{1}{c}([\mathbf{j}'_{\perp}, \mathbf{B}_0] + [\mathbf{j}^{(0)}, \mathbf{B}'])_{\perp}. \tag{12.13}$$

Here we have used the relations $n_e = n_i$ (the condition of electrical quasineutrality) and $\mathbf{R}'_e = \mathbf{R}'_i = 0$. In these equations, $\mathbf{j}^{(0)}$ and $\mathbf{j}'$ are the equilibrium and the perturbed currents:

$$\mathbf{j}^{(0)} = n_0(e_e \mathbf{V}_e^{(0)} + e_i \mathbf{V}_i^{(0)}), \tag{12.14}$$

$$\mathbf{j}'_{\perp} = e_i n_0 (\mathbf{V}'_{\perp i} - \mathbf{V}'_{\perp i}) + e_i n' (\mathbf{V}^{(0)}_{\perp i} - \mathbf{V}^{(0)}_{\perp e}). \tag{12.15}$$

It is assumed that $\mathbf{V}'_{\perp i}$ and $\mathbf{V}'_{\perp e}$ in (12.15) contain not only terms of zeroth order (12.12) but also corrections of order ω/ω_{Bi}.

Single-Fluid Description of a Plasma. It follows from (12.11) and (12.13) that for the motion at right angles to the magnetic field the plasma behaves as a single-component, perfectly conducting fluid with mass density ρ_0 and total pressure tensor $\overleftrightarrow{p'}$ equal to

$$\left.\begin{aligned} &\rho_0 = m_i n_0, \\ &\overleftrightarrow{p'} = \overleftrightarrow{I}(p'_i + p'_e) + \overleftrightarrow{\pi'_i} + \overleftrightarrow{\pi'_e}. \end{aligned}\right\} \tag{12.16}$$

Here, $\overleftrightarrow{I}$ is the unit tensor.

From this point of view, Eq. (12.13) determines the change in the perturbed velocity $\mathbf{V}'_\perp$ for the given currents $\mathbf{j}'$ and $\mathbf{j}^{(0)}$, pressure tensor $\overleftrightarrow{p'}$, fields $\mathbf{B}'$ and $\mathbf{B}_0$, and mass density ρ_0,

$$\rho_0 \frac{\partial \mathbf{V}'}{\partial t} = -\operatorname{div} \overleftrightarrow{p'} + \frac{1}{c}[\mathbf{j}'_\perp, \mathbf{B}_0] + \frac{1}{c}[\mathbf{j}^{(0)}, \mathbf{B}']_\perp. \tag{12.17}$$

A closed system of equations for the field and the plasma corresponding to this method of description can be obtained by expressing $\overleftrightarrow{p'}$, $\mathbf{j}'_\perp$, and $\mathbf{B}'$ in terms of $\mathbf{V}'$. The relationship between $\overleftrightarrow{p'}$ and $\mathbf{V}'_\perp$ can be found by solving the Boltzmann equation (this will be done in §12.3). The current $\mathbf{j}'$ is related to the field $\mathbf{B}'$ by the Maxwell equation

$$\mathbf{j}' = \frac{c}{4\pi} \operatorname{rot} \mathbf{B}', \tag{12.18}$$

and $\mathbf{B}'$ can be expressed in terms of $\mathbf{E}'$ by means of the equation $\operatorname{rot} \mathbf{E}' = -\frac{1}{c} \cdot \frac{\partial \mathbf{B}'}{\partial t}$. If we assume that $\mathbf{E}'\mathbf{B} = 0$ (the condition of applicability of this assumption will be discussed in the next subsection), the required relationship between $\mathbf{E}'$ and $\mathbf{V}'_\perp$ will be given by Eq. (12.11). Replacing $\mathbf{V}'$ by the perturbed displacement $\boldsymbol{\xi}$ of the plasma defined by the equation

$$\frac{\partial \boldsymbol{\xi}}{\partial t} = \mathbf{V}', \tag{12.19}$$

we then find in the manner described

$$\left.\begin{aligned} &\mathbf{E}' = -\frac{1}{c}\left[\frac{\partial \boldsymbol{\xi}}{\partial t}, \mathbf{B}_0\right], \\ &\mathbf{B}' = \operatorname{rot}[\boldsymbol{\xi}, \mathbf{B}_0]. \end{aligned}\right\} \tag{12.20}$$

Using (12.18) and (12.20) to express the current **j**' and the perturbed field **B**' in terms of ξ and substituting the result into Eq. (12.17), we arrive at a description of plasma oscillations as a hydrodynamic process characterized by the following equation for the displacement ξ:

$$\rho_0 \frac{\partial^2 \xi}{\partial t^2} = -\operatorname{div} \overleftrightarrow{p}'(\xi) + \frac{1}{4\pi}[\operatorname{rot}\mathbf{Q}, \mathbf{B}_0] + \frac{1}{4\pi}[\operatorname{rot}\mathbf{B}_0, \mathbf{Q}]_\perp. \qquad (12.21)$$

Here $\mathbf{Q} \equiv \operatorname{rot}[\xi, \mathbf{B}_0]$.

This hydrodynamic approach contrasts with the electrodynamic approach employed above, in which all the perturbed parameters of the plasma are expressed in terms of the electric field and the problem of the plasma oscillations reduces to an investigation of Maxwell's equations with known permittivity (see §7.1 in Volume 1).

2. The Longitudinal Component of Equation (12.6). Projecting the vector equation (12.6) onto the direction of $\mathbf{B}_0$, neglecting terms of order ω/ω_B [using the estimates (12.10) and (12.9)], and assuming $V_\parallel^{(0)}/L \ll \omega$, we obtain

$$mn_0 \frac{\partial V'_\parallel}{\partial t} = en_0 E'_\parallel + en' E_\parallel^{(0)} - \nabla_\parallel p' - (\operatorname{div} \overleftrightarrow{\pi}')_\parallel + R'_\parallel, \qquad (12.22)$$

(L is the characteristic longitudinal scale).

As in the treatment of the transverse motion of the plasma, we shall assume that the two equations (12.22) – the electron and ion equations – determine the corresponding components of the electric field **E**' and the mass velocity $\mathbf{V}'_m \approx \mathbf{V}'_i$. The field $E'_\parallel$ is determined by the electron equation

$$E'_\parallel = -\frac{R'_{\parallel e}}{e_e n_0} - \frac{n'}{n_0} E_\parallel^{(0)} - \frac{1}{e_e n_0}(\nabla_\parallel p'_e + (\operatorname{div} \overleftrightarrow{\pi}'_e)_\parallel) + \frac{m_e}{e_e} \cdot \frac{\partial V'_{\parallel e}}{\partial t}, \qquad (12.23)$$

and the mass velocity by the sum of the electron and ion equations:

$$\rho_0 \frac{\partial V'_{\parallel m}}{\partial t} = -\nabla_\parallel p'. \qquad (12.24)$$

Collisionless Plasma, $R_\parallel = 0$. In accordance with (12.7), the equilibrium electric field $E_\parallel^{(0)}$ when $R_\parallel = 0$ can be maintained only by a longitudinal gradient of the pressure tensor, and therefore $E_{\parallel}^{(0)} \simeq T/eL$. Taking into account this relation and using (12.9) and (12.11), we find that the part of $E'_{\parallel}$ due to the field $E_{\parallel}^{(0)}$ is approximately

$$\frac{E'_{\parallel}}{E'_{\perp}} \simeq \frac{\omega}{\omega_{B_i}} \cdot \frac{a}{L}. \tag{12.25}$$

In the same way we arrive at the same estimate for the part of $E'_{\parallel}$ due to $\nabla_{\parallel} p'_e$ and $(\mathrm{div}\, \overset{\leftrightarrow}{\pi}'_e)_{\parallel}$.

The contribution to the right-hand side of (12.23) from the inertial term $\frac{m_e}{e_e} \cdot \frac{\partial V'_{\parallel e}}{\partial t}$ can be written approximately as

$$\frac{m_e}{e_e} \cdot \frac{\partial V'_{\parallel e}}{\partial t} \simeq \frac{m_e}{n_0 e^2} \omega j'_{\parallel}. \tag{12.26}$$

We obtain an estimate for $j'_{\parallel}$ by means of the Maxwell equation $(\mathrm{rot\, rot}\, \mathbf{E}')_{\parallel} = 4\pi i \omega j'_{\parallel}/c^2$

$$j'_{\parallel} \simeq \frac{c^2}{4\pi\omega a^2} E'_{\parallel}. \tag{12.27}$$

Taking into account (12.26) and (12.27), we conclude that the inertial term in (12.23) can be neglected if

$$a^2 > c^2/\omega_{pe}^2. \tag{12.28}$$

This is the condition we have already encountered for the perturbations to be nonelectrostatic. Its meaning is: if (12.28) is satisfied, the component $E'_{\parallel}$ can be neglected even if the longitudinal wave number is nonvanishing. However, it is clear that in the investigation of fast instabilities the condition (12.28) may not be necessary (for example, in the case of electrostatic perturbations of flute type; in this case, the field $E'_{\parallel}$ does not occur in the problem at all).

Collisional Plasma, $\mathbf{R}_{\parallel e} \neq 0$. If collisions are fairly frequent, the main contribution to the right-hand side of Eq. (12.23) arises from the term with $R'_{\parallel e}$, so that it reduces to

$$E'_{\parallel} = (j_{\parallel}/\sigma_{\parallel})', \tag{12.29}$$

where

$$\left.\begin{aligned} j_{\parallel} &= e_e n (V_{e\parallel} - V_{i\parallel}), \\ \sigma_{\parallel} &= 1.96 e^2 n \tau_e / m_e. \end{aligned}\right\} \tag{12.30}$$

Here we have used the expression for R_e given in §6.2 of Volume 1 ($R_{\|e}$ does not depend on B_0).

Taking into account the approximate relationship (12.27) between the current and the field, we find that when $j_{\|}^{(0)}=0$ the right-hand side of (12.28) is small compared with the left-hand side if

$$\omega \gg \nu_{ei}\,(c/a\omega_{pe})^2. \tag{12.31}$$

In this case, (12.29) reduces to the condition of perfect longitudinal conductivity, $E'_{\|}=0$. The condition (12.31) has a transparent meaning – the right-hand side is the characteristic decay rate of diffusion damping of the solenoidal field.

Thus, we have shown that for fast perturbations with not too short wavelength it can be assumed that

$$E'_{\|}=0. \tag{12.32}$$

This justifies the approximation of perfect conductivity made in the foregoing subsection.

3. Relationship between the Tensor $\overleftrightarrow{p}'$ and the Longitudinal and Transverse Perturbations of the Pressure. To obtain a closed system of hydrodynamic equations, we must calculate the tensor of the perturbed pressure. For fast perturbations, the off-diagonal components in the viscosity tensor are small, so that $\overleftrightarrow{p}$ is diagonal and can be represented in the form

$$\overleftrightarrow{p} \equiv p_{ik} = p_{\perp}\delta_{ik} + (p_{\|} - p_{\perp})\, e_i e_k. \tag{12.33}$$

As yet we have not linearized, so that $p_{\perp} = p_{\perp 0} + p'_{\perp}$, $p_{\|} = p_{\|0} + p'_{\|}$, $e_i \equiv (\mathbf{B}/B)_i = (\mathbf{B}_0/B_0)_i + (\mathbf{B}'_{\perp}/B_0)_i$. Linearizing (12.33), we obtain

$$\overleftrightarrow{p}' = p'_{\perp}\delta_{ik} + (p'_{\|} - p'_{\perp})\,(\mathbf{e}_0)_i\,(\mathbf{e}_0)_k + (p_{\|0} - p_{\perp 0})\,[(\mathbf{e}_0)_i\,(\mathbf{e}')_k + (\mathbf{e}')_i\,(\mathbf{e}_0)_k], \tag{12.34}$$

where

$$\mathbf{e}_0 \equiv \mathbf{B}_0/B_0, \qquad \mathbf{e}' \equiv \mathbf{B}'_{\perp}/B_0.$$

In the special case of a scalar pressure,

$$p_{\|0} = p_{\perp 0} \equiv p_0,$$

so that

$$\overleftrightarrow{p}' = p'_{\perp}\delta_{ik} + (p'_{\parallel} - p'_{\perp})(\mathbf{e}_0)_i(\mathbf{e}_0)_k. \tag{12.35}$$

The components $p'_{\perp}$ and $p'_{\parallel}$ will be calculated in § 12.3.

§ 12.3. Pressure Tensor

We now find the relationship between the pressure tensor $\overleftrightarrow{p}'$ and the plasma displacement $\boldsymbol{\xi}$ needed if we are to use the hydrodynamic equations of §12.2.

1. Collisional Plasma. In the approximation $\nu_i \gg \omega$, the transport equation reduces to a system of equations for moments. In this case, $\overleftrightarrow{p}'$ is related very simply to $\boldsymbol{\xi}$. Collisions tend to isotropize the distribution function [see §6.2 of Volume 1], so that to small terms of order ρ/a and ω/ν the pressure tensor $\overleftrightarrow{p}'$ can be expressed in terms of the scalar pressure:

$$\overleftrightarrow{p}' = \overleftrightarrow{I}p'. \tag{12.36}$$

In this approximation we find, using equations for the moments of the type (4.52), that p is related to the mass density ρ and the velocity V by the equation of an adiabatic curve:

$$\left(\frac{\partial}{\partial t} + \mathbf{V}\nabla\right)(p/\rho^{5/3}) = 0 \tag{12.37}$$

Taking into account the continuity equation,

$$\frac{\partial\rho}{\partial t} + \operatorname{div}(\rho\mathbf{V}) = 0, \tag{12.38}$$

and the definition of $\boldsymbol{\xi}$ [Eq. (2.19)], we obtain the desired relation between p' and $\boldsymbol{\xi}$:

$$p' = -\boldsymbol{\xi}\nabla p_0 - \frac{5}{3}p_0 \operatorname{div}\boldsymbol{\xi}. \tag{12.39}$$

The applicability of (12.39) is restricted by the condition of frequent collisions, $\omega < \nu_i$, which for $\omega \simeq v_{T_i}/a$ reduces to

$$\lambda_{\text{coll}} < a. \tag{12.40}$$

Thus, (12.39) holds if the mean free path of the particles is small compared with the size of the plasma.

2. Collisionless Plasma. Suppose that the condition opposite to (12.40), $\lambda_{coll} > a$, is satisfied. In the transport equation we can then omit the collision term and write its solution in the form of an integral over the particle trajectory [cf. Eq. (1.95)]. To calculate the tensor of the perturbed pressure it is sufficient to know the mean value $\bar{f}_1$ of f_1 over the Larmor oscillations. The expression for $\bar{f}_1$ can be reduced to the form

$$\bar{f}_1 = -\left(\xi_\perp \nabla F + \mu \frac{\partial F}{\partial \mu} \cdot \frac{B'_{||}}{B_0}\right) - i\omega \frac{\partial F}{\partial \varepsilon} \int_{-\infty}^{t} \left\{ \left(v_{||}^2 - \frac{v_\perp^2}{2}\right) \mathbf{e}_0 \nabla_0 \xi_\perp + \frac{v_\perp^2}{2} \operatorname{div} \xi_\perp \right\} dt'. \tag{12.41}$$

Here $\varepsilon = v^2/2$ and $\mu = v_\perp^2/2B_0$ are the energy and the magnetic moment of the particle; $F = F(\varepsilon, \mu, r_\perp)$ is the equilibrium distribution function with the normalization

$$\int (B_0/v_{||}) F \, d\mu \, d\varepsilon = n_0, \tag{12.42}$$

$v_\perp$ and $v_{||}$ are the transverse and longitudinal velocities of the particle; and $\nabla_0 \equiv (\mathbf{e}_0 \nabla)$. The perturbed pressure can be expressed in terms of $\bar{f}_1$ by the equations

$$\left.\begin{aligned} p'_\perp &= \int m (B_0^2/v_{||}) \bar{f}_1 \mu \, d\mu \, d\varepsilon, \\ p'_{||} &= \int m (B_0/v_{||}) \bar{f}_1 v_{||}^2 \, d\mu \, d\varepsilon. \end{aligned}\right\} \tag{12.43}$$

The integral over the time in (12.41) is calculated for various limiting cases in the next section.

§12.4. Energy Method

In §6.1 we have shown that the flute instability of a plasma in a gravitational field can be investigated by the energy method. The same method can be used to study the fast gradient perturbations of a plasma in a curved magnetic field described by Eq. (12.21). This can be seen by considering the structure of Eq. (12.21).

Comparing this equation with (6.13), we find that they differ only in the actual form of the operator $\hat{K}$. It can be shown that in the case of a curved field $\hat{K}$ is self-adjoint, as in the case of a gravitational force. We can therefore once more construct an expression for the potential energy and reduce the stability problem to an analysis of the sign of the potential energy.

We obtain an expression for the energy by using the definition (6.15) in conjunction with (6.13) and (12.17):

$$W_{pot} = \frac{1}{2}\int \xi \left\{ \operatorname{div} \overleftrightarrow{p'} - \frac{1}{4\pi}[\operatorname{rot}\mathbf{Q}, \mathbf{B}_0] - \frac{1}{4\pi}[\operatorname{rot}\mathbf{B}_0, \mathbf{Q}]\right\} d\mathbf{r}, \quad (12.44)$$

where $\mathbf{Q} = \operatorname{rot}[\xi, \mathbf{B}_0]$. The contribution from the second term in the curly brackets can be transformed by integration by parts. We shall assume that the plasma is bounded by a solid conducting wall, so that at the surface $\xi ds = 0$, $\mathbf{B}_0 ds = 0$. Equation (12.44) is then replaced by

$$W_{pot} = \frac{1}{2}\int \left\{ \xi \operatorname{div} \overleftrightarrow{p'} + \frac{\mathbf{Q}^2}{4\pi} - \frac{1}{4\pi}\mathbf{Q}[\xi, \operatorname{rot}\mathbf{B}_0]\right\} d\mathbf{r}. \quad (12.45)$$

Into this equation we must substitute $\overleftrightarrow{p'}$ found in §12.3.

1. <u>Collisional Plasma.</u> The tensor of the perturbed pressure $\overleftrightarrow{p'}$ of a collisional plasma is determined by Eqs. (12.6) and (12.9). If we substitute the latter into (12.45), we obtain

$$W_{pot} = \frac{1}{2}\int \left\{ \gamma_0 p_0 (\operatorname{div}\xi)^2 + \xi\nabla p_0 \operatorname{div}\xi + \frac{\mathbf{Q}^2}{4\pi} - \frac{\mathbf{Q}}{4\pi}[\xi, \operatorname{rot}\mathbf{B}_0]\right\} d\mathbf{r}, \qquad \gamma_0 = 5/3. \quad (12.46)$$

Bernstein et al. have proved that $\hat{K}$ is self-adjoint for this case.

2. <u>Collisionless Plasma.</u> If collisions are ignored, the tensor $\overleftrightarrow{p'}$ is anisotropic, $p'_{\parallel} \neq p'_{\perp}$. Then, in accordance with (12.34),

$$\operatorname{div}\overleftrightarrow{p'} = \nabla_{\perp} p'_{\perp} + (p'_{\parallel} - p'_{\perp})(\mathbf{e}_0\nabla)\mathbf{e}_0 + $$
$$+ (p_{\parallel 0} - p_{\perp 0})\left\{\left(\frac{\mathbf{Q}_{\perp}}{B_0}\nabla\right)\mathbf{e}_0 + (\mathbf{e}_0\nabla)\frac{\mathbf{Q}_{\perp}}{B_0}\right\} + \frac{\mathbf{Q}_{\perp}}{B_0}\operatorname{div}[\mathbf{e}_0(p_{\parallel 0} - p_{\perp 0})]. \quad (12.47)$$

Substituting (12.47) into (12,45), assuming that the perturbations are localized only in the region of the plasma, and integrating by parts, we obtain

$$W_{pot} = \frac{1}{2}\int \left\{ \frac{\mathbf{Q}^2}{4\pi} - \frac{\mathbf{Q}}{4\pi}[\xi, \operatorname{rot}\mathbf{B}_0] - p'_{\perp}(\operatorname{div}\xi_{\perp} + \xi_{\perp}(\mathbf{e}_0\nabla)\mathbf{e}_0) + p'_{\parallel}\xi_{\perp}(\mathbf{e}_0\nabla)\mathbf{e}_0 + \right.$$
$$\left. + \frac{p_{\parallel 0} - p_{\perp 0}}{B_0}(\xi(\mathbf{Q}\nabla)\mathbf{e}_0 - \mathbf{Q}(\mathbf{e}_0\nabla)\xi)\right\} d\mathbf{r}. \quad (12.48)$$

Into this equation we must substitute $p'_{\perp}$ and $p'_{\parallel}$ from Eqs. (12.43), so that W_{pot} is expressed in terms of the integral over the time in the right-hand side of Eq. (12.41). We shall calculate this integral, assuming that either: a) during a time of order $1/\omega$ the longitudinal and transverse velocities of a particle do not change appreciably and the perturbation is strongly drawn out along the lines of force [a condition of the type $\omega \gg k_{\|}v_{\|}$ (long plasma)] or b) both $v_\perp$ and $v_\|$ change appreciably during the time $1/\omega$ (short plasma).

Long Plasma (Chew-Goldberger-Low Approximation). In this approximation, the integration over t' on the right-hand side of Eq. (12.41) reduces simply to multiplication by i/ω. The upshot is

$$W_{\text{pot}} = \frac{1}{2}\int \Big\{\frac{\mathbf{Q}^2}{4\pi} - \frac{\mathbf{Q}}{4\pi}[\xi, \operatorname{rot}\mathbf{B}_0] + \\ + (\xi_\perp \nabla p_{\perp 0})\operatorname{div}\xi_\perp + (\mathbf{Q}_\perp \nabla_0 \xi - \xi(\mathbf{Q}_\perp\nabla)\mathbf{e}_0) \times \\ \times (p_{\perp 0} - p_{\|0})/B_0 + 2p_{\perp 0}(\operatorname{div}\xi_\perp)^2 + \\ + (3p_{\perp 0} - p_{\|0})(\xi_\perp \nabla_0 \mathbf{e}_0)\operatorname{div}\xi + (\xi_\perp \nabla_0 e_0)^2 (p_{\perp 0} + 2p_{\|0}) - \\ - (\xi_\perp \nabla_0 \mathbf{e}_0)(\xi_\perp \nabla)(p_{\|0} - p_{\perp 0})\Big\}\, dr. \qquad (12.49)$$

If $p_{\perp 0} = p_{\|0} \equiv p_0$, then

$$W_{\text{pot}} = \frac{1}{2}\int \Big\{\frac{\mathbf{Q}^2}{4\pi} - \frac{\mathbf{Q}}{4\pi}[\xi, \operatorname{rot}\mathbf{B}_0] + (\xi_\perp \nabla p_0) \times \\ \times \operatorname{div}\xi_\perp + 2p_0[(\operatorname{div}\xi_\perp)^2 + (\xi_\perp\nabla_0\mathbf{e}_0)\operatorname{div}\xi_\perp + \frac{3}{2}(\xi_\perp\nabla_0\mathbf{e}_0)^2]\Big\}\, dr. \qquad (12.50)$$

For electrostatic perturbations of a zero-pressure plasma, (12.49) reduces to

$$W_{\text{pot}} = \frac{1}{2}\int \{-(\xi_\perp \nabla \ln B_0)\, \xi_\perp \nabla (p_{\|0} + p_{\perp 0}) + (4p_{\|0} + 3p_{\perp 0})(\xi\nabla \ln B_0)^2\}\, dr. \qquad (12.51)$$

Using either (12.50) or (12.51), we find that the energy of the electrostatic perturbations of a zero-pressure plasma with $p_{\|0} = p_{\perp 0} \equiv p_0$ is

$$W_{\text{pot}} = \frac{1}{2}\int \{-2(\xi_\perp \nabla p_0)(\xi_\perp \nabla \ln B_0) + 7p_0(\xi_\perp \nabla \ln B_0)^2\}\, dr. \qquad (12.52)$$

Short Plasma. General Expression for the Distribution Function. Suppose that during a time $\sim 1/\omega$ a particle repeats its trajectory many times, being reflected from stoppers or moving along a closed path without reflections. In this case, the integral (12.41) can be represented as a sum of integrals each of which is taken over the time of a single period of motion of the particle, so that

$$\int_{-\infty}^{t}\{\ldots\}\,dt'=\int_{t-\tau}^{t}\{\ldots\}\,dt'+\int_{t-2\tau}^{t-\tau}\{\ldots\}\,dt'+\int_{t-(n+1)\tau}^{t-n\tau}\{\ldots\}\,dt'+\ldots, \tag{12.53}$$

where $\tau=\int dl/v_{\|}$ is the period and the symbol $\{\cdot\cdot\cdot\}$ stands for the curly brackets in (12.41).

In each n-th interval, we make a change of the variable of integration, $t'=t''-n\tau$. All the functions in the curly brackets in (12.41) are unchanged except ξ, which transforms in accordance with the law

$$\xi(t''-n\tau)=e^{i\omega\tau n}\xi(t''). \tag{12.54}$$

As a result, (12.53) becomes

$$\int_{-\infty}^{t}\{\ldots\}\,dt'=(1+e^{i\omega\tau}+e^{i2\omega\tau}+\ldots)\int_{t-\tau}^{t}\{\ldots\}\,dt'\equiv\frac{1}{1-e^{i\omega\tau}}\int_{t-\tau}^{t}\{\ldots\}\,dt'. \tag{12.55}$$

Since we assume that $\omega\tau\ll 1$, the denominator in (12.55) is approximately equal to

$$1-e^{i\omega\tau}\approx -i\omega\tau. \tag{12.56}$$

In this manner we reduce (12.41) to

$$\bar{f}_1=-\left(\xi_{\perp}\nabla F+\mu\frac{\partial F}{\partial\mu}\cdot\frac{B_{\|}}{B_0}\right)+$$

$$+\frac{\partial F}{\partial\varepsilon}\cdot\frac{1}{\tau}\int_{t-\tau}^{t}\left\{\left(\frac{v_{\perp}^2}{2}-v_{\|}^2\right)\xi_{\perp}\nabla_0 e_0+\frac{v_{\perp}^2}{2}\operatorname{div}\xi_{\perp}\right\}dt'. \tag{12.57}$$

Since the particle time is uniquely determined by the position of the latter on the line of force, the integration over t' in (12.57) can be replaced by integration along the length of the line of force by setting $dt'=dl/v_{\|}$.

Zero-Pressure Plasma. If $\mathbf{E}'=-\nabla\psi$ and $\beta\to 0$, Eq. (12.57) can be reduced to

$$\bar{f}_1=-\xi_{\perp}\nabla F+\frac{1}{\tau}\cdot\frac{\partial F}{\partial\varepsilon}(\xi_{\perp}\nabla)J_{\|}, \tag{12.58}$$

where $J_{\|}=\int v_{\|}\,dl$ is the longitudinal adiabatic invariant.

The potential energy (12.48) in this approximation is

$$W_{\text{pot}} = \frac{1}{2}\sum_{i,e}\int d\Phi\left\{\int\frac{dl}{B_0}\Big[-(\xi_\perp\nabla\ln B_0)\,\xi_\perp\nabla(p_{\parallel 0}+p_{\perp 0})+(\xi_\perp\nabla\ln B_0)^2\times\right.$$

$$\left.\times\Big(p_{\parallel 0}+p_{\perp 0}+\int\frac{mB_0^3}{v_\parallel}\mu^2\frac{\partial F}{\partial\varepsilon}\,d\mu\,d\varepsilon\Big)\Big]-m\int d\varepsilon\,d\mu\frac{1}{\tau}\cdot\frac{\partial F}{\partial\varepsilon}(\xi_\perp\nabla J_\parallel)^2\right\}. \tag{12.59}$$

Here, as in §11.2, $d\Phi = B_0 dS$.

If the canonical variables α and β (see the appendix to this chapter) are employed, the potential energy (12.59) can be represented in the form

$$W'_{\text{pot}} = -\frac{m}{2}\int d\alpha\,d\beta\,d\mu\,dJ_\parallel(\xi_\perp\nabla\varepsilon)^2(\partial F/\partial\varepsilon)_{\mu,\,J_\parallel}. \tag{12.60}$$

If $p_{\parallel 0} = p_{\perp 0} \equiv p_0$, Eq. (12.59) reduces to

$$W_{\text{pot}} = \frac{1}{2}\int d\Phi\,(\xi_\perp\nabla p_0)(\xi_\perp\nabla U)+\frac{15}{4}\int p_0\frac{d\Phi}{K}\left(\xi_\perp\nabla\int\sqrt{1-\lambda B_0}\,dl\right)^2, \tag{12.61}$$

where

$$\lambda = \mu/\varepsilon,\quad U = \int dl/B_0,\quad K(\mathbf{r}_\perp,\lambda) = \oint dl'/\sqrt{1-\lambda B_0(l')}. \tag{12.62}$$

Finite-Pressure Plasma with $p_{\perp 0} = p_{\parallel 0}$. If the distribution of the particle velocities is isotropic, $\partial F/\partial\mu = 0$, the function $\bar{f}_1$ [see Eq. (12.57)] can be expressed in the form

$$\bar{f}_1 = -\xi\nabla F + 2\varepsilon\frac{\partial F}{\partial\varepsilon}\cdot\frac{J(\mathbf{r}_\perp,\lambda)}{K(\mathbf{r}_\perp,\lambda)}, \tag{12.63}$$

where

$$J(\mathbf{r}_\perp,\lambda) = \oint\left\{\Big[1-\frac{3}{2}\lambda B_0(l')\Big]\mathbf{e}_0\nabla_0\xi+\frac{\lambda B_0(l')}{2}\operatorname{div}\xi\right\}\frac{dl'}{\sqrt{1-\lambda B_0(l')}}.$$

Using this and Eq. (12.43), we can reduce the expression (12.48) to

$$W = W_1 + W_2, \tag{12.64}$$

where

$$W_1 = \frac{1}{2}\int\left\{\frac{\mathbf{Q}^2}{4\pi}-\frac{\mathbf{Q}}{4\pi}[\xi,\operatorname{rot}B_0]+\xi\nabla p_0\operatorname{div}\xi\right\}d\mathbf{r}, \tag{12.65}$$

$$W_2 = \frac{15}{4}\int p_0\,d\Phi\int d\lambda\,[J(\Phi,\lambda)]^2/K(\Phi,\lambda). \tag{12.66}$$

Finite-Pressure Plasma with $p_{\perp 0} \neq p_{\parallel 0}$. In this case the potential energy can be represented in the form (12.64) with W_1 and W_2 given by

$$\left.\begin{aligned} W_1 &= \frac{1}{2}\int\left\{\frac{Q^2}{4\pi} - \frac{Q}{4\pi}[\xi, \operatorname{rot} B_0] + \xi\nabla p_{\perp 0}\operatorname{div}\xi + \right.\\ &+ \frac{p_{\parallel 0}-p_{\perp 0}}{B_0}(\xi(Q_\perp\nabla)e_0 - Q_\perp(e_0\nabla)\xi) + \\ &+ (\xi_\perp\nabla_0 e_0)[(p_{\parallel 0}-p_{\perp 0})(e_0\nabla_0\xi - \operatorname{div}\xi) - \\ &- \xi\nabla(p_{\parallel 0}-p_{\perp 0})] + (e_0\nabla_0\xi - \operatorname{div}\xi)^2 \times \\ &\times\left.\left(2p_{\perp 0} + m\int\frac{B_0^3}{v_\parallel}\mu^2\frac{\partial F}{\partial\varepsilon}\,d\mu\,d\varepsilon\right)\right\} dr. \\ W_2 &= -\frac{m}{2}\int d\Phi\,d\mu\,d\varepsilon\frac{\partial F}{\partial\varepsilon}\left\{\int\frac{dl}{v_\parallel}\left[\mu B_0\operatorname{div}\xi_\perp + \right.\right.\\ &+\left.\left.\left(\frac{v_\perp^2}{2}-v_\parallel^2\right)\xi_\perp\nabla_0 e_0\right]\right\}^2\left[\int\frac{dl}{v_\parallel}\right]^{-1}. \end{aligned}\right\} \tag{12.67}$$

§12.5. Comparison Theorems

Using the Schwarz inequality

$$\left(\int fg\,dx\right)^2 \leqslant \int f^2\,dx\int g^2\,dx, \tag{12.68}$$

we shall show that the potential energy in the approximation of a short plasma does not exceed the energy in the approximation of a long plasma. For f, we substitute into (12.68) the expression in the square brackets on the right-hand side of (12.67) divided by $\sqrt{v_\parallel}$; for g, we substitute $\frac{1}{\sqrt{v_\parallel}}$.

Then

$$W_2 \leqslant -\frac{m}{2}\int d\Phi\,d\mu\,d\varepsilon\frac{\partial F}{\partial\varepsilon}\int\frac{dl}{v_\parallel}\left[\mu B_0\operatorname{div}\xi_\perp + \right.$$

$$+\left.\left(\frac{v_\perp^2}{2}-v_\parallel^2\right)\xi_\perp\nabla_0 e_0\right]^2 = \frac{1}{2}\int\left\{-2p_{\perp 0}(\operatorname{div}\xi_\perp + \xi_\perp\nabla_0 e_0)\times\right.$$

$$\times\,\xi_\perp\nabla_0 e_0 + 3(\xi_\perp\nabla_0 e_0)^2 p_{\parallel 0} - (\operatorname{div}\xi_\perp + \xi_\perp\nabla_0 e_0)^2\int\frac{\mu^2 B_0^3}{v_\parallel}\cdot\frac{\partial F}{\partial\varepsilon}\,d\mu\,d\varepsilon\left.\right\} dr. \tag{12.69}$$

Adding this to W_1 and comparing the result with (12.49), we arrive at the inequality postulated above:

$$W_{\text{short}} \leqslant W_{\text{long}} \tag{12.70}$$

If the velocities of the particles are distributed isotropically, it can be shown that the energy in the approximation of a short plasma is not less than the energy minimized with respect to $\xi_{||}$ in the approximation of a collisional plasma. To see this, we take the inequality (12.68) to obtain a lower bound for the integral over λ in (12.66), setting $f = J/\sqrt{K}$, $g = \sqrt{K}$:

$$\int d\lambda \frac{J^2}{K} \geqslant \frac{(\int J\, d\lambda)^2}{\int K\, d\lambda}. \tag{12.71}$$

In accordance with (12.62) and (12.63), the integrals on the right-hand side of (12.71) are

$$\left.\begin{aligned} \int J\, d\lambda &= \frac{2}{3} \int \frac{dl}{B} \operatorname{div} \xi, \\ \int K\, d\lambda &= 2 \int \frac{dl}{B}. \end{aligned}\right\} \tag{12.72}$$

As a result, we obtain a lower bound for (12.66):

$$W_2 \geqslant \frac{5}{6} \int p_0\, d\Phi \left[\int \frac{dl}{B} \operatorname{div} \xi_\perp \right]^2 \left(\int \frac{dl}{B} \right)^{-1}. \tag{12.73}$$

We now minimize the expression (12.46) for the energy of a collisional plasma with respect to $\xi_{||}$. We can minimize with respect to the function $y \equiv \operatorname{div} \xi_{||}$, but this must be done subject to the condition

$$\int \operatorname{div} \xi_{||}\, d\mathbf{r} = 0. \tag{12.74}$$

This equation plays the role of an additional condition which the varied function must satisfy. The problem therefore reduces to minimization of the integral

$$I = \int d\mathbf{r}\, \{\gamma_0 p_0 (\operatorname{div} \xi_\perp + y)^2 - \lambda y\}, \tag{12.75}$$

where λ is a Lagrangian multiplier. Varying (12.75) with respect to y, we obtain

$$y = -\operatorname{div} \xi + \frac{\lambda}{2\gamma p_0}. \tag{12.76}$$

We find the expression for λ by means of (12.74):

$$\lambda = 2\gamma_0 p_0 \int \frac{dl}{B} \operatorname{div} \xi_\perp \left(\int \frac{dl}{B} \right)^{-1}. \tag{12.77}$$

The result is

$$\left\{\frac{1}{2}\int \gamma_0 p_0 (\operatorname{div} \xi)^2\, dr\right\}_{\min} = \frac{1}{2}\int \gamma_0 p_0\, d\Phi \left[\int \frac{dl}{B} \operatorname{div} \xi_\perp\right]^2 \left(\int \frac{dl}{B}\right)^{-1}. \quad (12.78)$$

This, in conjunction with (12.73), proves the assertion.

Thus, if the particle velocities are distributed isotropically, $p_{\parallel 0} = p_{\perp 0} \equiv p_0$, the energy of the perturbations of a collisionless plasma defined in accordance with (12.64) is never less than the energy of the collisional approximation (12.46):

$$W_{\text{collisionless short}} \geqslant W_{\text{coll}} \quad (12.79)$$

This means that the stability conditions obtained by means of the simpler hydrodynamic formula (12.46) are sufficient for a collisionless plasma. Therefore, if it is desired to derive *sufficient* stability conditions of a plasma with $p_{\parallel 0} = p_{\perp 0}$, it is sufficient to use the hydrodynamic description and not have recourse to the kinetic description.

Comparing (12.64) and (12.46), we can estimate the error associated with the use of the collisional approximation. The difference between these two expressions is due to the terms of order $\partial \ln B/\partial \ln p$. This can be seen especially clearly in the example of formula (12.61) – in the essentially kinetic term the magnetic field is differentiated twice with respect to the coordinates, whereas in the "hydrodynamic" term the field is differentiated once and the pressure once. Therefore, if $\partial \ln B/\partial \ln p \ll 1$, the collisional and collisionless approximations give the same result. Since the derivatives of the magnetic field are related to the field curvature and the finiteness of β, the condition $\partial \ln B/\partial \ln p \ll 1$ entails qualitatively

$$\left(\beta, \frac{a}{R}\right) \ll 1. \quad (12.80)$$

In this case, $W_{\text{collisionless}} \approx W_{\text{coll}}$.

§12.6. Review of the Main Results of the Theory of Plasma Instabilities in a Curved Magnetic Field and Outline of the Subsequent Exposition

The most distinctive characteristic of an inhomogeneous plasma in a curved magnetic field is the magnetic drift of its particles. It is responsible for the fast gradient (magnetohydrodynamic) instabilities. The latter play a decisive role in the problem of plasma containment. Magnetic drift also affects the slow gradient (nonmagnetohydrodynamic) instabilities and is the cause of some of them.

The structure of the equations that describe the magnetohydrodynamic instabilities is relatively simple, and these equations do not contain many terms that are important in the equations of nonmagnetohydrodynamic instabilities. The investigation of magnetohydrodynamic instabilities reduces to establishing the sign of the minimum of the potential energy.

Despite its fundamental simplicity, the problem of magnetohydrodynamic instabilities of a plasma in a curved magnetic field entails laborious calculations in practice. This is due to the appearance of a large number of new parameters that characterize the configuration of the magnetic field. The theory of magnetohydrodynamic instabilities would be far less extensive if a restriction could be made to some definite configuration. However, the class of physically distinct steady states of a plasma and a magnetic field is very large. In principle, instability theory can give a description of the instabilities of each of these states. Thus, the manifold of parameters characteristic for instability theory of a plasma in a rectilinear magnetic field is replaced by the manifold of steady states in the case of a curved field. The content of the instability theory of a plasma in a curved magnetic field can be stated as follows: to find the complete set of magnetohydrodynamic instabilities or the stability boundary for all the physically distinct states and, if the plasma is magnetohydrodynamically stable, to augment this description by an analysis of the nonmagnetohydrodynamic instabilities.

In the modern theory of instabilities, the most important classes of steady states of a plasma and magnetic field are the following:

1. Plasma column in a cylindrically symmetric magnetic field.
2. Plasmas in adiabatic traps.
3. Plasmas in multipole traps.

4. Plasmas in closed traps with magnetic surfaces and no current.
5. Plasmas in closed traps with magnetic surfaces and a current.

We have discussed some aspects of the stability of a cylindrical plasma in §§11.1 and 11.6. Adiabatic traps have been discussed in §§11.3-11.5. A more systematic exposition of the instability theory of these steady states will be given below: in Chapter 13 we shall discuss the instabilities of a cylindrical column (both magnetohydrodynamic and nonmagnetohydrodynamic instabilities) and then the instabilities of a plasma in adiabatic traps (magnetohydrodynamic in Chapter 14 and nonmagnetohydrodynamic in Chapter 15). Multipole traps differ from a rod-stabilized probkotron (see §11.4) by the absence of an axisymmetric magnetic field. (The configuration of the field of a multipole trap is shown in Fig. 16.1; see below.) The instabilities of a plasma in such a system will be considered in Chapter 16.

Closed traps with magnetic surfaces and no current differ from the corresponding traps with a current in that their magnetic field is produced by currents that flow outside the plasma. We have already considered some aspects of the instabilities of traps with a current in §11.8, where we have shown that the main characteristics of such traps are the second derivative of the volume with respect to the longitudinal magnetic flux, which is related to the mean magnetic drift, and the derivative of the transverse flux with respect to the longitudinal flux – the shear. An example of a trap without a current is a plasma cylinder twisted into a toroid in a helical magnetic field (see §11.7). Chapter 17 will be devoted to a systematic analysis of the instabilities of such closed traps.

The simplest example of a closed trap with magnetic surfaces and a current is a cylindrically symmetric plasma column twisted into a torus with a current in a longitudinal magnetic field (see §11.6). The magnetic field of the current has an important influence on the equilibrium and stability of the plasma. This difference between traps with and without currents necessitates different approaches to the stability analysis for the corresponding systems. The stability of systems with a current will be analyzed in Chapter 18.

APPENDIX

Mean Particle Drift in the Space between Two Mirrors. Longitudinal Adiabatic Invariant

Suppose a particle moves between two magnetic mirrors. Because of the magnetic drift, it is displaced relative to a line of force with the velocity (11.1). This velocity varies along the line of force and is not therefore a sufficiently convenient characteristic. However, one can average the transverse drift over a period of the oscillations of the particle between the mirrors and in this way calculate the effective drift velocity of a particle on the corresponding line of force. We shall do this by the method proposed by Kadomtsev.

We introduce curvilinear coordinates x^1, x^2, x^3 with x^3 axis along the line of force. In these variables, the drift equation $d\mathbf{r}/dt = V_{dr}$ in conjunction with the relation

$$\frac{d\mathbf{r}}{dt}=\frac{\partial \mathbf{r}}{\partial x^i}\cdot\frac{dx^i}{dt} \tag{A.12.1}$$

and the laws of conservation of the energy and magnetic moment can be rewritten in the form

$$\frac{dx^i}{dt}=\frac{v_{||}}{\omega_B\sqrt{g}}\,e^{ijk}\,\frac{\partial}{\partial x^j}\left(v_{||}\,\frac{g_{k3}}{\sqrt{g_{33}}}\right). \tag{A.12.2}$$

We average this equation over a period of the oscillations of the particle between the mirrors. We shall denote the averaging by the symbol $\langle\ldots\rangle$, where

$$\langle\ldots\rangle\equiv\frac{1}{\tau}\oint(\ldots)\,dt=\frac{1}{\tau}\oint(\ldots)\,\frac{dl}{v_{||}}. \tag{A.12.3}$$

Here τ is the period of the oscillations:

$$\tau=\oint dt=\oint\frac{dl}{v_{||}}, \tag{A.12.4}$$

and the averaging is along a closed trajectory of the particle. As a result

$$\left\langle\frac{dx^i}{dt}\right\rangle=\frac{1}{\tau}\cdot\frac{\sqrt{g_{33}}}{\omega_B\sqrt{g}}\,e^{ij3}\,\frac{\partial J_{||}}{\partial x^j}, \tag{A.12.5}$$

where

$$J_{||}=\oint v_{||}\sqrt{g_{33}}\,dx^3\equiv\oint v_{||}\,dl. \tag{A.12.6}$$

Here, dl is an element of the line of force on which the particle is situated.

Equations (A.12.5) give the desired law of the mean displacement (that is, over the period of oscillations between the mirrors) of the particle at right angles to the lines of force. It can be shown that the particle is displaced along transverse trajectories on which the quantity $J_{||}=J_{||}(x^1,\ x^2)$ remains constant; for substituting into the right-hand side of the equation

$$\frac{dJ_{||}}{dt}=\frac{\partial J_{||}}{\partial x^1}\left\langle\frac{dx^1}{dt}\right\rangle+\frac{\partial J_{||}}{\partial x^2}\left\langle\frac{dx^2}{dt}\right\rangle \tag{A.12.7}$$

the expressions for $\langle dx^i/dt \rangle$ from (A.12.5) we find that

$$\frac{dJ_{\parallel}}{dt} = 0. \qquad (A.12.8)$$

The quantity $J_{\parallel}$ is known as the longitudinal adiabatic invariant of the particle.

As the variables x^1 and x^2, we can take any function of the ordinary coordinates x, y, and z that remain constant along the line of force. In particular, these variables can be chosen such that

$$\mathbf{B} = [\nabla x^1, \nabla x^2]. \qquad (A.12.9)$$

Taking here the contravariant component, we find that in these variables

$$B\sqrt{g}/\sqrt{g_{33}} = 1. \qquad (A.12.10)$$

Coordinates x^1 and x^2 satisfying the conditions (A.12.9) are frequently denoted by α and β.

In the variables $x^1 = \alpha$ and $x^2 = \beta$, Eqs. (A.12.5) can be written especially simply:

$$\left.\begin{aligned} \frac{d\alpha}{dt} &= \frac{mc}{e\tau} \cdot \frac{\partial J_{\parallel}}{\partial \beta}, \\ \frac{d\beta}{dt} &= -\frac{mc}{e\tau} \cdot \frac{\partial J_{\parallel}}{\partial \alpha}. \end{aligned}\right\} \qquad (A.12.11)$$

The right-hand sides of these equations are expressed in terms of the variables α, β, μ, and ε. Instead of these, we can use the set of variables α, β, μ, and $J_{\parallel}$, assuming that the energy ε is a function of these variables that satisfies the relation

$$J_{\parallel} = \oint \sqrt{2(\varepsilon - \mu B)}\, dl. \qquad (A.12.12)$$

In the new variables, the derivatives with respect to α and β transform in accordance with the law

$$\frac{\partial}{\partial x^i} \to \frac{\partial}{\partial x^i} - \frac{\partial \varepsilon/\partial x^i}{\partial \varepsilon/\partial J_{\parallel}} \cdot \frac{\partial}{\partial J_{\parallel}}, \qquad (A.12.13)$$

where, in accordance with (A.12.12),

$$\frac{\partial \varepsilon}{\partial J_{\parallel}} = \frac{1}{\tau}. \qquad (A.12.14)$$

As a result, Eqs. (A.12.11) take the simple form

$$\frac{d\alpha}{dt} = -\frac{mc}{e} \cdot \frac{\partial \varepsilon}{\partial \beta}, \quad \frac{d\beta}{dt} = \frac{mc}{e} \cdot \frac{\partial \varepsilon}{\partial \alpha}. \qquad (A.12.15)$$

The form of these equations is completely analogous to the usual equations of motion of a particle with the Hamiltonian $(mc/e)\varepsilon$. They are called canonical equations.

Bibliography

1. S. Lundquist, Phys. Rev., 83:307 (1951). One of the results of this paper, which is devoted to the "dynamo" problem, is the statement of a variational principle for investigating the stability of a conducting liquid in a magnetic field.
2. G. F. Chew, M. L. Goldberger, and F. E. Low, Proc. Roy. Soc., A236:112 (1956). Derivation of the equations of one-fluid hydrodynamics with anisotropic pressure tensor, $p_{\parallel} \neq p_{\perp}$.
3. I. B. Bernstein et al., Proc. Roy. Soc., A244:17 (1958).
4. S. I. Braginskii and B. B. Kadomtsev, in: Plasma Physics and the Problem of Controlled Thermonuclear Reactions, Vol. 3, Pergamon Press, Oxford (1959), p. 356.
5. K. Hain, R. Lüst, and A. Schlüter, Z. Naturforsch., 12a:833 (1957). The energy method for investigating magnetohydrodynamic instabilities is developed in [3-5]. The most general expression for the potential energy is given in [3]. Besides the single-fluid model with isotropic pressure, a study is also made of the Chew–Goldberger–Low approximation.
6. B. B. Kadomtsev, in: Plasma Physics and the Problem of Controlled Thermonuclear Reactions, Vol. 3, Pergamon Press, Oxford (1959), p. 340. The conservation of $J_{\parallel} = \int v_{\parallel} dl$ is proved in the appendix to Chapter 12.
7. B. B. Kadomtsev, in: Plasma Physics and the Problem of Controlled Thermonuclear Reactions, Vol. 4, Pergamon Press, Oxford (1960), p. 17. The potential energy of the perturbations of a zero-pressure plasma is calculated in the hydrodynamic approximation.
8. B. B. Kadomtsev, ibid. (in the articles beginning on p. 417). The transport equation is solved for a zero-pressure plasma. The perturbed pressure and potential energy of such a plasma are found.
9. M. D. Kruskal and C. R. Oberman, Phys. Fluids, 1:275 (1958).
10. M. N. Rosenbluth and N. Rostoker, Phys. Fluids, 2:23 (1959). In [9-10] a kinetic expression is obtained for the potential energy of a Maxwellian plasma with arbitrary β. The investigation in [9] is based on a functional of the energy and not the transport equation. A plasma with anisotropic pressure is also considered in [9]. The comparison theorems of §12.5 are proved in [9, 10].
11. T. G. Northrop and E. Teller, Phys. Rev., 117:215 (1960). Canonical variables (see the appendix to this chapter) are introduced.
12. J. Andreoletti, Compt. Rend. Acad. Sci., 256:1251 (1963). A lucid derivation is given of a general kinetic expression for the potential energy which agrees with the result of Kruskal and Oberman [9]. The limiting case of a zero-pressure plasma is discussed. Canonical variables are used.
13. J. Andreoletti, Compt. Rend. Acad. Sci., 256:1469 (1963). A kinetic expression is given for the potential energy of a zero-pressure plasma.
14. J. B. Taylor, Phys. Fluids, 7:767 (1964). An expression is derived for the potential energy of a zero-pressure plasma in canonical variables [Eq. (12.60)].
15. A. A. Vedenov, E. P. Velikhov, and R. Z. Sagdeev, Usp. Fiz. Nauk, 73:701 (1961) [Sov. Phys. – Uspekhi, 4:332 (1961)].

16. B. B. Kadomtsev, in: Reviews of Plasma Physics, Vol. 2, Consultants Bureau, New York (1966), p. 153.
17. M. D. Kruskal, in: La Theorie des Gaz Neutres et Ionizés, Hermann, Paris (1960), p. 251. The review articles [15-17] contain a more rigorous derivation and detailed exposition of the energy method than the present chapter.
18. V. F. Aleksin and V. N. Yashin, Zh. Eksp. Teor. Fiz., 39:822 (1960) [Sov. Phys. – JETP, 12:572 (1961)]. Development of a generalized energy principle that takes into account charge neutrality. In such an approach the sign of the potential energy enables one to establish stability of the plasma even against some nonmagnetohydrodynamic perturbations; for example, those whose growth rate is determined by the interaction between resonant particles and the wave [see, for example, the paper by the same authors in Zh. Eksp. Teor. Fiz., 40:1115 (1961)] [Sov. Phys. – JETP, 13:787 (1961)].
19. B. A. Trubnikov, Phys. Fluids, 5:184 (1962). Proposal to investigate the stability of the plasma by inspecting the sign of the kinetic energy of perturbations (dynamical principle).
20. R. Kulsrud, Phys. Fluids, 5:192 (1962).
21. R. Kulsrud, in: Advanced Plasma Theory, Academic Press, New York and London, (1964), p. 54. In [20-21] an attempt is made to include more effects than those covered by the Kulsrud–Oberman approximation [9].
22. H. Grad, Phys. Fluids, 7:1283 (1964). A further development of the magnetohydrodynamic variational method.
23. H. Grad, Phys. Fluids, 9:225 (1966). Critical analysis of the earlier papers in which variational methods had been developed. Formulation of a general approach to the investigation of plasma stability in the guiding center approximation.

Chapter 13

Cylindrical Plasma Column

§13.1. Magnetohydrodynamic Instabilities of a Column without a Longitudinal Magnetic Field

Suppose that a plasma is cylindrically symmetric and is in a magnetic field that has only an azimuthal component:

$$\mathbf{B} = (0,\ B_\varphi,\ 0). \tag{13.1}$$

A field of this kind can be produced by a longitudinal current flowing along the plasma or a current that flows along a rod inside the plasma. The first case corresponds to the ordinary pinch; the second, to the inverse pinch.

An elementary perturbation of a cylindrical plasma has a coordinate dependence of the form $\xi(r)\exp(il\varphi + ik_z z)$, where $l = 0, 1, 2, \ldots$. Perturbations with $l = 0$ correspond to flutes oriented along the lines of force (in the geometry under consideration these are sausage-type perturbations). Perturbations of this nature were investigated in §11.1. Using the results then obtained, we conclude that the ordinary pinch is unstable if the plasma pressure decreases sufficiently rapidly along the radius. In the case of an inverse pinch, the pressure may increase with the radius (with increasing distance from the central conductor). In accordance with §11.1, the flute instability does not then arise.

If the pressure does not decrease too rapidly along the radius, the flute instability may not develop. Using the expression for the potential energy (12.46) and employing the minimization procedure, we obtain the following condition for stabilization of the flute instability:

$$-\frac{d \ln p}{d \ln r} < \frac{4\gamma_0}{2+\gamma_0\beta}. \tag{13.2}$$

Here γ_0 is the adiabatic exponent; in a collisional plasma $\gamma_0 = 5/3$; in a collisionless plasma, $\gamma_0 = 2$.

Since the pressure is high, $\beta \geq 1$, in the internal region of an ordinary pinch, it is necessary to investigate the possible growth of nonflute-type perturbations (see §6.4.2). In the case of the field geometry under consideration, this corresponds to bending (a kink) of the filament, $l \neq 0$ and $k_z \neq 0$. Using (12.46), we obtain a condition for stability against such perturbations:

$$-\frac{d \ln p}{d \ln r} < \frac{l^2}{\beta}. \tag{13.3}$$

This is a more stringent condition than (13.2) if $l = 1$ and $\beta > 2\gamma_0/3$. However, it can also be satisfied if the pressure decreases sufficiently smoothly.

Thus, an ordinary pinch with a fairly smoothly decreasing pressure may be magnetohydrodynamically stable. However, if the pressure decreases abruptly at some point in the column, the plasma is unstable.

§13.2. Magnetohydrodynamic Instabilities of a Column with a Longitudinal Magnetic Field

We shall now assume that, in addition to the azimuthal field, there is a longitudinal magnetic field in the plasma:

$$\mathbf{B} = (0,\ B_\varphi,\ B_z). \tag{13.4}$$

The stability of such a plasma was investigated in §11.6 on the basis of a qualitative treatment of the shear and magnetic drift of the particles. In this manner, a condition was obtained for the stability of a plasma against perturbations of flute type localized near a surface on which $k_{\parallel} = 0$, i.e., Suydam's condition, which states that the pressure gradient must not be a too large negative number:

$$-\frac{8\pi\, dp/dr}{rB_z^2} < \frac{1}{4}\left(\frac{d \ln \mu}{dr}\right)^2, \tag{13.5}$$

where

$$\mu \equiv B_\varphi / r B_z.$$

The same result can be obtained by the energy method. Using the expression for the potential energy, one can also investigate the possibility of the growth of perturbations with a large localization region of the order of the column radius. The instability condition that is then obtained is similar to (8.103). A plasma is stable or unstable depending on the distribution of the current over the cross section. This kind of instability is known as the kink instability.

§13.3. General Remarks on the Equations That Describe Nonmagnetohydrodynamic Instabilities of a Cylindrical Column

If a plasma is magnetohydrodynamically stable, it is necessary to investigate the possible excitation of nonmagnetohydrodynamic perturbations. In a cylindrical column, the latter can be analyzed in the same general manner as in the case of a field with rectilinear lines of force.

If collisions are not important, then, in accordance with Chapter 1, it is necessary to integrate the linearized transport equation along the unperturbed trajectories (or solve it by some other method) and find the perturbed charge density and, in the case of nonelectrostatic perturbations, the perturbed currents as well. As a result, one obtains an equation for the scalar potential or a system of equations for the scalar and vector potentials, these equations forming the basis of the instability analysis. If one is concerned with only fairly small-scale perturbations, the system of equations can be reduced to a local dispersion equation.

In the case of a collisional plasma, the method of derivation of the basic equations is somewhat different. In this case the transport equation must be solved by making an expansion in the reciprocal of the collision frequency. If the plasma is strongly magnetized, one can also use an expansion in $1/\omega_B$. In the approximation of rectilinear lines of force, this procedure was outlined in §§4.2-4.4. In this manner, one obtains a system of equations for the perturbations of the density, longitudinal velocity, and temperature; this system then forms the starting point for the investigation of the stability.

The instabilities of a collisional plasma can also be investigated by the method typical for a collisionless plasma if one dispenses with an exact description of the collision process, describing the latter by a model. This approach was outlined in §4.7.

A helical form of the lines of force of a cylindrically symmetric magnetic field makes it necessary to include in the treatment of each of these cases the following two factors: magnetic drift of the particles due to the curvature and inhomogeneity of the magnetic field (see §11.1) and the coordinate dependence of the longitudinal wave number due to the shear (see §11.6).

Magnetic drift leads to much the same rearrangement of the equations as the gravitational drift discussed in Chapter 6. It must, however, be borne in mind that magnetic and gravitational drifts do not lead to exactly the same effects. Gravitational drift leads to not only a motion of the guiding centers of the individual particles but also a macroscopic displacement of the whole of the corresponding component of the plasma. It follows that gravitational drift $\mathbf{V}_g$ can be taken into account by the simple substitution $\omega \rightarrow \omega - \mathbf{k}\mathbf{V}_g$. Magnetic drift is characterized by a spatial displacement of the guiding centers of the particles, but this is by no means equivalent to a macroscopic velocity of the corresponding component. The macroscopic velocity is determined by the expression $\mathbf{V} = -c\,[\nabla p, \mathbf{e}_0]/en_0B$, which does not contain the radius of curvature.

Despite this difference, the introduction of a gravitational drift does indeed simulate qualitatively a number of effects due to magnetic drift. This is confirmed by the results that follow from the exact equations, as we have already seen, for example, in §11.1.

One can, however, show that a simulation of this kind is justified only if $\omega \gg \mathbf{k}\mathbf{V}_g$. Bearing this in mind, we did not consider perturbations with $\omega \lesssim \mathbf{k}\mathbf{V}_g$ in Chapters 6 and 9, since very incorrect results would otherwise have been obtained.

In the case of a zero-pressure plasma, the magnetic drift due to curvature and inhomogeneity of the magnetic field does not depend on the gradient of the plasma pressure but is uniquely determined by the radius of curvature:

$$\mathbf{V}_{dr} = \frac{1}{R\omega_B}\left(\frac{v_\perp^2}{2} + v_\parallel^2\right)[\mathbf{e}_r, \mathbf{e}_0], \qquad R \equiv r\,(B/B_\varphi)^2. \tag{13.6}$$

If $\beta \neq 0$, this equation is replaced by

$$\mathbf{V}_{\mathrm{dr}} = \frac{1}{\omega_B}\left\{\frac{1}{R}\left(\frac{v_\perp^2}{2} + v_\parallel^2\right) + \frac{\beta}{2}\cdot\frac{\partial \ln p}{\partial r}\cdot\frac{v_\perp^2}{2}\right\}[\mathbf{e}_r, \mathbf{e}_0]. \quad (13.7)$$

If it is assumed that the entire drift is similar to gravitational drift, a result is obtained that does not agree with the result of exact analysis for the flute instability in the limit $R \to \infty$. In reality, however, we have the following unusual situation: under the conditions when the term with β is important, i.e., when

$$\beta > a/R, \quad (13.8)$$

it is also important to take into account the fact that perturbations are not electrostatic; the latter effect leads to a contribution to the dispersion equation that exactly compensates the contribution of the term with β. Therefore, in the investigation of the instabilities of a plasma with $\beta \geq a/R$, it is in general necessary to take into account the fact that the perturbations are not electrostatic.

The shear of a helical magnetic field can be taken into account in the same way as in §8.1 (see also §11.6).

§13.4. Nonmagnetohydrodynamic Instabilities of a Column without a Longitudinal Magnetic Field

Suppose that a plasma column in a field of the form (13.1) is magnetohydrodynamically stable, i.e., suppose that fast gradient (magnetohydrodynamic) instabilities cannot develop in the plasma. Let us consider whether slow gradient (nonmagnetohydrodynamic) instabilities are possible.

1. Ordinary Pinch. In this case one must distinguish a peripheral region, in which $\beta < 1$, and a central region, in which $\beta \geq 1$. A picture of the instabilities in the peripheral region can be obtained by using the results of the first part of this volume. The central region requires an additional analysis.

A. Instabilities in the Peripheral Region, $\beta < 1$. If the ratio of the Larmor radius ρ of the particles to the characteristic inhomogeneity scale a of the plasma is not too small, i.e., we have a plasma with finite ρ/a, then, in accordance with §§2.1 and 2.2, the development of ion-cyclotron and high-frequency instabilities is to be expected.

If ρ/a is small, the main role must be played by the low-frequency instabilities considered in §§3.1-3.3, or the instabilities discussed in Chapter 5, depending on the extent to which the plasma is collisional. The results of §§6.5 and 6.6 indicate that the curvature of the lines of force must affect all these kinds of instability. The effect must be destabilizing, since the field curvature is directed along ∇p if the pressure decreases along the radius and is therefore unfavorable from the point of view of stability.

If the model treatment given in Chapter 6 is replaced by a more rigorous treatment, which takes into account the real curvature and the temperature gradient, an additional class of instabilities is found with frequencies of the order of the magnetic drift $\omega \simeq kV_{dr}$ (see §11.1); these develop at small values of $\partial \ln T/\partial \ln n_0$.

B. Instabilities in the Central Region, $\beta \geq 1$. In the investigation of perturbations of a plasma with $\beta \geq 1$, it is not possible to make the assumption adopted almost everywhere in the first part of this volume that the perturbations are electrostatic. Instead one must take into account the solenoidal part of the electric field and the perturbations of the magnetic field. An analysis of this nature shows that, on the whole, a plasma with a high β is more stable than a plasma with low values of β. However, for large β as well there are growing perturbations if ρ/a is not too small or if $\partial \ln T/\partial \ln n_0 \geq 1$.

2. Inverse Pinch. If the pressure increases along the radius, the stabilizing effect of a favorable curvature must play a role. This may make it possible to suppress the ion-cyclotron and high-frequency instabilities with longest wavelengths to which a plasma with finite ρ/a is subject (see §§2.1 and 2.2). The low-frequency instabilities (see §§6.5 and 6.6) with the longest wavelengths can be suppressed in the same manner.

Thus, the field configuration in an inverse pinch is favorable both as regards magnetohydrodynamic and nonmagnetohydrodynamic perturbations.

§13.5. Nonmagnetohydrodynamic Instabilities of a Column with a Longitudinal Magnetic Field

Our entire discussion in §13.4 can also be applied to a column with a longitudinal magnetic field if we consider the region of a plasma in which the shear of the magnetic field is sufficiently small. The presence of a shear appreciably changes the picture of non-magnetohydrodynamic instabilities. If the shear is sufficiently large, the existence of many types of instability that are excited in the absence of a shear is impossible. These suppression effects were discussed in Chapters 8 and 9.

Our treatment of the role of a shear in Chapters 8 and 9 was restricted to a model of a planar layer. We found that the shear is uniquely related to the longitudinal current that flows at the given position in the plasma. Therefore, as follows from this treatment, shear effects are important only if there is a sufficiently strong longitudinal current.

This model also gives a qualitatively correct description of the real situation in the case of a cylindrical plasma if the shear is due to a current that flows in the plasma and not along a conductor which can be situated within the plasma. However, if we have the latter situation, then, using the estimates of Chapters 8 and 9, we can ignore the estimates in which the longitudinal current or the dimensionless particle density number play a role. Instead of this kind of relationship between the shear and the local parameters of the plasma, the shear is related to the current that flows within the plasma. This relation has been discussed in §11.6. In particular, if the shear is produced by means of a conductor situated within the plasma, the estimate of Θ in the relations of Chapters 8 and 9 has the form

$$\Theta \simeq 2I/crB_0, \tag{13.9}$$

where I is the total current flowing along the conductor.

If a central conductor is present, the pressure may increase along the radius. If the formulas of Chapter 9 are used in this case, the fictitious force of gravity **g** must be assumed to be directed along the density gradient, $\mathbf{g}\nabla n_0 > 0$. This means that, in addition to the stabilizing effect of the shear, the stabilizing effect of the favorable curvature discussed in §§ 6.6 and 6.7 must also play a role (in this connection see also the comments made in §13.4).

§13.6. Plasma Cylinder with Finite β in a Rectilinear Magnetic Field

Both the flute instability and all the other forms of magnetohydrodynamic instability can arise only if the lines of force of the magnetic field are curved. There must therefore be magnetohydrodynamic stability if a plasma cylinder is situated in a magnetic field that has only a longitudinal component:

$$\mathbf{B} = (0,\ 0,\ B_z). \tag{13.10}$$

If the particles have a nearly Maxwellian velocity distribution and the radial electric field is sufficiently weak, only slow gradient instabilities due to the inhomogeneity of the density and temperature can develop. If only the density is inhomogeneous but not the temperature $\nabla n_0 \neq 0,\ \nabla T = 0,$ the plasma will be more stable if β is finite than if the plasma has zero pressure. In practice, this will mean that a plasma with finite β will decay over much longer times than a low-pressure plasma.

If the collisions are not very frequent, $\nu_{ii} < \omega^*$, which is characteristic of a fairly hot plasma, and β is small, all the main types of instability due to a density gradient – low-frequency (see §3.1) and ion-cyclotron (see §2.1) – are revealed by a theoretical analysis. It was shown in §3.3 that for even a relatively small β, $\beta \simeq (m_e/m_i)$, the maximal growth rate of the long-wavelength part of the spectrum of the low-frequency instability, $k_\perp \rho_i \leq 1$, is smaller than when $\beta < m_e/m_i$. If β is appreciably greater than m_e/m_i, there is another stabilizing effect – the damping of perturbations as a result of interaction with the ions. This effect stabilizes both the long-wavelength, $k_\perp \rho_i < 1$, and the short-wavelength, $k_\perp \rho_i > 1$, perturbations. If $T_e = T_i$, the low-frequency instability is completely suppressed if

$$\beta > 0.13. \tag{13.11}$$

In accordance with §2.1, the ion-cyclotron instability of a plasma with small β develops if $\rho_i/a > 2\,(m_e/m_i)^{1/2}$. This instability condition also remains in force if $\beta \simeq 1$. However, as β increases, there is a decrease in the growth rate of the perturbations. The decrease of the growth rate with increasing β begins when

$$\beta > (\rho_i/a)^{1/2}, \tag{13.12}$$

and for these values of β

$$\gamma \sim 1/\beta. \tag{13.13}$$

Thus, the effects of a finite β lead to stabilization, complete or partial, of both forms of instability of a plasma with a density gradient.

A finite-pressure plasma with $\nabla T \neq 0$ is less stable than one with $\nabla T = 0$. If $\partial \ln T/\partial \ln n_0 \geq 1$, then, as in the limit $\beta \rightarrow 0$, instabilities with $\omega \simeq \omega^*$ and $k_\perp \simeq 1/a$ can develop. Estimates show that such instabilities must lead to diffusion of the order of the Bohm diffusion.

Bibliography

Aspects of magnetohydrodynamic stability

Early papers on the stability of a plasma column with a current are cited in the review:

1. B. B. Kadomtsev, Reviews of Plasma Physics, Vol. 2, Consultants Bureau, New York (1966), p. 153.

Of these, we mention the following, from which results were quoted in §§13.1-13.2:

2. W. A. Newcomb, Ann. Phys., 10:232 (1960). Transformation of the expression for the potential energy.
3. B. B. Kadomtsev, Zh. Eksp. Teor. Fiz., 37:1096 (1959) [Sov. Phys. – JETP, 37:1096 (1959)]. Investigation of the perturbations of a column with $B_z = 0$ (§13.1).
4. B. R. Suydam, in: Proceedings of the Second United Nations International Conference on the Peaceful Uses of Atomic Energy, Geneva, (1958), Vol. 31, published by U.N., Geneva (1958), p. 157. Derivation of a condition of local stability of a cylindrical column (§13.2).
5. B. Coppi, J. M. Greene, and J. L. Johnson, Nucl. Fusion, 6:101 (1966). Analysis of nonlocal perturbations of a plasma column with a current in a longitudinal magnetic field; discussion of the solutions corresponding to the kink instability under conditions when a singular point is situated in the plasma (§13.2).
6. S. A. Colgate and H. P. Furth, Phys. Fluids, 3:982 (1960). Discussion, in particular, of the stability of a column with a solid core (inverse pinch). Extensive bibliography on the theory of the magnetohydrodynamic stability of plasma columns.

Investigation of nonmagnetohydrodynamic instabilities

7. Yu. A. Tserkovnikov, Zh. Eksp. Teor. Fiz., 32:67 (1957) [Sov. Phys. – JETP, 5:58 (1957)].
8. L. I. Rudakov and R. Z. Sagdeev, Zh. Eksp. Teor. Fiz., 37:1337 (1959) [Sov. Phys. – JETP, 10:952 (1960)].
9. B. B. Kadomtsev, Zh. Eksp. Teor. Fiz., 37:1096 (1959) [Sov. Phys. – JETP, 10:780 (1960)].
10. L. V. Mikhailovskaya and A. B. Mikhailovskii, Nucl. Fusion, 3:276 (1963). In [7-10] investigations are made of instabilities with frequencies of the order of the frequency of magnetic drift $\omega \simeq kV_{dr}$, the shear being neglected. Instabilities of this kind develop in a plasma with $|\partial \ln T/\partial \ln n_0| \geq 1$ (§13.4).

11. L. V. Mikhailovskaya and A. B. Mikhailovskii, Zh. Eksp. Teor. Fiz., 45:1566 (1963) [Sov. Phys. – JETP, 18:1077 (1964)]. The instability boundary of a plasma with $\nabla n_0 \neq 0$ $\nabla T = 0$ at large β is found (§13.6). The dispersion equation is given for nonelectrostatic low-frequency oscillations of a Maxwellian plasma.

12. A. B. Mikhailovskii, Nucl. Fusion, 5:125 (1965). Study of the cyclotron instability of a plasma with finite β (§13.6). Some of the results of the investigation of nonmagnetohydrodynamic instabilities of a cylindrical column can also be found in the papers mentioned in the bibliographies to Chapters 3-10.

Chapter 14

Energy Method Investigation of Plasma Stability in Adiabatic Traps

§14.1. Scalar-Pressure Plasma in a Slightly Curved Field

The conclusions we have drawn in §§11.3 and 11.4 concerning the stability of a plasma in adiabatic traps, which are important from the experimental point of view, lack rigor in several respects. Above all, the very method of obtaining the general stability condition (11.21), $\nabla p \nabla U > 0$, is not rigorous. This lack of rigor can be eliminated if the energy method is employed. In the present section we shall derive the condition (11.21) in this manner. In addition, it is not justified to assume that the particles have a Maxwellian velocity distribution, since there is a loss cone in an adiabatic trap, and the plasma in the trap is certainly anisotropic. The condition (11.21) is generalized to the case of an anisotropic plasma in §14.2. In the same section we shall show that inclusion of anisotropy does not affect the qualitative conclusions that follow from (11.21).

In deriving (11.21) we have made a further restrictive assumption – the inhomogeneity of the magnetic field is small compared with the plasma inhomogeneity. We shall consider the case when $\partial \ln B/\partial \ln p \simeq 1$ in §14.3. In this case a stabilizing role can be played by the "finiteness of the specific volume." In §14.4 we dispense with the simplifying assumptions that the pressure is a scalar and that a/R is small and obtain rigorous conditions for the stability of a collisionless plasma in an adiabatic trap. Finally, in §14.5 we take into account a finite value of β. In the following calculations we omit the index 0 from the steady-state values of the magnetic field and the pressure.

We shall investigate the sign of the potential energy of fast gradient perturbations of a plasma under the assumption that the steady-state pressure of the plasma is a scalar, $p_{\parallel} = p_{\perp}$, the field curvature is small, $a \ll R$, the plasma pressure is such that $8\pi p/B^2 \ll a/R$, and the perturbations themselves are electrostatic, rot $\mathbf{E}' \to 0$. The crudest approximation is that the pressure is a scalar and, strictly speaking, it is not true in the case now under consideration – a plasma confined in an adiabatic trap. However, the results obtained in this approximation can be justified by a more rigorous treatment.

In the approximation of a scalar pressure, the potential energy is determined by the expression (12.46). If $\beta \to 0$, rot $\mathbf{E}' \to 0$, and the curvature is slight, $a/R \to 0$, this expression reduces to

$$W_{pot} = \frac{1}{2} \int \xi_{\perp} \nabla p \operatorname{div} \xi_{\perp} d\mathbf{r}. \tag{14.1}$$

Noting that $\operatorname{div} \xi_{\perp} = \nabla \ln B 2 \xi_{\perp}$, and representing $d\mathbf{r}$ in the form $d\Phi dl/B$ (cf. §11.2), we find

$$W_{pot} = \frac{1}{2} \int d\Phi (\xi_{\perp} \nabla p)(\xi_{\perp} \nabla U), \tag{14.2}$$

where $U = \int dl/B$ [cf. (11.20)].

A necessary and sufficient condition for stability is that W_{pot} be nonnegative for all ξ. It can be seen that this is the case if

$$\nabla p \nabla U > 0, \tag{14.3}$$

which agrees with the condition (11.21) obtained heuristically.

The integral U is calculated in §11.3 for an axisymmetric trap. It increases with the radius. Therefore, if the pressure decreases along the radius, the condition (14.3) cannot be satisfied, i.e., the plasma in unstable. In §11.4 we have shown that in the case of axially asymmetric traps the integral U can decrease with the radius; in accordance with §14.3, a plasma in such traps must be stable.

Thus, we have shown that the energy method gives the same picture of the stability of a plasma in adiabatic traps as the qualitative method of §11.2, which employs the notion of magnetic drift of the particles.

§14.2. Plasma with Nonscalar Pressure in a Slightly Curved Field

We now take into account the anisotropy of the particle velocity distribution, $p_{\parallel} \neq p_{\perp}$. All the other basic assumptions made in §14.1 remain in force. Let us consider how anisotropy affects the conclusions drawn in §§11.2-11.4 and §14.1 about the plasma stability.

In accordance with Eq. (12.59), the potential energy of the perturbations of an anisotropic plasma in the limit $\beta \to 0$ and $\frac{\partial \ln B}{\partial \ln p} \ll 1$ has the form

$$W_{\text{pot}} = -\frac{1}{2}\int d\Phi \int \frac{dl}{B}(\xi_{\perp}\nabla \ln B)(\xi_{\perp}\nabla)(p_{\parallel} + p_{\perp}). \tag{14.4}$$

1. Axisymmetric Trap. The assumption $\partial \ln B/\partial \ln p \ll 1$ means, in particular, that the trap is long compared with its radius, so that $\nabla \ln B \equiv \mathbf{n}/R$ and $\nabla(p_{\parallel} + p_{\perp})$ have essentially radial components: $\mathbf{n} \approx -\mathbf{e}_r$, $\nabla(p_{\parallel} + p_{\perp}) \approx \mathbf{e}_r \frac{\partial}{\partial r}(p_{\parallel} + p_{\perp})$. From (14.4) we can obtain a condition for the stability of an axisymmetric trap:

$$\int \frac{dl}{B}\cdot\frac{1}{R}\cdot\frac{\partial}{\partial r}(p_{\parallel} + p_{\perp}) > 0. \tag{14.5}$$

If the pressure decreases along the radius, this condition cannot be satisfied, which corresponds to instability. This agrees with the result obtained in §11.3 by means of the stability condition $\nabla p \nabla U > 0$ for a plasma with scalar pressure.

2. Axially Asymmetric Trap. If the field is not axisymmetric, the gradients of the plasma pressure and the magnetic field satisfy the integral condition

$$\oint \mathbf{e}_0\left[\nabla_{\perp}(p_{\perp} + p_{\parallel}),\ \nabla_{\perp}\frac{dl}{B_0}\right] = 0. \tag{14.6}$$

In particular, the condition (14.6) is satisfied by a distribution for which

$$[\nabla_{\perp}(p_{\perp} + p_{\parallel}),\ \nabla_{\perp}B] = 0 \tag{14.7}$$

at every point along the line of force. In this condition we have used $\nabla_{\perp} B \| \nabla_{\perp} dl$ (cf. §11.2). For a distribution of this kind, (14.4) yields the stability condition

$$\oint \frac{\partial}{\partial x^1}(p_{\parallel} + p_{\perp}) \frac{\partial}{\partial x^1}\left(\frac{dl}{B}\right) \geqslant 0, \tag{14.8}$$

where x^1 is the direction of the transverse gradient of B. This condition can be regarded as a generalization of (14.3) to the case of a plasma with a nonscalar pressure.

To prove that a plasma can be confined stably in an axially asymmetric trap, we must show that the inequality (14.8) can be satisfied for at least some $p_{\parallel}(r)$ and $p_{\perp}(r)$, where $p_{\parallel}(r)$ and $p_{\perp}(r)$ are the corresponding moments of the steady-state distribution function. However, we can make an even cruder assumption by setting

$$\partial (p_{\parallel} + p_{\perp})/\partial x_1 = -C (p_{\parallel} + p_{\perp}), \tag{14.9}$$

where C > 0 does not depend on x^3. Then the stability condition (14.8) reduces to

$$\oint (p_{\parallel} + p_{\perp}) \frac{\partial}{\partial x^1}\left(\frac{dl}{B}\right) < 0. \tag{14.10}$$

The main difference between (14.10) and the condition $\partial U/\partial x^1 < 0$ is that we have now taken into account more correctly the limits of integration along the line of force and the relative contribution of the curvature on each element of length.

An investigation of plasma stability by means of the condition (14.10) (Trubnikov) confirms that a plasma in a trap with a minimum-B configuration can be confined stably.

§14.3. High-Curvature Field. Effect of the Finite Value of $\partial \ln U/\partial \ln p$

We now obtain a condition for the stability of a zero-pressure plasma in an adiabatic trap at finite values of $\partial \ln B/\partial \ln p$.

1. Hydrodynamic Stability Condition. We assume that the plasma can be described hydrodynamically. Then the potential energy of the perturbations is given by Eq. (12.46). We minimize this expression with respect to the longitudinal displacement $\xi_{\parallel}$

[the latter occurs in the first term of the right-hand side of (12.46)]. If (12.78) and the equation $\int (dl/B)\,\mathrm{div}\,\xi_\perp = \xi_\perp \nabla U$ are taken into account,

$$W_{\min} = \frac{1}{2}\int d\Phi \left\{\frac{\gamma p}{U}(\xi_\perp \nabla U)^2 + (\xi_\perp \nabla U)(\xi_\perp \nabla p). \right. \tag{14.11}$$

From this we obtain the stability condition [cf. (14.3)]:

$$\nabla p\, \nabla U + \gamma p\, (\nabla U)^2/U \geqslant 0. \tag{14.12}$$

The second term of the left-hand side is always positive. This term is associated with a stabilization effect that is important if

$$|\partial \ln U/\partial \ln p| > 1, \tag{14.13}$$

which entails approximately $|\partial \ln B/\partial \ln p| \geq 1$.

2. Kinetic Stability Condition for a Plasma with Isotropic Pressure. This condition is obtained by means of formula (12.61):

$$\nabla p\, \nabla U + \frac{15}{2} p \int \left(\nabla \oint \sqrt{1-\lambda B}\, dl\right)^2 \frac{d\lambda}{\oint dl/\sqrt{1-\lambda B}} > 0. \tag{14.14}$$

This means approximately the same as (14.12). If we have recourse to the comparison theorem (12.79), we find that the integral term in (14.14) is not less than $\gamma p\,(\nabla U)^2/U$ in (14.12) if $\gamma = 5/3$.

§14.4. Exact Stability Conditions for a Plasma in an Adiabatic Trap

1. Sufficient Condition for Stability. If collisions are ignored, the potential energy of a zero-pressure plasma obtained in the kinetic approach has the form (12.60). It is not negative at least when

$$(\partial F/\partial \varepsilon)_{\mu,\, J_\parallel} < 0. \tag{14.15}$$

This inequality is a sufficient condition for stability [it is assumed that (14.15) holds for all ε, μ and $J_\parallel$].

The inequality (14.15) has a clear meaning, which is the following. Consider particles with given μ and $J_\parallel$. Such particles move on surfaces which are contained one within another and their position, for a given magnetic field, is uniquely determined by the value of ε (drift surfaces). One can show that the surfaces with larger values of ε are further from the center if the magnetic field has a minimum-B configuration. In this case, the condition (14.15) means simply that the number of particles decreases on the drift surfaces with increasing distance from the center. This condition is satisfied automatically if the plasma density decreases toward the periphery.

A special case of a spatially localized distribution in a minimum-B field is one to which there corresponds an F that is independent of the coordinates α and β and is therefore independent of $J_{\|}$:

$$F = F(\varepsilon, \mu). \tag{14.16}$$

A sufficient condition for stability in this case has the form

$$(\partial F/\partial \varepsilon)_{\mu} < 0. \tag{14.17}$$

As an example of a distribution of the type (14.16), we may mention the following:

$$f(\mu, \varepsilon) = (\mu B^{(0)} - \varepsilon)^n g(\mu), \quad \varepsilon \leqslant \mu B^{(0)}. \tag{14.18}$$

where $g(\mu)$ is an arbitrary function. For a distribution of the form (14.18), the condition (14.17) means simply

$$n > 0. \tag{14.19}$$

The functions (14.18) correspond to the longitudinal and transverse pressure:

$$\left.\begin{aligned} p_{\|} &= CB(B^{(0)} - B)^{n+3/2}, \\ p_{\perp} &= \left(n + \frac{3}{2}\right) C(B^{(0)} - B)^{(n+1/2)}, \end{aligned}\right\} \tag{14.20}$$

which are nonvanishing only for $B < B^{(0)}$. This last condition means that a plasma with the function (14.18) is localized in the region $B < B^{(0)}$.

This example shows that it is in principle possible for there to be hydrodynamically stable containment of a plasma in minimum-B traps.

2. Necessary and Sufficient Condition for Stability. The condition (14.15) need not be satisfied, but the plasma may still be stable if, as follows from (12.60),

$$\lambda_{\alpha\alpha} < 0, \ \lambda_{\beta\beta} < 0, \ (\lambda_{\alpha\beta})^2 < \lambda_{\alpha\alpha}\lambda_{\beta\beta}, \tag{14.21}$$

where

$$\lambda_{xy} = \int d\mu dJ_{\parallel} \frac{\partial \varepsilon}{\partial x} \cdot \frac{\partial \varepsilon}{\partial y} \cdot \frac{\partial F}{\partial \varepsilon}, \quad (x, y) = (\alpha, \beta). \tag{14.22}$$

This is the desired necessary and sufficient condition for the stability of a collisionless plasma in an adiabatic trap. It can be seen that the stability is determined by the configuration of the magnetic field and the nature of the energy distribution of the particles.

§14.5. Limiting Pressure of a Plasma That Is Stably Confined in an Adiabatic Trap with a Minimum-B Configuration

Introduction. In an adiabatically confined plasma the particles have an anisotropic velocity distribution, so that $T_\perp > T_\parallel$, where $T_\perp$ and $T_\parallel$ are certain effective temperatures. The velocity anisotropy may be the cause of instabilities of the plasma. We have already considered some of the anisotropic instabilities, which are frequently known as Harris type instabilities, in Chapter 15 in Volume 1. These are electrostatic instabilities. They are the only possible instabilities in a plasma with $T_\perp > T_\parallel$ if $\beta_\perp$ has very small values. If $\beta_\perp$ is not very small, one must take into account the possibility that nonelectrostatic perturbations are excited as a result of the anisotropy. In the approximation of a homogeneous plasma for nonelectrostatic perturbations with $\mathbf{B}'_\perp = 0$ and $B'_\parallel \neq 0$ ($\mathbf{B}'_\perp$ and $B'_\parallel$ are the perturbed magnetic fields at right angles to and along $\mathbf{B}_0$) at frequencies $\omega \ll (\omega_{Bi}, k_\parallel v_{T_\parallel i})$, we can obtain the dispersion equation

$$N^2 - \varepsilon_{22} = 0, \tag{14.23}$$

where $N^2 \equiv c^2k^2/\omega^2$ is the square of the refractive index; ε_{22}, a component of the permittivitv tensor, is equal to

$$\varepsilon_{22} = -\frac{4k_\perp^2 T_\perp}{m_i\omega^2}\left(\frac{\omega_{p_i}}{\omega_{B_i}}\right)^2\left[1-\frac{T_\perp}{T_\parallel}\left(1+\frac{i\sqrt{\pi}\,\omega}{2\,|k_\parallel|\,v_{T_{\parallel i}}}\right)\right], \tag{14.24}$$

$$(T_{\parallel e}=T_{\parallel i},\ T_{\perp e}=T_{\perp i}).$$

It follows from (14.23) and (14.24) that the plasma is unstable if

$$p_\perp - p_\parallel > \frac{p_\parallel}{p_\perp}\cdot\frac{B_0^2}{8\pi}, \tag{14.25}$$

the growth rate of the perturbations being due to the interaction between the resonant ions and the wave; it is given approximately by

$$\gamma \simeq k_\parallel v_{T_{\perp i}}\beta_\perp^{1/2}. \tag{14.26}$$

The condition (14.25) can be interpreted from the energy point of view: if one writes down the expression for the potential energy in the magnetohydrodynamic approximation, assumes that the plasma is homogeneous, and considers perturbations with $\mathbf{Q}_\perp = 0$ and $\mathbf{Q}_\parallel \neq 0$, the potential energy will be found to be negative if the condition (14.25) is satisfied. The approximation of a homogeneous plasma is a special case of the Chew-Goldberger–Low approximation. The potential energy in this case therefore has the form (12.49). Proceeding from the expression (12.49) and making appropriate simplifications, we can verify directly this assertion.

2. Electromagnetic Anisotropic Instability of an Inhomogeneous Plasma Confined in an Adiabatic Trap. We shall now assume that the plasma has a restricted length and that the frequency of the perturbations is so low that the particles succeed in traversing the system many times. The potential energy of such perturbations is determined by the expression (12.67). We shall use this expression to study the stability of the plasma with a Taylor type distribution (14.20) (in the approximation of electrostatic perturbations this corresponds to a very stable plasma; see §14.4).

In accordance with §14.5.1, we consider perturbations with $Q_\perp = 0$ localized in a region that is small compared with the length of the plasma (large values of $k_\parallel$ correspond to balloon-type perturbations). Because the localization region is small, the quantities averaged over the length of the trap can be neglected in W_2, and we then obtain the value for the maximal $\beta_\perp$ of a stably confined plasma:

$$\beta_{\perp\max} \equiv 8\pi p_{\max}/B_0^2 = (B_{\max}^2 - B_0^2)/B_0^2, \tag{14.27}$$

where $p_{\perp\max}$ and B_0 are the transverse pressure and magnetic field in the center and $B_{\max}$ is the magnetic field at the point where $p_\perp = 0$.

The right-hand side of (14.27) is the depth of the magnetic well, so that the maximal attainable β are determined by this depth.

Bibliography

1. M. N. Rosenbluth and C. Longmire, Ann. Phys., 1:120 (1957). It is shown that an axisymmetric trap is unstable. The cases $p_\perp = p_\parallel$ and $p_\perp \neq p_\parallel$ (§14.2) are studied. The instability condition is interpreted thermodynamically.
2. B. B. Kadomtsev, in: Plasma Physics and the Problem of Controlled Thermonuclear Reactions, Vol. 4, Pergamon Press, Oxford (1960), p. 17. A study is made in the hydrodynamic approximation of the stability of a zero-pressure plasma. The effect of a finite $\partial \ln U/\partial \ln p$ (§14.3) is studied.
3. I. B. Bernstein et al., Proc. Roy. Soc., A244:17 (1958). The stability theory of axisymmetric traps is developed in the hydrodynamic approximation. The effect of a finite $\partial \ln U/\partial \ln p$ is considered. It is shown that balloon-type perturbations can develop (see also §11.5).
4. B. B. Kadomtsev, in: Plasma Physics and the Problem of Controlled Thermonuclear Reactions, Vol. 4, Pergamon Press, Oxford (1960) (in the articles beginning on p. 417). A thermodynamic interpretation of the quantity $\int dl/B$ and the effect of a finite $\partial \ln U/\partial \ln p$ is given. The stability of various types of magnetic trap is discussed.
5. B. B. Kadomtsev, ibid. Discussion of the effect of finite $\partial \ln U/\partial \ln p$ in the kinetic approach (§14.3).
6. J. Berkowitz, H. Grad, and H. Rubin, in: Proceedings of the Second United Nations International Conference on the Peaceful Uses of Atomic Energy, Geneva 1958, Vol. 31, Published by U.N., Geneva (1958), p. 177.
7. R. F. Post et al., Phys. Rev. Lett., 4:166 (1960).
8. B. B. Kadomtsev and V. E. Rokotyan, Dokl. Akad. Nauk SSSR, 133:68 (1960) [Sov. Phys. – Doklady, 5:747 (1960)].
9. B. B. Kadomtsev, Zh. Eksp. Teor. Fiz., 40:328 (1961) [Sov. Phys. – JETP, 13:223 (1961)]. The papers [6-9] contain, in particular, a discussion of the role of the freezing in of lines of force at ends (see also §§6.4 and 11.5 and the bibliography to Chapters 6 and 11).
10. B. B. Kadomtsev, Nucl. Fusion, 1:286 (1961).

11. B B. Kadomtsev, in: Reviews of Plasma Physcis, Vol. 2, Consultants Bureau, New York (1966), p. 153. In [10, 11] some aspects of the magnetohydrodynamic stability of adiabatic traps are considered.
12. J. Andreoletti, Compt. Rend. Acad. Sci., 257:1235 (1963).
13. J. B. Taylor, Phys. Fluids, 6:1529 (1963). In [12, 13] it is shown that in an axially asymmetric trap a plasma may be stable; this is the minimum-B effect.
14. J. Andreoletti, Compt. Rend. Acad. Sci., 256:1469 (1963).
15. J. Andreoletti, Compt. Rend. Acad. Sci., 257:1033 (1963).
16. J. B. Taylor, Phys. Fluids, 7:767 (1964). In [14-16] the necessary and sufficient condition (14.21) for stability is obtained. In [16] the sufficient condition (14.15) is also obtained and given a very lucid interpretation. This paper also contains a calculation of the stability of magnetic traps using the concept of drift surfaces. In §14.4 we have essentially followed Taylor [16]. The papers [14, 15] also reproduce the previously known results of Rosenbluth and Longmire [1] and Kadomtsev [5].
17. B. A. Trubnikov, At. Energ., 19:415 (1965).
18. B. A. Trubnikov, in: Plasma Physics and Controlled Nuclear Fusion Research, Vol. 1, IAEA, Vienna (1966), p. 83. The papers [17, 18] contain detailed treatments of various specific examples of traps with stabilizing rods. The stability is calculated by means of the approximate condition (14.10) obtained in [17].
19. R. A. Hastie and J. B. Taylor, Phys. Rev. Lett., 13:123 (1964).
20. J. Andreoletti, Compt. Rend. Acad. Sci., 258:5183 (1964).
21. J. Andreoletti, Compt. Rend. Acad. Sci., 259:2392 (1964).
22. J. B. Taylor and R. A. Hastie, Phys. Fluids, 8:323 (1965).
23. H. Grad, Phys. Fluids, 9:499 (1966).
24. A. Kadish, Phys. Fluids, 9:514 (1966). In [19-24] the maximal pressure of a plasma that can be stably confined in an adiabatic trap is calculated.
25. G. Schmidt, Phys. Fluids, 8:754 (1965). Calculation of the maximum possible energy of perturbations in the case where the condition $(\partial F/\partial \varepsilon)_{\mu, J_{\parallel}} < 0$ is not satisfied for all μ and $J_{\parallel}$ (see §14.4).
26. J. B. Taylor, in: Plasma Physics, IAEA, Vienna (1965), p. 449.
27. J. B. Taylor, Proc. Roy. Soc., A304:335 (1968). The review papers [26, 27] are devoted to aspects of the stability theory of adiabatic traps.
28. B. A. Trubnikov, Introduction to Plasma Theory [in Russian], Izd. MIFI, Moscow (1969). This book contains a good review of the main ideas and various applications of the stability theory of a low-pressure plasma in adiabatic traps.

Chapter 15

Nonmagnetohydrodynamic Instabilities of Plasmas Confined in Adiabatic Traps

§15.1. Introductory Comments on Instabilities of Plasmas in Adiabatic Traps

A plasma may be unstable against two kinds of perturbation – gradient instabilities, with growth rates that depend on the spatial gradients of the steady-state plasma parameters and the magnetic field, and nongradient instabilities, which are usually investigated in the approximation of a homogeneous plasma and are due to streaming or velocity anisotropy. In their turn, the gradient instabilities can be divided into two groups – fast gradient (magnetohydrodynamic) and slow gradient (nonmagnetohydrodynamic) instabilities. This classification was explained in §12.1. It is applicable only to the case of a curved magnetic field since there are no magnetohydrodynamic instabilities in the approximation of a rectilinear field.

A plasma in an adiabatic trap is one example of a plasma in a curved magnetic field. If one wishes to investigate the problem of the prolonged confinement of a fairly dense plasma in such a trap, the general stability problem must be formulated as follows. It is first of all necessary to establish whether magnetohydrodynamic instabilities can arise; for if so the plasma will rapidly leave the containment volume. If such an analysis predicts magnetohydrodynamic stability, one must then study the other types of instability – nongradient and slow gradient. These may be responsible for slower plasma losses, which, however, exceed the classical losses. Since the nongradient (ordinary) instabilities have, in general, larger growth rates than the slow gradient instabilities, the former are to be considered before the latter.

Thus, keeping to our applied point of view, we must establish the following hierarchy of instabilities: fast gradient (magnetohydrodynamic), nongradient, slow gradient.

We shall now briefly summarize the knowledge we have already gained about these three groups of instabilities and outline the further investigation of these questions.

1. Fast Gradient (Magnetohydrodynamic) Instabilities. These instabilities are due to inhomogeneity of the plasma and an unfavorable curvature of the lines of force. The most effective way to eliminate these instabilities is therefore to produce an appropriate configuration of the magnetic field (minimum-B configurations).

The most typical representative of the fast gradient instabilities is the flute instability. At not too small β, the balloon-mode instability may also play a role.

The flute instability was discussed in §§6.1-6.4, where the curvature was simulated by a gravitational force, and in §§11.1-11.5, where an analogous discussion was given for various cases of a curved field on the basis of the notion of magnetic drift of particles; it was also considered in Chapters 13 and 14, which contain the results of the magnetohydrodynamic approach to this instability. The balloon mode was discussed in §11.5.

We saw that, besides the stabilizing effect, other effects of a similar kind could be important: finiteness of the ion Larmor radius, freezing in of the lines of force at the ends, finite values of $\partial U/\partial p$, etc. We did not discuss all the possible stabilization effects. Information concerning them can be found in the original papers cited at the ends of the corresponding chapters.

2. Nongradient Instabilities. The instabilities of this class are due either to streaming or velocity anisotropy but not the presence of a spatial inhomogeneity, although, of course, the latter may affect the dynamic development of the instabilities. In general, streaming and anisotropy of the plasma are a necessary consequence of the very method by which a plasma is confined in

adiabatic traps. In the case of anisotropy, this assertion is obvious. We have shown in §14.5 of Volume 1 how a beam-type velocity distribution of a plasma arises in an adiabatic trap.

Streaming and anisotropy lead to the excitation of cyclotron and high-frequency oscillations and, in a number of cases, low-frequency oscillations (for the electrons or ions, respectively). The instabilities due to a nonequilibrium velocity state of the electrons were discussed in Chapters 10 and 11 in Volume 1; the instabilities due to one of the ion velocity distribution, in Chapters 14-17 of Volume 1. To use the results obtained in Volume 1 to analyze the stability of a plasma in adiabatic traps we must, in general, take into account the inhomogeneity of the plasma and the magnetic field. This is done to a certain extent in §14.6 of Volume 1. This question is analyzed further in papers cited at the end of the present chapter.

3. Slow Gradient Instabilities. In their turn, these can be split into two subclasses: ion-cyclotron and high-frequency, $\omega \gtrsim \omega_{B_i}$; low-frequency, $\omega \ll \omega_{B_i}$. We shall discuss these subclasses separately.

A. Ion-Cyclotron and High-Frequency Instabilities. These are due to spatial inhomogeneity of the plasma. At the same time, transverse streaming of the ions due to the loss cone can also play a destabilizing role.

The instabilities of this subclass can develop only if the ratio of the Larmor radius of the particles to the transverse dimension of the plasma is not too small (system with finite ρ/a. The most important of these is the gradient-cone instability considered in §2.3, although in a number of cases some of the other instabilities considered in Chapter 2 may be important.

The effect of longitudinal inhomogeneity of the plasma and inhomogeneity of the magnetic field on these instabilities is discussed in the papers cited at the end of the present chapter.

B. Low-Frequency Instabilities. Like the instabilities of the subclass A, they are due to spatial inhomogeneity of the plasma. In the approximation of a homogeneous magnetic field and longitudinally homogeneous plasma, the main forms of these instabilities have been considered in §§3.1-3.3 (see also §3.6 and Chapter 5). In §6.5 we have considered how they are affected by curvature of the lines of force in the model that employs a force of gravity.

It follows from this simplified treatment that the low-frequency instabilities are very sensitive to the ratio of the longitudinal to the transverse dimension of the system, to the nature of the longitudinal velocity distribution of the particles, and also to the curvature of the lines of force. Therefore, to study the importance of these instabilities in the problem of plasma containment in adiabatic traps, it is necessary to make a further analysis. This will be done in the following sections.

§15.2. Basic Equations for Low-Frequency Perturbations of a Plasma in a Field of Complicated Geometry

1. General Expression for the Perturbed Charge Density. At a sufficiently low plasma density, the perturbations can be assumed to be electrostatic, $\mathbf{E} = -\nabla\psi$. Such perturbations are described by Poisson's equation $\Delta\psi + 4\pi\Sigma n' = 0$. We find the perturbed charge density n' by the trajectory integral method. We proceed from the expression for f' in the form (1.95) and recall that f_0 depends on ε, μ_0, and $\mathbf{r}_\perp$ ($\mu_0 = \mu - \delta\mu$, where $\mu = v_\perp^2/2B$ is the magnetic moment of the particle and $\delta\mu$ is the part of the magnetic moment that oscillates in time) and express $\partial f_0/\partial \mathbf{v}$ in terms of the integrals of the motion. Then

$$f_1 = \frac{e}{m}\int_{-\infty}^{t} \nabla\psi\left(\mathbf{v}\frac{\partial F}{\partial\varepsilon} + \frac{\mathbf{v}_\perp}{B}\cdot\frac{\partial F}{\partial\mu_0} + \frac{1}{\omega_B}[\mathbf{e}_0, \nabla F]\right)dt'. \qquad (15.1)$$

As everywhere hitherto, we assume that the perturbations have the time dependence $\exp(-i\omega t)$. We represent the dependence of ψ on the transverse coordinates in the form

$$\psi \sim \psi_0 \exp\left(i\int \mathbf{k}_\perp d\mathbf{r}_\perp\right), \qquad (15.2)$$

assuming that $k_\perp$ is large compared with $(\nabla_\perp \ln F)^{-1}$ (the approximation of small-scale perturbations). We note that $\mathbf{r}_\perp$ has a part that oscillates with the time and we expand $\exp(i\mathbf{k}_\perp\delta\mathbf{r}_\perp)$ in a series in Bessel functions. We then obtain an expression for f_1 as an infinite series analogous to (1.103). In the approximation of low-frequency perturbations, $\omega \ll \omega_B$, summation can be carried out in this series, and for the part of f_1 that is averaged over the angle in the velocity space we obtain

$$\bar{f}_1 = \frac{e\psi}{m}\left[\frac{\partial F}{\partial \varepsilon} + (1 - J_0^2)\frac{\partial F}{\partial \mu}\right] + \frac{ie}{m}\left(\omega\frac{\partial F}{\partial \varepsilon} + \frac{\mathbf{k}_\perp}{\omega_B}[\mathbf{e}_0, \nabla F]\right) J_0 \times$$

$$\times \int_{-\infty}^{t} \psi_0 \exp\left[-i\int_{t}^{t'} (\omega - \omega_{dr})\, dt''\right] J_0 dt'. \tag{15.3}$$

Here $\omega_{dr} \equiv \mathbf{k}_\perp \mathbf{V}_{dr}$, where $\mathbf{V}_{dr}$ is the magnetic drift velocity of the particles defined by Eq. (11.1). In deriving (15.3) we have taken into account the indentity

$$\frac{\mathbf{k}_\perp}{\omega_B}[\mathbf{e}_0, \nabla F] = -\frac{i\sqrt{g_{33}}}{\omega_B\sqrt{g}} e^{3jk}\frac{\partial F}{\partial x^j}\frac{\partial \ln \psi}{\partial x^k}, \qquad (j, k) = (1, 2) \tag{15.4}$$

and the fact that $\sqrt{g_{33}}/B\sqrt{g}$ is constant along a line of force. As a result of this, the expression $(\mathbf{k}_\perp/\omega_B)[\mathbf{e}_0, \nabla F)$ could be taken in front of the integral over t'.

The expression (15.3) is similar to (12.41), and in the limiting case ($\omega \gg \omega_{dr}$, $k_\perp v_\perp \ll \omega_B$) these expressions are identical if the fact that the perturbations are not electrostatic is ignored in (12.41), $B'_\parallel = 0$, $\mathbf{E} = -\nabla\psi$.

Assuming that the particles oscillate between stoppers (mirrors) with period $\tau = \oint dl/v_\parallel$, we can go over from the infinite integral in (15.3) to a finite integral, using a transformation of the type of (12.55):

$$\int_{-\infty}^{t} \{\ldots\}\, dt' = \left[1 - \exp i\int_0^{\tau} (\omega - \omega_{dr})\, dt\right]^{-1} \int_{t-\tau}^{t} \{\ldots\}\, dt'. \tag{15.5}$$

The integral over t' on the right-hand side of the equation can be transformed as follows:

$$\int_{t-\tau}^{t} \{\ldots\}\, dt' = \int_{t-\tau}^{-\tau/2} \{\ldots\}\, dt' + \int_{-\tau/2}^{0} \{\ldots\}\, dt' +$$

$$+ \int_0^{t} \{\ldots\}\, dt' = i\left\{\exp[-i\nu M(l_1, l)]\frac{1}{\sin M(l_1, l_2)}\int_{l_1}^{l_2} J_0\psi \cos M(l_2, l')\frac{dl'}{v_\parallel} + \right.$$

$$\left. + i\nu\int_{l_1}^{l} J_0\psi \exp[i\nu M(l', l)]\, dl'/v_\parallel\right\}\left[1 - \exp i\int_0^{\tau} (\omega - \omega_{dr})\, dt\right]. \tag{15.6}$$

Here $M(a, b) \equiv \int_a^b (\omega - \omega_{dr})\, dl/v_{\parallel}$. In deriving this last equation we have used the fact that, because the number of particles moving in one direction equals the number moving in the other,

$$\psi[l(t)] = \psi[l(\tau - t)]. \tag{15.7}$$

In (15.6) we have introduced the index ν, which has the value + 1 for particles moving in the positive direction along the line of force and −1 for those moving in the opposite direction. In calculating the charge density, we must sum over ν. We then obtain an expression for the density of each species of charge:

$$\rho^{(\alpha)} = \frac{e^2}{m} \int \frac{B}{v_{\parallel}}\, d\mu\, d\varepsilon \left\{ \psi \left[\frac{\partial F}{\partial \varepsilon} + (1 - J_0^2) \frac{1}{B} \cdot \frac{\partial F}{\partial \mu} \right] - J_0 \left(\omega \frac{\partial F}{\partial \varepsilon} + \right.\right.$$

$$\left. + \frac{\mathbf{k}_{\perp}}{\omega_B} [\mathbf{e}_0, \nabla F] \right) \left[\frac{\cos M(l_1, l)}{\sin M(l_1, l_2)} \int_{l_1}^{l_2} J_0 \psi \cos M(l', l_2) \frac{dl'}{v_{\parallel}} - \int_{l_1}^{l} J_0 \psi \times$$

$$\left. \left. \times \sin M(l, l') \frac{dl'}{v_{\parallel}} \right] \right\}. \tag{15.8}$$

Substitution of this expression into Poisson's equation gives the basic equation which must be used to investigate the stability of a plasma in a field of complicated geometry.

2. Integral Relation. It can be seen from (15.8) that the perturbation of the density is related to the perturbation of the potential integrally and not locally as in the case of a longitudinally homogeneous magnetic field and perturbations of the potential of the form $\exp(ik_z z)$. It is therefore now quite difficult to find the eigenfrequencies of the oscillations. However, to establish stability and instability conditions one does not need to know the frequency but only the sign of its imaginary part. In a number of cases this can be found by means of integral relations that follow from Poisson's equation and are to a certain extent analogous to those used in the analysis of magnetohydrodynamic instabilities in Chapters 13 and 14. In the magnetohydrodynamic approximation we were concerned with a real functional representing the potential energy of the perturbations. We shall now take into account the processes of resonant interaction between the particles and the wave, and we must therefore introduce a complex functional. In certain cases, the real part of this functional corresponds to the total energy of the

oscillations, and the imaginary part characterizes the energy balance between the wave and the resonant particles.

We substitute the charge density (15.8) into Poisson's equation, multiply the latter by ψ^*, and integrate over the space. As transverse coordinates, we employ the variables α and β introduced in the appendix to Chapter 12. Then $d\mathbf{r} = d\alpha\, d\beta\, dl/B$. After some transformations, we obtain the integral relation

$$Q \equiv \int d\alpha\, d\beta \Big\{ \int \frac{|\nabla\psi|^2 dl}{B} - \sum \frac{4\pi e^2}{m} \int d\mu\, d\varepsilon \times$$

$$\times \Big[\int_{l_1}^{l_2} \frac{|\psi|^2 dl}{v_\parallel} \Big(\frac{\partial F}{\partial \varepsilon} + (1 - J_0^2) \frac{\partial F}{B \partial \mu} \Big) - \Big(\omega \frac{\partial F}{\partial \varepsilon} +$$

$$+ \frac{\mathbf{k}_\perp}{\omega_B} [\mathbf{e}_0, \nabla F] \Big) \Big[\frac{\cos M(l_1, l)}{\sin M(l_1, l_2)} \Big| \int_{l_1}^{l_2} J_0 \psi \cos M(l, l_2) \frac{dl}{v_\parallel} \Big|^2 +$$

$$+ \int_{l_1}^{l_2} \frac{dl}{v_\parallel(l)} \int_{l}^{l_2} \frac{dl'}{v_\parallel(l')} \sin M(l, l_2) \cos M(l', l_2) J_0(l) J_0(l') \times$$

$$\times \psi^*(l)\psi(l') + \psi(l)\psi^*(l')) \Big]\Big] \Big\} = 0. \qquad (15.9)$$

For real ω and $\mathbf{k}_\perp$, all the terms in the curly brackets are real provided the function $\sin M(l_1, l_2)$ does not vanish for any values of ε and μ. Otherwise there is resonance between the wave and the particles with corresponding ε and μ and the term with $1/\sin M$ must then be written in a complex form of the type $\mathcal{P}(1/\sin M) - i\pi a \delta(\sin M)$, where $\mathcal{P}$ is the principal value and a is a constant. It is the term with the δ-function that makes a contribution to Im Q.

3. Some Special Cases. A. Short-Plasma Approximation, $\omega\tau \ll 1$. Suppose that over a time of order $1/\omega$ a particle can make many oscillations between the stoppers. Then on the right-hand side of (15.8) the functions cos M and sin M can be expanded in the small argument M. Neglecting terms of order $\omega\tau$, we have

$$\rho^{(\alpha)} = \frac{e^2}{m} \int \frac{B}{v_\parallel} d\mu\, d\varepsilon \Big\{ \Big(\psi - \frac{\omega J_0 (\overline{\psi J_0})}{\omega - \bar{\omega}_{dr}} \Big) \frac{\partial F}{\partial \varepsilon} +$$

$$+ \psi(1 - J_0^2) \frac{1}{B} \cdot \frac{\partial F}{\partial \mu} - \frac{J_0 (\overline{\psi J_0})}{\omega - \bar{\omega}_{dr}} \frac{\mathbf{k}_\perp}{\omega_B} [\mathbf{e}_0, \nabla F] \Big\}, \qquad (15.10)$$

where

$$\overline{(\psi J_0)} = \Big(\int_{l_1}^{l_2} \psi J_0 dl / v_{\parallel} \Big) \Big/ \Big(\int_{l_1}^{l_2} dl / v_{\parallel} \Big).$$

If the approximation $\omega\tau \ll 1$ is satisfied for both the electrons and the ions, the expression (15.9) for Q has the form

$$Q = \int d\alpha\, d\beta \Big\{ \int \frac{|\nabla\psi|^2 dl}{B} - \sum \frac{4\pi e^2}{m} \int d\mu\, d\varepsilon \frac{\tau}{2} \Big[\Big(\overline{|\psi|^2} - \frac{\omega\, |\overline{(J_0\psi)}|^2}{\omega - \bar{\omega}_{dr}} \Big) \frac{\partial F}{\partial \varepsilon} +$$

$$+ \Big(\overline{|\psi|^2} - |\overline{(J_0\psi)}|^2 \Big) \frac{1}{B} \cdot \frac{\partial F}{\partial \mu} - \frac{|\overline{(J_0\psi)}|^2}{\omega - \bar{\omega}_{dr}} \cdot \frac{\mathbf{k}_\perp}{\omega_B} [\mathbf{e}_0, \nabla F] \Big] \Big\} = 0. \qquad (15.11)$$

B. Long-Plasma Approximation, $\omega\tau \gg 1$. In this case, $\cos M(l', l_2)$ and $\sin M(l, l')$ in the integrals on the right-hand side of (15.8) are rapidly varying functions of l'. Using this fact, we reduce (15.8) to the form

$$\rho^{(\alpha)} = \frac{e^2\psi}{m} \int \frac{B}{v_\parallel} d\mu\, d\varepsilon \Big\{ \Big(1 - \frac{\omega J_0^2}{\omega - \omega_{dr}} \Big) \frac{\partial F}{\partial \varepsilon} +$$

$$+ (1 - J_0^2) \frac{\partial F}{\partial \mu} - \frac{J_0^2 \mathbf{k}_\perp}{(\omega - \omega_{dr})\, \omega_B} [\mathbf{e}_0, \nabla F] \Big\} +$$

$$+ \frac{e^2}{m} \int \frac{B}{v_\parallel} d\mu\, d\varepsilon \Big(\omega \frac{\partial F}{\partial \varepsilon} + \frac{\mathbf{k}_\perp}{\omega_B} [\mathbf{e}_0, \nabla F] \Big) \times$$

$$\times \frac{J_0 v_\parallel}{\omega - \omega_{dr}} \cdot \frac{\partial}{\partial l} \cdot \frac{v_\parallel}{\omega - \omega_{dr}} \cdot \frac{\partial}{\partial l} \Big(\frac{\psi J_0}{\omega - \omega_{dr}} \Big). \qquad (15.12)$$

In contrast to (15.10), the perturbed density is determined in this case by the local characteristics of the field and plasma.

C. Approximation of Finite $\omega\tau$. In this approximation, allowance is made for the longitudinal resonance between the particles and the wave, which is not present in the limiting cases $\omega\tau = 0$ and $\omega\tau = \infty$. We shall consider this effect under the additional assumptions $\omega \gg \omega_{dr}$, $k_\perp v_\perp \ll \omega_B$, $\omega\tau \ll 1$. From (15.8) we then obtain

$$\rho^{(\alpha)} = \frac{e^2}{m} \int \frac{B}{v_\parallel} d\mu\, d\varepsilon \Big\{ \tilde{\psi} \frac{\partial F}{\partial \varepsilon} - \frac{\mathbf{k}_\perp}{\omega\omega_B} [\mathbf{e}_0, \nabla F]\, \bar{\psi} + \frac{i\pi\tau}{2} \Big(\omega \frac{\partial F}{\partial \varepsilon} + \frac{\mathbf{k}_\perp}{\omega_B} \times$$

$$\times [\mathbf{e}_0, \nabla F] \Big) \delta [\sin(\omega\tau/2)] \cos\omega\tau \int_{t_1}^{t_2} \psi \cos\omega t' dt' \Big\}. \qquad (15.13)$$

Here $\tilde{\psi} = \psi - \bar{\psi}$, $t = \int_{l_1}^{l} dl/v_\parallel$. In the expression (15.13) we have omitted small real terms of order ω_{dr}/ω and $(k_\perp v_\perp/\omega_B)^2$, which can be readily recovered by referring to (15.10).

§15.3. Stability of a Thermodynamically Quasiequilibrium Plasma in a Minimum-B Configuration

In Chapter 3 we have shown that if the curvature of the magnetic field is negligible, the plasma may be unstable if its density or temperature depends on the transverse coordinates. The instabilities considered in Chapter 3 are the main types of nonmagnetohydrodynamic instabilities of a plasma that is collisionless, an adjective that can be applied with certain reservations to a plasma in an adiabatic trap. However, the results of Chapter 3 cannot be automatically transferred to the case of an adiabatic trap. For this there are at least two reasons.

First, the low-frequency instabilities are sensitive to curvature of the lines of force. This follows from the analysis of Chapter 6, in which we simulated curvature by the introduction of a force of gravity. Using the results of Chapter 6 and the analogy between a force of gravity and curvature, we conclude that in the case of an ordinary adiabatic trap there must be additional destabilization of nonmagnetohydrodynamic perturbations, whereas in the case of a minimum-B configuration there may be partial or complete stabilization of these perturbations. We shall investigate this question quantitatively in §15.6.

Another reason why a further analysis of nonmagnetohydrodynamic perturbations of a plasma in an adiabatic trap is necessary is that an inhomogeneous plasma in a minimum-B configuration may be closer to thermodynamic equilibrium than in the case of a rectilinear field. This is due to a distinctive stabilizing effect which we shall discuss below.

We recall that the steady-state distribution function of each species of charge can be expressed as a function of the integrals of the energy, magnetic moment, and transverse coordinates:

$$f_0(\mathbf{r},\ \mathbf{v}) = F(\varepsilon,\ \mu,\ \mathbf{r}_\perp). \tag{15.14}$$

Here $\varepsilon = v^2/2$, $= \mu = v_\perp^2/2B$. On the transition to the approximation of a rectilinear homogeneous field, the coordinate dependence of f_0 is determined solely by the dependence of F on the integral $\mathbf{r}_\perp$.

In the case of a curved field, a coordinate dependence is also contained in μ. Moreover, one can conceive of a spatially inhomogeneous distribution with an F that does not depend on $\mathbf{r}_\perp$ at all:

$$f_0(\mathbf{r}, \mathbf{v}) = F(\varepsilon, \mu). \tag{15.15}$$

In §14.4 we have shown that in the case of a minimum-B field an f_0 of this kind can correspond to a spatial distribution of particles localized with respect to all three coordinates. A plasma with such an F is magnetohydrodynamically stable in accordance with §14.4 if

$$\partial F/\partial \varepsilon < 0. \tag{15.16}$$

We shall now show that if this condition is satisfied a plasma with f_0 of the form (15.15) is stable against all kinds of low-frequency perturbations for a very large class of distributions $F(\mu)$.

The equations that describe the low-frequency perturbations of a plasma in a curved field were derived in §15.2. We shall use the integral relation (15.9) derived in that section, assuming in it $\nabla F = 0$. [We recall that ∇F in (15.9) is taken with constant μ, so that $\nabla F = 0$ does not mean that the coordinate dependence of f_0 is ignored.]

We shall establish the conditions under which only solutions with $\operatorname{Im}\omega \le 0$ can satisfy this relation. Proceeding from the Nyquist condition (see §2.7 in Volume 1) and without as yet specifying the form of Q, we conclude that there is no instability, $\operatorname{Im}\omega \le 0$, if for all real ω

$$\operatorname{Im}\omega Q < 0 \tag{15.17}$$

and in addition

$$\lim_{(\omega\to\infty)} Q(\omega) > 0. \tag{15.18}$$

The imaginary part of Q is related to the poles of the function $1/\sin M(l_1, l_2)$. These poles correspond to resonance between the particles and the wave. The resonance condition, $\sin M(l_1, l_2) = 0$, entails

$$M(l_1, l_2) \equiv \int_{l_1}^{l_2} (\omega - \mathbf{k}\mathbf{V}_{\text{dr}})\, dl/v_\parallel = \pi n, \quad n = 0, 1, 2\ldots \tag{15.19}$$

In view of what we have said and using (15.9), we find that if f_0 has the form (15.15), then

$$\operatorname{Im} \omega Q = \pi\omega^2 \sum \frac{4\pi e^2}{m} \int d\alpha\, d\beta \int d\mu\, d\varepsilon \times$$

$$\times \left| \int_{l_1}^{l_2} J_0 \psi \cos M(l_1, l_2)\, dl/v_{\|} \right|^2 \delta[\sin M(l_1, l_2)]\, \partial F/\partial\varepsilon. \qquad (15.20)$$

It can be seen that the inequality (15.17) can be satisfied if (15.16) holds.

The second necessary condition for stability (15.18) when Q has the form (15.9) entails

$$\int d\alpha\, d\beta \left[\int \frac{|\nabla\psi|^2}{B}\, dl - \sum \frac{4\pi e^2}{m} \int d\varepsilon\, d\mu\, \frac{|\psi|^2 (1 - J_0^2)\, dl}{v_{\|}} \left(\frac{\partial F}{\partial\varepsilon} + \frac{\partial F}{B\, \partial\mu} \right) \right] > 0. \qquad (15.21)$$

It is certainly satisfied if

$$\frac{\partial F}{\partial\varepsilon} + \frac{1}{B} \cdot \frac{\partial F}{\partial\mu} < 0. \qquad (15.22)$$

In the approximation of a homogeneous magnetic field, the conditions (15.16) and (15.22) have a simple meaning. The first shows that F must decrease monotonically with increasing longitudinal velocity, $\partial F/\partial \ln v_{\|} < 0$. It is clear that otherwise beam instabilities associated with a resonance of the type $\omega = k_{\|} v_{\|}$ could develop. The second condition means that the distribution function must decrease monotonically with the transverse velocity, $\partial f_0/\partial v_{\perp}^2 < 0$. If this is not the case, the loss-cone effect may be manifested. In the approximation adopted above, we are concerned with the effect of nonmonotonicity of f_0 on the low-frequency instabilities, $\omega \ll \omega_B$, which include the Dory–Guest–Harris instability considered in §10.4 of Volume 1. As we then saw, it does not develop for every nonmonotonic distribution but only one such that $\langle (1 - J_0^2) \partial F/\partial\varepsilon_{\perp} \rangle > 0$. This necessary condition for instability also follows from (15.21).

§15.4. Some General Stability Conditions for Low-Frequency Perturbations

We shall now consider a plasma with a distribution function of the form (15.14). A plasma with such a distribution is more unstable than in the case (15.15). Nevertheless, if the distribution with respect to ε and μ is sufficiently favorable, stability can be achieved.

As in §15.3, we shall here be concerned with stability of the plasma only against low-frequency perturbations, $\omega \ll \omega_{B_i}$. The frequency of such perturbations satisfies the integral relation (15.9). We shall investigate this relation in the same manner as in §15.3, assuming, however, that we now have $\nabla F \neq 0$. In this case the expression for Im ωQ is similar to (15.20), except that in the integrand we must make the substitution

$$\frac{\partial F}{\partial \varepsilon} \to \frac{\partial F}{\partial \varepsilon} + \frac{\mathbf{k}_\perp}{\omega \omega_B} [\mathbf{e}_0, \nabla F]. \tag{15.23}$$

We go over in the expression for Im ωQ from the variables ε, μ, α, β to μ, α, β, $J_{||}$ – the variables introduced in the appendix to Chapter 12. The particle energy ε is now a function of the new variables, and the distribution function will depend on μ, $J_{||}$, and $\varepsilon(\mu, J_{||}, \alpha, \beta)$. On the transition to the new variables, the right-hand side of (15.23) takes the form

$$\left(\frac{\partial F}{\partial \varepsilon}\right)_{\mu, r} + \frac{\mathbf{k}_\perp}{\omega \omega_B} [\mathbf{e}_0, (\nabla F)_{\varepsilon, \mu}] = \left(\frac{\partial F}{\partial \varepsilon}\right)_{\mu, J_{||}} + \left(\frac{\partial F}{\partial J_{||}}\right)_{\mu, \varepsilon} \left(\tau + \frac{\mathbf{k}_\perp [\mathbf{e}_0, \nabla J_{||}]}{\omega \omega_B}\right), \tag{15.24}$$

and the expression (15.19) for $M(l_1, l_2)$ becomes

$$M(l_1, l_2) = \omega \left(\tau + \frac{\mathbf{k}_\perp [\mathbf{e}_0, \nabla J_{||}]}{\omega \omega_B}\right). \tag{15.25}$$

The expressions in the brackets in (15.24) and (15.25) are identical. The second term in these brackets bears the same relation to the first as the characteristic frequency of magnetic drift does to the oscillation frequency, $\mathbf{k}V_{dr}/\omega$.

We shall now consider separately the cases of perturbations when this term is small or of the order of the first.

<u>1. Perturbations with $\omega/k_\perp \gg V_{dr}$.</u> If the perturbation frequencies are such that $\omega \gg \mathbf{k}V_{dr}$, the second term in the brackets we have been discussing can be neglected and the sign of Im ωQ is then determined by the sign of the sum $(\partial F/\partial \varepsilon)_{\mu, J_{||}} + \tau(\partial F/\partial J_{||})_{\mu, \varepsilon} \equiv (\partial F/\partial \varepsilon)_{\mu, \alpha, \beta}$. Stability corresponds to a distribution for which

$$\left(\frac{\partial F}{\partial \varepsilon}\right)_{\mu, \alpha, \beta} < 0. \tag{15.26}$$

This stability condition is a generalization of (15.16) to the case of distribution functions that depend on the transverse coordinates. The resonance condition (15.19) is satisfied when $\omega > kV_{dr}$ only if $n \neq 0$, i.e., for particles for which

$$\tau = \pi n/\omega, \qquad n = 1, 2. \tag{15.27}$$

This kind of resonance is characteristic of gradient instabilities in the approximation of a rectilinear field in which the role of $\tau/\pi n$ is played by $k_z v_z$. Therefore, (15.26) can be interpreted as a condition of stability against gradient perturbations of the type of oblique waves.

2. Perturbations with $\omega/k_\perp \leq V_{dr}$. In this case the resonance $n = 0$ is also important. If, in addition, $k_\perp V_{dr} < 1/\tau$, i.e., $k_\perp \rho < R/L$, where R is the characteristic radius of curvature and L is the length of the system, this is the only resonance possible. Then the expression in the last bracket in (15.24) vanishes, so that the stability condition reduces to the form

$$\left(\frac{\partial F}{\partial \varepsilon}\right)_{\mu,\, J_{||}} < 0. \tag{15.28}$$

It is identical with the condition of magnetohydrodynamic stability (14.15).

§15.5. Flute Instabilities of a Plasma with a Finite Ion Larmor Radius

We now turn to the investigation of the various concrete types of perturbation of a plasma in an adiabatic trap. Some of them have already been considered using the energy method (see Chapter 14). These are the flute-type perturbations of a plasma with vanishing ion Larmor radius and belong to the class of magnetohydrodynamic instabilities. In this section we shall also consider flute-type perturbations but in the approximation of a finite ion Larmor radius.

We shall assume that during a time $\sim 1/\omega$ the ions can traverse the trap many times, $\omega \tau_i \gg 1$. For flute-type perturbations with $\omega \simeq v_{T_i}/(aR)^{1/2}$, this means $L < (aR)^{1/2}$, where R is the mean radius of curvature, L is the length of the trap, and a is the characteristic transverse dimension. To investigate such perturbations, we can use the integral relation (15.11). We assume $k_\perp \rho_i \ll 1$ but retain the terms of order $(k_\perp \rho_i)^2$. We represent the potential in the form $\psi = \bar{\psi} + \tilde{\psi}$. Using Poisson's equation and the expression (15.10) for the charge densities, we obtain an estimate for $\tilde{\psi}$:

$$\tilde{\psi} \simeq (k_\perp \rho_i)^2 \bar{\psi}. \qquad (15.29)$$

The contribution of terms with $\tilde{\psi}$ to the relation (15.11) is of the order $(k_\perp \rho_i)^4$, so that it can be ignored. As a result, (15.11) reduces to

$$Q = -\int d\alpha\, d\beta\, |\bar{\psi}|^2 \left\{ \frac{4\pi e^2}{m_i} \int d\mu\, d\varepsilon \left(\frac{\partial F_i}{\partial \varepsilon} + \right.\right.$$
$$\left. + \frac{1}{B}\frac{\partial F_i}{\partial \mu} + \frac{\mathbf{k}_\perp}{\omega\omega_{B_i}} [\mathbf{e}_0, \nabla F_i] + \frac{\mathbf{k}_\perp \bar{\mathbf{V}}_{\mathrm{dr}}^{(i)}}{\omega} \cdot \frac{\partial F_i}{\partial \varepsilon} \right) \int \frac{(k_\perp v_\perp)^2}{2\omega_B^2} \cdot \frac{dl}{v_\parallel} -$$
$$- \frac{1}{\omega^2} \sum_{i,\,e} \frac{4\pi e^2}{m} \int d\varepsilon\, d\mu \int \frac{dl}{v_\parallel} \mathbf{k}_\perp \bar{\mathbf{V}}_{\mathrm{dr}} \times$$
$$\left. \times \left(\frac{\mathbf{k}_\perp}{\omega_B} [\mathbf{e}_0, \nabla F] + \frac{\mathbf{k}\mathbf{V}_{\mathrm{dr}}}{\omega} \cdot \frac{\partial F}{\partial \varepsilon} \right) \right\} = 0. \qquad (15.30)$$

If the term with $1/\omega$ is ignored, we arrive at the condition of magnetohydrodynamic stability (14.15) obtained by the energy method in §14.4 and the identical condition (15.28). The term with $1/\omega$ describes the effect of the finite ion Larmor radius discussed for the case of a plasma in a gravitational field in §6.2.

Thus, the problem of the flute instability in a curved magnetic field reduces qualitatively to the problem of a plasma in a gravitational field. In such an analysis, however, we obtain additional information about the nature of the averaging of the steady-state parameters over the velocities and the space.

§15.6. Nonflute Perturbations

We shall now consider long-wavelength perturbations, $k_\perp \rho_i \ll 1$, with frequency satisfying the conditions

$$(1/\tau_i, \omega_{\mathrm{dr}}) \ll \omega \ll 1/\tau_e. \qquad (15.31)$$

Using (15.10) and (15.12) and taking into account the quasineutrality condition, we obtain the following integral relation in the zeroth approximation in the given small parameters [cf. (15.9)]:

$$\int d\alpha\, d\beta\, d\mu\, d\varepsilon \frac{\tau}{2} \left\{ |\tilde{\psi}|^2 \left(\frac{\partial F_e}{\partial \varepsilon} + \frac{\mathbf{k}_\perp}{\omega\omega_{Be}} [\mathbf{e}_0, \nabla F_i] \right) \right\} = 0. \qquad (15.32)$$

From this we find that, if $\tilde{\psi}$ is not identically equal to zero, the plasma can sustain a branch of oscillations with the frequency

$$\omega = -\frac{\int \tau \frac{\mathbf{k}_\perp}{\omega_{B_e}} [\mathbf{e}_0, \nabla F_i] \, |\tilde{\psi}|^2 \, d\alpha \, d\beta \, d\mu \, d\varepsilon}{\int \tau \frac{\partial F_e}{\partial \varepsilon} |\tilde{\psi}|^2 \, d\alpha \, d\beta \, d\mu \, d\varepsilon}. \tag{15.33}$$

This expression satisfies the approximation $\omega \gg \omega_{dr}$ adopted above if

$$\left(\frac{\partial \ln F}{\partial r_\perp}\right)_{\mu,\,\varepsilon} \gg \frac{\partial \ln B}{\partial r_\perp}, \tag{15.34}$$

i.e., only in situations that are very different from those discussed in §15.3, in which it was assumed that $(\partial \ln F/\partial r_\perp)_{\mu,\,\varepsilon} = 0$.

The branch (15.33) is the analog of the branch (3.13), which exists in the case of a homogeneous magnetic field. Oscillations of the type (3.13) grow or are damped when they interact with resonant particles. The same is true in the case of oscillations of the type (15.33). The instability condition obtained when allowance is made for the imaginary term with (15.13) does not have such a transparent form as in the case of a homogeneous magnetic field. Moreover, it can only be written down explicitly when the dependence $\psi(l)$ is known.

Bibliography

Effect of longitudinal inhomogeneity of the plasma and magnetic field on the high-frequency and cyclotron instabilities due to the loss cone

1. M. N. Rosenbluth and R. F. Post, Phys. Fluids, 8:547 (1965).
2. R. F. Post and M. N. Rosenbluth, Phys. Fluids, 9:730 (1966). In [1, 2] an estimate is made of the critical length of a trap in which the high-frequency cone instability can arise. The reflection of waves due to longitudinal inhomogeneity of the plasma is ignored.
3. R. E. Aamodt and D. L. Book, Phys. Fluids, 9:143 (1966). Like [1, 2], this paper is concerned with the critical length of the trap, but allowance is made in this case for the reflection of waves.
4. J. G. Cordey, Phys. Lett., 23:228 (1966).
5. C. O. Beasley, Phys. Fluids, 10:466 (1967).
6. C. O. Beasley and J. G. Cordey, Plasma Physics, 10:411 (1968).
7. G. E. Guest and R. A. Dory, Phys. Fluids, 11:1775 (1968).
8. C. O. Beasley et al., in: Plasma Physics and Controlled Nuclear Fusion Research, Vol. 2, IAEA, Vienna (1969), p. 141. It is shown in [4-8] that the ion-cyclotron oscillations at a sufficiently high plasma density are unstable absolutely and can therefore develop in a trap with shorter length than is the case for the high-frequency oscillations.
9. H. L. Berk et al., Phys. Fluids, 11:365 (1968).

10. H. L. Berk et al., in: Plasma Physics and Controlled Nuclear Fusion Research, Vol. 2, IAEA, Vienna (1969), p. 151. In [9, 10] the convective high-frequency instabilities are discussed with allowance for the effect of nonlocal reflection, and absolute cyclotron instabilities are considered with allowance for longitudinal inhomogeneity of the plasma.
11. A. B. Mikhailovskii and É. A. Pashitskii, Zh. Tekh. Fiz., 35:1960 (1965) [Sov. Phys. – Tech. Phys., 10:1507 (1966)]. Study of the effect of longitudinal inhomogeneity of the magnetic field on the excitation of ion-cyclotron oscillations by ions with a δ-function distribution.
12. J. G. Cordey, L. G. Kuo-Petravic, and M. Petravic, Nucl. Fusion, 8:153 (1968). Study of the effect of magnetic drift on the gradient-cone instability when $\omega \simeq \omega_{Bi}$.
13. L. V. Mikhailovskaya, ZhETF Pis. Red., 5:339 (1967) [JETP Letters, 5:279 (1967)]. It is shown that if allowance is made for magnetic drift there are growing cyclotron oscillations of flute type (with $k_{\parallel} = 0$) even if $\nabla n_0 = 0$.
14. B. B. Kadomtsev and O. P. Pogutse, in: Plasma Physics and Controlled Nuclear Fusion Research, Vol. 2, IAEA, Vienna (1969), p. 125.
15. J. F. Clarke et al., ibid., Vol. 2, p. 291.
16. J. F. Clarke and G. G. Kelley, Phys. Rev. Lett., 21:1041 (1968). It is shown in [14-16] that in an anisotropic plasma in a longitudinally inhomogeneous magnetic field ion-cyclotron oscillations with $k_{\parallel} = 0$ can develop.
17. V. V. Arsenin, Zh. Tekh. Fiz., 37:614 (1967) [Sov. Phys. – Tech. Phys., 12:442 (1967)].
18. M. Cotsaftis, Phys. Lett., 25A:170 (1967).
19. M. Cotsaftis, Nucl. Fusion, 7:3 (1967). In [17-19] a study is made of the influence of end effects (reflection of particles from the ends of the trap) on the excitation of ion-cyclotron and high-frequency oscillations.

Low-frequency gradient instabilities

20. J. B. Taylor, Phys. Fluids, 6:1529 (1963). It is shown that a plasma with a distribution function of the form $f(\mu, \varepsilon)$ is stable against all perturbations in which the magnetic moment μ is conserved if $\partial f/\partial \varepsilon < 0$.
21. M. N. Rosenbluth and N. A. Krall, Phys. Fluids., 8:1004 (1965).
22. A. A. Ivanov, Dokl. Akad. Nauk SSSR, 166:1084 (1966) [Sov. Phys. – Doklady, 11:145 (1966)].
23. A. A. Ivanov, Zh. Tekh. Fiz., 37:229 (1967) [Sov. Phys. – Tech. Phys., 37:162 (1967)]. In [21-23] concrete examples demonstrate the stability of a plasma with a distribution function $f(\mu, \varepsilon)$ having a negative derivative $\partial f/\partial \varepsilon$ in a minimum-B field.
24. H. K. Wimmel and R. Saison, Phys. Lett., 23:449 (1966).
25. R. Saison and H. K. Wimmel, Z. Naturforsch., 22a:281 (1967). In [24-25] it is shown that a plasma with $\partial f(\mu, \varepsilon)/\partial \varepsilon < 0$ may be unstable against low-frequency perturbations if the condition $\frac{\partial f}{\partial \varepsilon} + \frac{\partial f}{B \partial \mu} < 0$ is not satisfied.
26. J. B. Taylor, Phys. Fluids, 10:1357 (1967). A direct calculation shows that in short-wavelength low-frequency perturbations the magnetic moment is not conserved. This proves that there is no contradiction between the result of [24, 25] and the theorem of [20].

27. P. H. Rutherford and E. A. Frieman, Phys. Fluids, 11:252 (1968). Formulation of an energy principle applicable to perturbations with arbitrary transverse wavelength. The sufficient conditions for stability (15.16) and (15.22) are derived as a consequence.
28. P. H. Rutherford and E. A. Frieman, Phys. Fluids, 11:569 (1968).
29. J. B. Taylor and R. J. Hastie, Plasma Physics, 10:479 (1968). In [28, 29] the transport equation is solved in a field of complicated geometry. The corresponding results of these investigations (especially [29]) were used in §15.2. The sufficient conditions for stability (15.16) and (15.22) are obtained for a distribution of the form $f(\varepsilon, \mu)$. In [29] there is also an analysis of the stability of a plasma with a distribution function of a general form. This analysis is given in §15.4.
30. J. M. Greene and B. Coppi, Phys. Fluids, 8:1745 (1965).
31. V. V. Arsenin and A. V. Timofeev, Zh. Tekh. Fiz., 37:1244 (1967) [Sov. Phys. – Tech. Phys., 12:904 (1968)]. In [30-31] a study is made of the influence of dissipative effects on the growth of flute-type perturbations in adiabatic traps.

Chapter 16

Plasmas in Multipole Traps

§16.1. Equilibrium and Magnetohydrodynamic Stability of Plasmas in Multipole Traps

The geometry of a linear multipole trap is shown in Fig. 16.1. A magnetic field with components B_r and B_φ is produced by currents that flow in the same direction along conductors situated inside the plasma. Taking into account the relation div B = 0, one can characterize the magnetic field by a flux function $\Psi(r, \varphi)$ such that

$$B_r = (1/r)\, \partial\Psi/\partial\varphi, \qquad B_\varphi = -\partial\Psi/\partial r. \tag{16.1}$$

As can be seen in Fig. 16.1, each line of force that passes around the axis of the system is closed after one circuit. Therefore, the condition of equilibrium of a zero-pressure plasma in such a trap reduces to

$$[\nabla p, \nabla U] = 0, \tag{16.2}$$

where $U = \int dl/B$ is an integral taken along the length of the line of force. If instead of cylindrical coordinates (r, φ) we use orthogonal coordinates (Ψ, χ), where Ψ is the magnetic flux passing through the separatrix and the corresponding line of force and χ has the meaning of a generalized azimuth, $\nabla\chi\nabla\Psi = 0$ (see Fig. 16.1), it follows from geometrical considerations that U must depend only

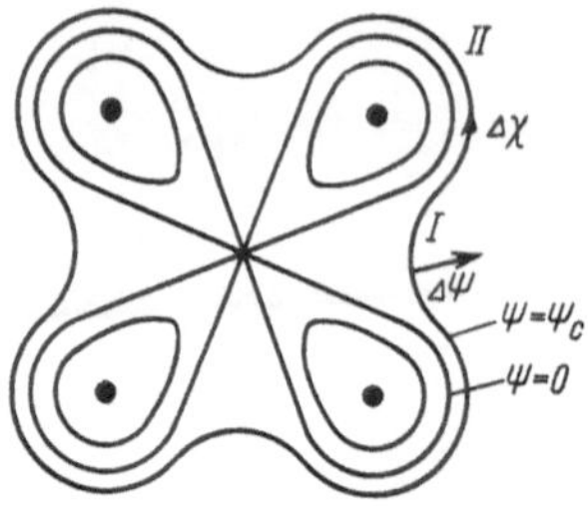

Fig. 16.1. Magnetic field configuration in a multipole trap. The line $\Psi = 0$ corresponds to the separatrix.

on Ψ, $U = U(\Psi)$. Therefore, $\nabla U \parallel \nabla\Psi$, and we then find from (16.2) that the pressure must also depend only on Ψ:

$$p = p(\Psi). \tag{16.3}$$

The result (16.3) is also true in the case of finite β. In this case it can be obtained by forming the vector product of $\nabla\Psi$ and the equilibrium equation

$$4\pi\nabla p = [\mathrm{rot}\,\mathbf{B}, \mathbf{B}]. \tag{16.4}$$

1. Magnetohydrodynamic Stability of a Zero-Pressure Plasma in a Multipole Trap. As we have already noted in §11.2, the condition for the stability of a zero-pressure plasma with an isotropic particle velocity distribution against the flute instability has the form

$$\nabla p \nabla U \geqslant 0. \tag{16.5}$$

This result was justified by means of an energy treatment in §14.1. We shall show that for a multipole trap this condition is satisfied. Since $dp/d\Psi < 0$, it is sufficient to show that $dU/d\Psi < 0$.

The dependence $U = U(\Psi)$ can be established qualitatively by geometrical arguments. As can be seen from Fig. 16.1, the mag-

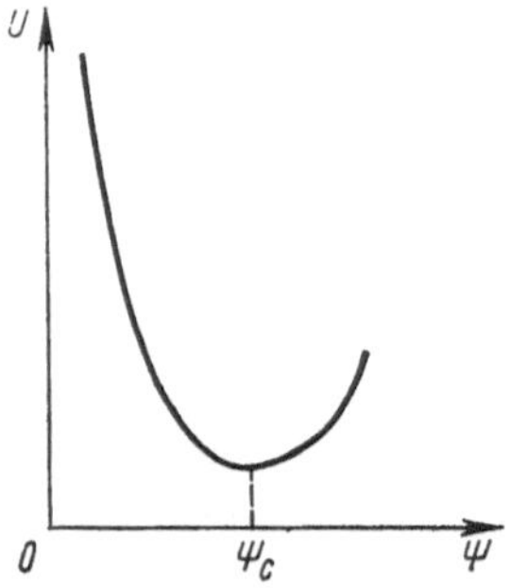

Fig. 16.2. Dependence of U on Ψ.

netic field vanishes at $\Psi = 0$, which corresponds to the lines of force of the separatrix. Therefore, $U(0) = \infty$ (Fig. 16.2). As one moves away from the separatrix, the magnetic field is nonvanishing and $U(\Psi)$ decreases. This decrease occurs until some $\Psi = \Psi_c$, after which $U(\Psi)$ again increases (see Fig. 16.2). If a vessel wall is placed at $\Psi = \Psi_c$ (see Fig. 16.1), then $dU/d\Psi < 0$ everywhere between the vessel wall and the separatrix, and this corresponds to stability.

2. Balloon-Mode Instability of a Finite-Pressure Plasma. As in the case of a stabilized probkotron, the field of a multipole trap does not increase at every point toward the periphery but only on the average. At the same time, only the region I in Fig. 16.1 makes a favorable contribution to $dU/d\Psi$, whereas the region II is unfavorable. It is therefore clear that if a perturbation is localized entirely in region II it "feels" only the unfavorable field curvature and an instability may therefore develop with a growth rate of order

$$\gamma \simeq v_{T_i}/(aR)^{1/2}, \tag{16.6}$$

where a is the characteristic scale of the pressure gradient and R is the radius of curvature of the field in region II. This is the balloon-mode instability. This kind of perturbation distorts the magnetic field, which may suppress the instability (see §§6.2.1 and 11.5.2) if

$$\gamma \leqslant c_A k_{\|} \simeq \pi c_A/L, \tag{16.7}$$

where L is the length of the region of unfavorable curvature. It follows from (16.6) and (16.7) that the balloon-mode instability does not develop if

$$\beta \leqslant \pi^2 aR/L^2. \tag{16.8}$$

This condition determines the maximal pressure of a magnetohydrodynamically stable plasma in a multipole trap.

§16.2. General Comments on the Possible Nonmagnetohydrodynamic Instabilities of a Plasma in Multipole Traps

In contrast to a probkotron plasma, the velocity distribution of a plasma confined in a multipole trap does not have a "deleted" cone. Such a plasma may therefore have a Maxwellian velocity distribution of its particles and therefore be free of the high-frequency and ion-cyclotron instabilities of cone type predicted for a plasma.

However, in a plasma confined in a multipole trap other types of nonmagnetohydrodynamic instability may arise – slow gradient instabilities due to density and temperature gradients. In the approximation of a rectilinear and homogeneous magnetic field such instabilities have been discussed in Chapters 2 and 3 (collisionless plasma) and in Chapter 5 (collisional plasma). In considering a model with a force of gravity in Chapter 6, we have noted that these instabilities can be either amplified or suppressed by magnetic drift. It is therefore to be expected that magnetic drift will play an important role in a multipole field that, in general, has a curvature comparable with the inhomogeneity scale of the density.

In accordance with Chapters 8 and 9, a shear of the magnetic field also has an important stabilizing effect. However, in the case considered here – multipoles without longitudinal magnetic field – there is no shear. In this sense, both multipole and adiabatic traps must be included in the class of systems with an incomplete set of stabilizing factors, in contrast to, for example, various types of closed traps with magnetic surfaces discussed in the following chapters.

The magnetic field of multipole traps is inhomogeneous along a line of force. Therefore, in accordance with §11.9, instabilities due to the presence of trapped particles may develop in such systems.

In the following sections we shall establish which of the possible types of instabilities mentioned above can actually occur in the specific magnetic field geometry of a multipole trap.

§16.3. Low-Frequency Instabilities of a Plasma Confined in a Multipole Trap for $\rho_i/a < a/L$

In the approximation of a longitudinally homogeneous magnetic field we have shown that gradient instabilities do not develop if $\omega^* < k_{\|} v_{T_i}$ (see Chapter 3). In the case of perturbations with

$k_\perp a \simeq 1$ this means that the length L of the device must be relatively short:

$$\rho_i/a < \pi a/L. \tag{16.9}$$

Let us now consider the problem of gradient instabilities of a plasma in a multipole trap assuming that the parameters of the field and the plasma satisfy this condition.

If the condition (16.9) is satisfied, perturbations with $\omega \leq \omega^*$ must be regarded as slow compared with the period of the motion of the particles along the lines of force, $\omega\tau_i < 1$. In this case, the perturbation of the potential satisfies the integral relation (15.11). Using the latter, we can obtain in the same manner as in §15.4 a sufficient condition for stability. Assuming $\partial F/\partial\varepsilon < 0$, which is true, for example, for a Maxwellian F and $\partial F/\partial\varepsilon + \partial F/B\partial\mu < 0$ (this excludes the class of strongly "cone-like" distributions) and remembering that F and $\bar{\omega}_{dr}$ depend solely on Ψ (see Fig. 16.1), we find that the plasma is stable if for all ε and μ

$$\frac{\partial F}{\partial \Psi} \cdot \frac{\partial J_{||}}{\partial \Psi} > 0, \tag{16.10}$$

where $J_{||} = \int v_{||}\, dl.$ Here we have used Eqs. (A.12.11), which relate the magnetic drift V_{dr} of a particle averaged along the line of force to its longitudinal adiabatic invariant $J_{||}$.

If F is Maxwellian and $\nabla T = 0$ the condition (16.10) means

$$\frac{\partial n_0}{\partial \Psi} \cdot \frac{\partial J_{||}}{\partial \Psi} > 0. \tag{16.11}$$

If the density decreases with Ψ, $\partial n_0/\partial\Psi < 0$, it follows from (16.11) that the plasma is stable if for all of its particles

$$\partial J_{||}/\partial \Psi < 0. \tag{16.12}$$

Devices in which this is so are called $\max J_{||}$ systems.

If the velocity distribution is Maxwellian but $\nabla T \neq 0$, Eq. (16.10) does not yield (16.11) but

$$\frac{\partial \ln n_0}{\partial \Psi} \cdot \frac{\partial J_{||}}{\partial \Psi} \left[1 - \left(\frac{3}{2} - \frac{mv}{2T}\right)^2 \eta\right] > 0, \tag{16.13}$$

where $\eta \equiv \partial \ln T/\partial \ln n_0$. This means that for

$$\eta > 2/3 \tag{16.14}$$

a $\max J_{||}$ system can, in general, be unstable.

Following Rosenbluth, we now establish whether the sufficient condition (16.22) for stability can be satisfied in the case of the multiopole field of the type shown in Fig. 16.1. Using (A.12.6), we make the following identical transformations:

$$\frac{1}{\sqrt{\varepsilon}}\cdot\frac{\partial J_{\|}}{\partial\Psi}=\int\frac{dl}{R}B\frac{d}{dB}\frac{\sqrt{1-\alpha B}}{B}=-\int dBB\frac{d\alpha}{dB}\frac{d}{dB}\frac{\sqrt{1-\lambda B}}{B}. \quad (16.15)$$

Here $\lambda \equiv \mu/\varepsilon$ and α is the angle between the line of force and some fixed direction on the xy plane. Referring to Fig. 16.1, we see that $d\alpha/dB \to -\infty$ near the points $B = B_{max}$ and $d\alpha/dB \to +\infty$ near $B = B_{min}$. The contribution to the integral from the region with $B \simeq B_{max}$ is greater than that from the region with $B \simeq B_{min}$ for particles for which

$$\lambda \simeq 1/B_{max}. \quad (16.16)$$

For such particles the condition opposite to (16.12) is satisfied and these particles may therefore lead to instability.

That instability may really occur can be seen from the following treatment. Assume that F_e and F_i are Maxwellian with $T_e = T_i \equiv T$ and that $\nabla T = 0$. Consider perturbations with $\omega_{dr} < \omega < \omega^*$ and $k_\perp\rho_i \ll 1$. It then follows from (15.11) that

$$\omega^2 = -\frac{\omega^*\int d\mu\, d\varepsilon\tau F\bar{\omega}_{dr}\,|\bar{\psi}|^2}{\int d\mu\, d\varepsilon\tau F\,(\overline{|\psi|^2}-|\bar{\psi}|^2)}. \quad (16.17)$$

Here the denominator is not negative, so that the sign of the square of the frequency is determined by the sign of the numerator. In the case of flute-type perturbations, when $\bar{\psi} \approx \psi$, the integration in the numerator reduces to averaging of the magnetic drift over all the particles and gives $\omega^2 > 0$ in agreement with §§16.1 and 11.2. However, we consider perturbations localized near the point $\chi = \chi_0$ at which $B(\chi_0) = B_{max}$. Assuming for simplicity $\psi \sim \delta(\chi-\chi_0)$, we obtain

$$\bar{\psi} \sim \int\frac{\psi\, dl}{v_{\|}} \sim \frac{1}{\sqrt{1-\lambda B_{max}}}. \quad (16.18)$$

It can be seen that the largest contribution to the integral of the numerator (16.17) arises from particles with $\lambda \equiv \mu/\varepsilon \simeq 1/B_{max}$.

But by what has been said above such particles have an unfavorable magnetic drift. We therefore now have $\omega^2 < 0$, which corresponds to instability. Approximately

$$\gamma \simeq (\omega_{dr}\omega^*)^{1/2} \simeq \omega^* (a/R)^{1/2}. \quad (16.19)$$

Thus, even if the condition (16.9) holds, a multipole trap is unstable. This instability is due to the presence of trapped particles that reach the region with $B = B_{max}$ and a small fraction of untrapped particles whose velocity at $\chi = \chi_0$ is small compared with the thermal velocity. The instability mechanism is similar to that discussed in §11.9.

§16.4. Low-Frequency Instabilities of a Plasma Confined in a Multipole Trap for $\rho_i/a > a/L$

In contrast to §16.3, we shall now assume that the plasma is fairly long so that

$$\rho_i/a > \pi a/L. \quad (16.20)$$

If this condition is satisfied, perturbations with $\omega \simeq \omega^*$ and $k_\perp a \simeq 1$ are fast compared with the ion circulating time, $\omega\tau_i > 1$. It is clear that the inequality $\omega\tau_i > 1$ can also be satisfied when the condition opposite to (16.20) holds if $k_\perp a \gg 1$. It follows that some of the results obtained in this section in the approximation $\omega\tau_i > 1$ are also applicable to the case of a short plasma (discussed in §16.3).

If $|\partial \ln T/\partial \ln n_0| \geq 1$, then in a plasma confined in a multipole trap, as also in a longitudinally homogeneous plasma, a gradient-temperature instability with $\gamma \simeq \mathrm{Re}\,\omega \simeq \omega^*$ (see §3.2) can develop; estimates show that this must lead to turbulent losses of the order of the Bohm loss.

It follows from the analysis of Chapters 3 and 6 that a plasma with $\nabla T = 0$ is more stable than if $\nabla T \neq 0$. However, instabilities can still arise if $\nabla T = 0$ with growth rate proportional to the small parameters $(k_\perp \rho_i)^2$ (see §3.1) or a/R (see §6.5). These results were obtained in the approximation of a longitudinally homogeneous magnetic field. Let us now consider what happens in a longitudinally inhomogeneous field, in particular, in the field of a multipole trap.

We shall assume that the perturbation frequency lies in the range

$$(\omega_{\mathrm{dr}},\ \tau_i^{-1}) < \omega < (\tau_e^{-1},\ \omega_{B_i}), \tag{16.21}$$

and that the wave number $k_\perp$ is small compared with $1/\rho_i$, $k_\perp \rho_i \ll 1$. We shall first establish which oscillation branches exist under these conditions and then take into account the interaction between them and the resonant electrons and estimate the growth rate due to this effect.

1. Oscillation Branches. Proceeding from the expressions (15.10) for the electron density and (15.12) for the ion density and expanding in these expressions with respect to the small parameters mentioned above, we write down the quadratic form Q [cf. (15.9) and (15.11)]:

$$Q = \int d\alpha\, d\beta \left\{ \frac{4\pi e^2}{T_e}\left(1 - \frac{\omega_{ne}}{\omega}\right) \int \frac{\tau}{2} F_e\, d\mu\, d\varepsilon \times \right.$$
$$\times \left[\overline{|\psi|^2} - |\bar{\psi}|^2 \left(1 + \frac{\bar{\omega}_{\mathrm{dr.e}}}{\omega}\right)\right] + \frac{4\pi e^2}{T_i}\left(1 - \frac{\omega_{ni}}{\omega}\right) \times$$
$$\left. \times \int \frac{dl}{B}\left[\left(k_\perp^2\rho_i^2 - 2\frac{\omega_{ni}}{\omega}\cdot\frac{\partial \ln B}{\partial \ln n_0}\right)|\psi|^2 - \frac{v_{T_i}^2}{2\omega^2}\left|\frac{\partial \psi}{\partial l}\right|^2\right]\right\} = 0. \tag{16.22}$$

Approximately, this equation can be satisfied when $\psi \approx \bar{\psi}$ or when $\omega \approx \omega_{ne}$. The first possibility corresponds to flute-type perturbations. In this case, (16.22) reduces to the dispersion equation

$$\int \frac{dl}{B}\left[\left(1 - \frac{\omega_{ni}}{\omega}\right)k_\perp^2\rho_i^2 + 2\left(\frac{\omega_{ne}}{\omega}\right)^2\left(1 + \frac{T_i}{T_e}\right)\frac{\partial \ln B}{\partial \ln n_0}\right] = 0. \tag{16.23}$$

The second possibility corresponds to nonflute type perturbations with a frequency that differs from ω_{ne} by a small quantity of order $(k_\perp \rho_i)^2$ or a/R:

$$1 - \frac{\omega_{ne}}{\omega} = -\frac{\left(1 + \frac{T_e}{T_i}\right)\int \frac{dl}{B}\left[\left(k_\perp^2\rho_i^2 + 2\frac{T_i}{T_e}\cdot\frac{\partial \ln B}{\partial \ln n_0}\right)|\psi|^2 - \frac{T_i}{m_i(\omega_{ne})^2}\left|\frac{\partial \psi}{\partial l}\right|\right]^2}{\int \frac{\tau}{2} F_e\, d\mu\, d\varepsilon\, \left(\overline{|\psi|^2} - |\bar{\psi}|^2\right)}. \tag{16.24}$$

This is the analog of the branch of oblique gradient perturbations described by the dispersion equation (3.12).

2. Interaction between Oscillations and Resonant Electrons. The resonant particles can be taken into

account by adding to (16.22) the imaginary terms obtained when a contour is taken around the singularity in the denominator of the perturbed distribution function (cf. §§15.3 and 15.4). Since $\omega \gg (\omega_{dr.i}, \tau_i^{-1})$, the number of resonant ions is exponentially small, as in the approximation of a homogeneous magnetic field (cf. §3.1). As for the electrons, they satisfy approximately $\omega_{dr.e} < \omega < \tau_e^{-1}$. It follows that the contribution to the growth rate resulting from the transverse electron resonance, $\omega = \bar{\omega}_{dr}$, is exponentially small like the ion contribution. Longitudinal resonance, $\omega = 2\pi n/\tau_e$, is realized by electrons whose energy is small compared with the thermal energy as

$$m_e \varepsilon / T \simeq (\omega L / v_{T_e})^2. \tag{16.25}$$

Since $f_e \sim 1/v_{T_e}^3$ for small ε, the number of such electrons is small as $(\omega L/v_{T_e})^3$. This is a qualitative difference from the case of a homogeneous magnetic field, in which only the longitudinal distribution function occurs, this being small as $1/v_{T_e}$ for small $v_{\parallel}$. Therefore, the contribution to the growth rate due to the resonant electrons now contains the factor

$$\gamma \sim (\omega L / v_{T_e})^3 \tag{16.26}$$

in contrast to

$$\gamma \sim \omega / k_z v_{T_e}, \tag{16.27}$$

which we had in the approximation of a longitudinally homogeneous field.

The interaction between the resonant electrons and the oscillations can lead to growth of the latter only if $1 - \omega_{ne}/\omega < 0$ (cf. §3.1). At not too small $k_\perp$, this condition is satisfied by one of the branches of flute-type perturbations (16.23). The nonflute branch (16.24) grows if the integrand of the numerator of the right-hand side is positive, i.e., if

$$\left(\frac{v_{T_i}}{\omega_{ne}}\right)^2 \int \frac{dl}{B} \left|\frac{\partial \psi}{\partial l}\right|^2 < \int \frac{dl}{B} |\psi|^2 \left(k_\perp^2 \rho_i^2 + 2\frac{T_i}{T_e} \cdot \frac{\partial \ln B}{\partial \ln n_0}\right). \tag{16.28}$$

Qualitatively, this corresponds to the same upper bounds on the plasma length as those found in §§3.1 and 6.2. An estimate for the growth rate of the nonflute perturbations can be obtained by using the results of §3.1 and taking into account the difference between the number of resonant particles characterized by the relations (16.26) and (16.27).

Bibliography

Magnetohydrodynamic stability of multipole traps

1. S. I. Braginskii and B. B. Kadomtsev, in: Plasma Physics and the Problem of Controlled Thermonuclear Reactions, Vol. 3, Pergamon Press, Oxford (1959), p. 356. It is suggested that guard conductors could have a stabilizing effect. A plasma with a sharp boundary is considered.
2. B. B. Kadomtsev, in: Plasma Physics and the Problem of Controlled Thermonuclear Reactions, Vol. 4, Pergamon Press, Oxford (1960) (in the articles beginning p. 417). Discussion of the stability of a low-pressure plasma.
3. T. Ohkawa and D. W. Kerst, Phys. Rev. Lett., 7:41 (1961).
4. T. Ohkawa and D. W. Kerst, Nuovo Cimento, 22:784 (1961).
5. T. Ohkawa and N. Rostoker, Phys. Today 20(12):49 (1967). In [3-5] a study is made of the critical β that is stable against balloon-type perturbations.

Nonmagnetohydrodynamic instabilities

6. T. Ohkawa and M. Yoshikawa, Phys. Rev. Lett., 17:685 (1966). It is pointed out that a magnetohydrodynamically stable plasma in a multipole trap could serve as a suitable object for the investigation of nonmagnetohydrodynamic instabilities. The gradient-cyclotron instability is discussed.
7. B. B. Kadomtsev, ZhETF Pis. Red., 4:15 (1966) [JETP Letters, 4:10 (1966)]. It is noted that in traps with variable curvature of the magnetic field a trapped-particle instability can develop if $J_{\parallel}$ increases as the periphery of the plasma is approached (in min $J_{\parallel}$ systems).
8. M. N. Rosenbluth, Phys. Fluids, 11:869 (1968). It is shown that multipole traps do not belong to the class of max $J_{\parallel}$ systems and they may therefore sustain a trapped-particle instability.
9. P. H. Rutherford and E. A. Frieman, Phys. Fluids, 11:252 (1968). The energy principle is formulated for low-frequency nonmagnetohydrodynamic perturbations and is used to obtain the sufficient condition for stability $\frac{\partial F}{\partial \psi}\frac{\partial J_{\parallel}}{\partial \psi} > 0$ [see condition (16.10)].
10. P. H. Rutherford and E. A. Frieman, Phys. Fluids, 11:569 (1968). Detailed analysis of the instability of a plasma in multipole traps with allowance for trapped particles and other effects associated with the longitudinal inhomogeneity of the magnetic field. It is assumed that $\nabla T = 0$.
11. B. Coppi, M. N. Rosenbluth, and P. Rutherford, Phys. Rev. Lett., 21:1055 (1968). Investigation of various types of instability with $\gamma \simeq \mathrm{Re}\,\omega$.
12. P. Rutherford et al., in: Plasma Physics and Controlled Nuclear Fusion Research, Vol. 1, IAEA, Vienna (1969), p. 367. Systematic exposition of the theory of low-frequency instabilities of a plasma in multipole traps.
13. B. Coppi, Phys. Lett., 28A:518 (1969).
14. B. Coppi, Phys. Lett., 28A:685 (1969).
15. O. P. Pogutse, Zh. Eksp. Teor. Fiz., 52:1536 (1967) [Sov. Phys. JETP, 25:1021 (1967)]. In [13-15] some specific types of instabilities of systems with a longitudinally inhomogeneous magnetic field are investigated.

16. A. B. Mikhailovskii, Nucl. Fusion, 5:125 (1965).
17. N. A. Krall and T. K. Fowler, Phys. Fluids, 10:1526 (1967).
18. D. K. Bhadra, Phys. Rev., 161:126 (1967).
19. D. K. Bhadra, Phys. Rev., 171:188 (1968). The gradient-cyclotron instability is discussed in [16-19].
20. T. Ohkawa and N. Rostoker, see [5]. Review of the results on nonmagnetohydrodynamic instabilities.

Chapter 17

Plasmas in Closed Traps with Magnetic Surfaces and No Current

§17.1. General Comments on Magnetohydrodynamic Stability

Closed traps with magnetic surfaces can be divided into two large classes. In one the plasma is confined by a magnetic field induced by currents in conductors situated outside or inside the plasma (systems without current). In the other an important role is played by the current that flows in the plasma (current systems). In this chapter we shall briefly discuss the stability of systems of the first type; in the next chapter, of those of the second type.

Magnetohydrodynamic stability of closed systems without current depends on the sign of the mean magnetic drift of the particles and the shear. As we have already noted in §11.8.1, the mean magnetic drift of particles in the case of systems with magnetic surfaces is characterized by $V''(\Phi)$, the second derivative of the volume with respect to the longitudinal magnetic flux. In accordance with §11.8.2, the shear is associated with the derivative $(d/dV)(\chi'/\Phi')$, where χ is the azimuthal magnetic flux.

If $V'' > 0$, magnetic drift leads to a flute instability. However, the latter cannot develop if the shear is sufficiently large. For a given shear and V'', a plasma can be confined stably in a trap with $V'' > 0$ only if its pressure is not too high. This conclusion follows from the investigation in §11.6 of the idealized case of a system with a shear and an unfavorable magnetic drift [see, for example, the condition (11.78)].

In systems with $V'' < 0$, a flute instability is impossible irrespective of the magnitude of the shear. In the case of a plasma with sufficiently low pressure this leads to magnetohydrodynamic stability as a whole. However, as the pressure increases, the magnetohydrodynamic perturbations of nonflute type – the balloon mode (see §11.5.2) – increase in importance. Perturbations of this kind are localized in regions with unfavorable magnetic drift – the cause of the instability. The instability condition can be characterized qualitatively by (11.44).

A great many theoretical investigations have been devoted to the magnetohydrodynamic stability of plasma systems without current (see the bibliography at the end of this chapter). The simplest of these results are discussed in §§17.2 and 17.3.

§17.2. Magnetohydrodynamic Instability of Classical Stellarators

It is not in every system with magnetic surfaces that the mean curvature of the lines of force has a favorable sign and the shear is sufficiently large. The simplest example of closed systems with magnetic surfaces is a plasma cylinder surrounded by helical windings and bent into a torus (see §11.7). The main function of the helical windings becomes clear when the cylinder is closed – the magnetic field of these windings, added to the longitudinal field, gives rise to a configuration with magnetic surface, needed for equilibrium of the plasma. (This configuration corresponds to one of the forms of a stellarator.) At the same time, the helical windings cause the magnetic field to be curved and, as can be seen from §11.7, the mean radius of curvature is directed into the plasma, $V'' > 0$. This curvature may give rise to a flute instability. From this point of view, the presence of the helical windings is not advantageous.

However, the helical windings lead not only to a curvature of the lines of force but also to a shear, which helps to suppress the flute instability. This suppression effect can be characterized by the generalized Suydam's condition (11.78). It can be seen from the latter, that, as in the case of an ordinary cylindrical plasma [cf. (13.5)], the flute instability does not develop if β is sufficiently small

$$\beta < \beta_{crit}. \tag{17.1}$$

In accordance with (11.78), the value of β_{crit} depends on the number of pairs of helical windings, their pitch, and the current that flows in them. The analysis of Johnson et al. shows that the largest value of β is obtained for $l = 3$ and that in this case

$$\beta_{crit} \simeq 0.1. \tag{17.2}$$

Among the closed systems with magnetic surfaces that do not have a shear we may mention the stellarator proposed by Spitzer in the form of a spatial figure of eight. The curvature of the lines of force in this form of stellarator is directed inward on the average, $V'' > 0$. Since there is no shear, a flute instability can develop at even very small values of β.

§17.3. Systems with V'' < 0

We have shown in §11.4 that the superposition of a multipole field on an axisymmetric field enables one to make an adiabatic trap that is free of the flute instability. The idea of using a multipole field also formed the basis of the method for making closed traps with $V'' < 0$ proposed by Furth and Rosenbluth. The scalar potential of the magnetic field Ψ is then taken in the form

$$\Psi = \int f\,dz - \frac{1}{4} f' r^2 + \sum_{l=1}^{\infty} \frac{1}{l} g_l r^l \cos(l\theta). \tag{17.3}$$

It is assumed that f and g are certain arbitary periodic functions of z. In the case of a quadrupole field, $g_l = 0$, $l \neq 2$, satisfying the periodicity conditions $f(z + \pi) = f(\pi - z)$ and $g_2(z) = -g_2(z + \pi) = -g_2(\pi - z)$, we can, using (17.3) and proceeding as in §§11.3 and 11.4, obtain the stability condition

$$\int_0^{\pi} \frac{dz}{f^4} \left(\frac{f'}{2} + g_2\right) \left(\frac{3}{2} f' - g_2\right) \exp\left(2 \int_0^z dz_1 \frac{g_2}{f}\right) < 0. \tag{17.4}$$

It follows that a stabilizing contribution arises from the regions where $g_2 \approx f'$. In addition to this region, there are regions in which $g_2 \approx -f'$. They have a destabilizing effect. The integral as a whole can be made negative by an appropriate choice of the weight factor $\exp\left(2 \int_0 dz_1 g_2/f\right)$. The numerical calculation of Furth and Rosenbluth shows that a depth of the magnetic well of the order of 1% can be achieved in this manner.

More promising methods for increasing $-V''$ are afforded by the use of magnetic fields with a curved magnetic axis. The most general results concerning the stability of systems with a curved magnetic axis have been obtained by Solov'ev and Shafranov. We give one of the simplest results of these authors, which applies to a two-turn stellarator with a circular magnetic axis. The scalar potential of such a field near the axis has the form

$$\Psi = B_0 s + \frac{1}{2} B_0 \varepsilon \delta' \rho^2 \sin 2(\omega - \delta' s), \tag{17.5}$$

where s is the path length along the axis; ρ and ω are polar coordinates in the plane s = const; δ' and ε are certain constants that characterize the pitch of the winding and the ellipticity of the magnetic surfaces. It is assumed that the magnetic axis has constant curvature k.

Calculation of the function $V''(\Phi)$, whose form cannot be given here for reasons of space, for Ψ of the form (17.5) yields

$$V''(\Phi) = \frac{1}{\pi B_0^2 \sqrt{1-\varepsilon^2}} \left(\varepsilon^2 \delta'^2 - \frac{11}{2} k^2 \right). \tag{17.6}$$

A low-pressure plasma must be stable in such a trap, $V'' < 0$, if

$$\varepsilon^2 < \frac{11}{2} \left(\frac{k}{\delta'} \right)^2. \tag{17.7}$$

According to Shafranov, the depth of the magnetic well may be increased when the magnetic trap is filled with plasma. A magnetic well may be formed even if there is no vacuum magnetic well. This shows that it may be simpler to achieve magnetohydrodynamic stability of a plasma with finite β in closed systems than in the case $\beta \to 0$.

§17.4. Nonmagnetohydrodynamic Instabilities

In magnetohydrodynamically stable closed systems (see §17.3), nonmagnetohydrodynamic instabilities of the type considered in §§16.1 and 16.2 can develop. There must also be some other forms of nonmagnetohydrodynamic instability peculiar to the class of fields with a longitudinal gradient, i.e., with $(\mathbf{B}\nabla) B \neq 0$. A typical representative of this kind of instability is the trapped-particle instability discussed in §§11.9 and 16.3.

The presence of trapped particles in a system leads not only to this instability but also to a whole class of additional instabilities. In contrast to the instabilities of a cylindrical plasma, some of them are weakly sensitive to the shear of the magnetic field and are not suppressed by a mean minimum of B (i.e., they remain when $V'' < 0$).

The presence of trapped particles may be manifested in two principal ways:

1. Their unfavorable magnetic drift may be important. It is this that is responsible for the instability mechanism discussed in §§11.9 and 16.3. As we have noted in §16.3, this effect is absent in $\max J_{\parallel}$ systems.
2. Because of their low longitudinal velocity, the trapped particles may collide relatively often with the remaining particles. (We recall that the collisional term contains the second derivative with respect to the velocity, $C \sim \partial^2/\partial v_{\parallel}^2$. As a result $\nu \sim 1/v_{\parallel}^3$.) This is the cause of a new class of dissipative instabilities that do not have an analog in a cylindrical plasma. These instabilities are not due to magnetic drift and may therefore arise in $\min J_{\parallel}$ or $\max J_{\parallel}$ systems.

The growth rates of the trapped-particle instabilities are larger, the larger is the relative fraction of trapped particles. This fraction is determined by the longitudinal inhomogeneity of the magnetic field. In a number of cases, strongly inhomogeneous magnetic fields are used to make traps with the maximum possible depth of the mean magnetic well. This improves the magnetohydrodynamic stability but, as is clear from the foregoing, must also favor the development of trapped-particle instabilities.

A longitudinal inhomogeneity of the magnetic field and the associated curvature of alternating sign of the lines of force somewhat modify the instability of a cylindrical plasma even if the effects of trapped particles are ignored. In particular, these factors make possible the existence of perturbations localized in regions of favorable or unfavorable curvature (cf. §16.4).

Bibliography

1. L. Spitzer, Phys. Fluids, 1:253 (1958).
2. J. Johnson et al., Phys. Fluids, 1:281 (1958). Questions of the stability of classical stellarators (§17.2) are discussed in [1, 2].
3. H. P. Furth and M. N. Rosenbluth, Phys. Fluids, 7:764 (1964). This paper contains the proposal to use multipole fields for producing closed configurations with $V'' < 0$.
4. L. S. Solov'ev and V. D. Shafranov, in: Plasma Physics and Controlled Nuclear Fusion Research, Vol. 1, IAEA, Vienna (1966), p. 169.
5. L. S. Solov'ev and V. D. Shafranov, in: Reviews of Plasma Physics, Vol. 5, Consultants Bureau, New York (1970), p. 1. In [4, 5] the methods and results of calculations for closed traps with $V'' < 0$ are discussed in detail.
6. H. Furth et al., in: Plasma Physics and Controlled Nuclear Fusion Research Vol. 1, IAEA, Vienna (1966), p. 103.
7. R. M. Kulsrud, ibid., p. 127.
8. V. D. Shafranov, ZhETF Pis. Red., 6:975 (1967) [JETP Letters 6:387 (1967)]. In [6-8] the magnetohydrodynamic theory of the stability of a finite-pressure plasma in traps with $V'' < 0$ is discussed. In [6,7] a model treatment indicates that a plasma is unstable if $\beta > (\pi a/L)^2 (\delta U/U)_*^{-1}$, where a is the characteristic transverse scale of the plasma; L is the length of the region with unfavorable curvature; and $(\delta U/U)_*$ is the relative height of the local magnetic peak. It is shown in [8] that a rigorous treatment taking into account the actual geometry leads in general to a larger value of the critical β.

Chapter 18

Plasmas in Closed Traps with Magnetic Surfaces and a Current

§18.1. Magnetohydrodynamic Stability of Current Systems

In contrast to systems without a current, the quantity $V''(\Phi)$ when a plasma is confined by means of a current system is not in general a unique characteristic of the degree of stability of a plasma. Physically, the explanation for this is that the mean magnetic drift of the particles in a solenoidal magnetic field is determined not only by the function $V''(\Phi)$ but also by more complicated parameters of the system that depend on the longitudinal current. This means that the analysis of the stability of current systems is more complicated than for current-free systems. If they are analyzed by the energy method, account must be taken of terms that are nonlinear in the current in the expression for the potential energy even if the plasma pressure is low. We shall summarize the results obtained by the energy method a little later; first, we shall discuss some qualitative aspects of the stability of current systems.

The problem of the stability of closed systems with a current is to a certain extent similar to the problem of the stability of a cylindrical plasma column in a longitudinal magnetic field (§13.2). The new aspects in this case are the curvature of the magnetic axis and some other parameters that characterize the magnetic axis and the field near the axis. It is to expected that if the curvature of the axis is very small the results concerning the stability of a cylindrical column will also be applicable to a closed current system.

The results mentioned in §13.2 concern two types of instability – the flute and kink instabilities. The flute instability is characterized by a growth rate of order $v_{T_i}/(aR)^{1/2}$ and Suydam's stability condition (13.5). In accordance with the model of ideal hydrodynamics, the kink instability has a growth rate of order $(B_\varphi/B_0)c_A/a$ and the stability condition depends on the distribution of the current over the radius of the column and the presence of a vacuum region. The flute instability is the most sensitive to a curvature of the column, and its very existence is due to curvature of the lines of force. This means that we must concentrate our attention on the flute and not the kink instability in our study of closed systems with a current.

The effect of curvature of the plasma column on the flute instability can be estimated by comparing the mean velocities of magnetic drift of the particles in cylindrical and curved columns. The particles of a cylindrical plasma drift with velocities of the order

$$V_{\mathrm{dr}} \simeq (\rho/R)\, v_T, \tag{18.1}$$

where v_T and ρ are the thermal velocity and the mean Larmor radius; $R = r(B_0/B_\varphi)^2$ is the radius of curvature of the lines of force.

In estimating the drift of particles in a curved field we shall for simplicity have in mind the case of a toroidal plasma with a circular magnetic axis (radius of curvature R_0). The local drift of the particles due to this curvature is $\sim (\rho/R_0)v_T$. However, this part of the drift depends on the azimuthal angle in the plane of the plasma cross section, so that it vanishes when averaged along a line of force. The next highest order in r/R_0 is therefore the principal order, so that

$$\bar{V}_{\mathrm{dr}} \simeq (\rho r/R_0^2)\, v_T. \tag{18.2}$$

A comparison of (18.1) and (18.2) shows that the effect of the curvature can be ignored if

$$(r/R_0)^2 < (B_\varphi/B_0)^2. \tag{18.3}$$

This condition is frequently written in the form

$$q < 1, \tag{18.4}$$

where $q \equiv rB_0/R_0B_\varphi$.

This makes it clear that for $q > 1$ the flute instability of a toroidal plasma must be described by relations that differ strongly from those of a cylindrical plasma. The general situation in this case is fairly complicated. Simplifying factors in the derivation of a stability condition for a curved configuration are as follows:

1. The fact that the perturbations have a small scale compared with the transverse dimension of the plasma. A similar assumption is used in the derivation of Suydam's condition for a cylindrical column.
2. The deviation of the plasma and magnetic field parameters from their mean values is small at all points of the corresponding magnetic surface.

Under these assumptions one can obtain a general geometric stability condition which generalizes the ordinary Suydam condition. In the simplest case of a plasma of circular cross section with a circular magnetic axis this condition reduces to

$$\frac{1}{4}\left(\frac{q'}{q}\right)^2 + \frac{8\pi p'}{rB_0^2}(1-q^2) \geqslant 0 \tag{18.5}$$

(the prime denotes the derivative with respect to the radius).

This inequality shows that if $q > 1$ there is shear-free stabilization of the flute instability.

More general results that follow from the general geometric stability condition can be found in the papers of Shafranov, Solov'ev, and Yurchenko cited at the end of the chapter.

§18.2. Two-Stream Instability in Current Systems

A plasma with a current is not a system in thermodynamic equilibrium even if it is spatially homogeneous. The presence of a longitudinal current corresponds to a relative motion of the electron and ion components or the presence of beams of fast electrons. Depending on the actual conditions, both these factors can lead to beam (two-stream) excitation of different branches of oscillations of the plasma. Chapter 12 in Volume 1 was devoted to a special discussion of the instabilities of a plasma with a longitudinal current. Using the results of that chapter, we can obtain the following picture of the possible beam instabilities in a closed system with a current.

If the current is produced by an electric field that exceeds the critical value, $E > E_{crit} \simeq \nu_{ei} v_{Te} m_e/e$, a plasma is formed with a current velocity that exceeds the thermal velocity of the electrons, $V > v_{T_e}$. In such a plasma, the Buneman instability must develop. It is characterized by

$$\left.\begin{aligned} &k_\perp \simeq k_\parallel \simeq \omega_{pe}/V, \\ &\gamma \simeq \mathrm{Re}\,\omega \simeq (m_e/m_i)^{1/3}\,\omega_{p_i}. \end{aligned}\right\} \quad (18.6)$$

If $E < E_{crit}$, only a small fraction of the electrons attain a high velocity. If the distribution of these electrons over the longitudinal velocities has a maximum, $\partial f/\partial v_\parallel > 0$, they must excite electron oscillation branches by means of the Cherenkov resonance, $\omega = k_\parallel v_\parallel$, with

$$k \simeq \omega_{pe}/v_1, \qquad \mathrm{Re}\,\omega \simeq \omega_{pe}, \qquad \gamma \simeq \alpha\omega_{pe}, \quad (18.7)$$

where α is the fraction of fast electrons and v_1 is their velocity. If $\partial f/\partial v_\parallel \ll 0$, this instability does not occur. Instead, cyclotron resonance $(\omega = k_\parallel v_\parallel - \omega_{Be})$ may excite oscillations with

$$k \simeq \omega_{Be}/v_1, \qquad \mathrm{Re}\,\omega \simeq \omega_{pe}\cos\theta, \qquad \gamma \simeq \alpha\,\mathrm{Re}\,\omega, \quad (18.8)$$

if $\omega_{p_e} < \omega_{B_e}$ and $v_1/v_{T_e} > \omega_{B_e}/\omega_{p_e}$. This instability is impossible if the velocities of the fast particles are not too high, $1 < v_1/v_{T_e} < \omega_{B_e}/\omega_{p_e}$. Instead of these oscillations, fast particles satisfying this inequality must excite high-frequency ion-acoustic oscillations if $T_e > T_i$. In this case

$$k \simeq 1/d_e, \qquad \mathrm{Re}\,\omega \simeq \omega_{p_i}, \qquad \gamma \simeq \alpha\omega_{p_i}(\omega_{pe}/\omega_{Be})^2. \quad (18.9)$$

Besides the directed motion of a small group of fast electrons when $E < E_{crit}$, there is also a directed motion of the majority of the electrons with velocity $V < v_{Te}$. If $V > v_{T_i}$ and $T_e > T_i$, the relative motion of the electrons and ions must excite high-frequency ion-acoustic oscillations. In this case

$$k \simeq 1/d_e, \qquad \mathrm{Re}\,\omega \simeq \omega_{p_i}, \qquad \gamma \simeq (m_e/m_i)^{1/2}\,\omega_{p_i}. \tag{18.10}$$

If $T_e \simeq T_i$, such an instability is impossible. Instead, we know from §6.3 of Volume 1 that in a sufficiently cold plasma, $\lambda_e \equiv v_{T_e}/\nu_{ei} < a$, there may be collisional excitation of low-frequency ion-acoustic oscillations:

$$k \simeq 1/\lambda_e, \qquad \mathrm{Re}\,\omega \simeq \nu_{ii}, \qquad \gamma \simeq \nu_{ie}. \tag{18.11}$$

In a plasma with $T_e \simeq T_i$ and $v_{T_i} < V < v_{Te}$, ion-cyclotron oscillations can be excited. In this case

$$\left.\begin{aligned} &k_\perp \simeq 1/\rho_i, \qquad k_\| \simeq \omega_{B_i}/V,\\ &\mathrm{Re}\,\omega \simeq \omega_{B_i}, \qquad \gamma \simeq 0.1\,(V/v_{Te})\,\omega_{B_i}. \end{aligned}\right\} \tag{18.12}$$

If the shear of the magnetic field is sufficiently small (this is the case in the central part of the plasma), then for $\beta > m_e/m_i$ and $c_A < V < v_{Te}$ a longitudinal current can excite Alfvén oscillations. This instability is characterized by

$$k_\perp \lesssim 1/\rho_i, \qquad k_\| \lesssim \omega_{p_i}/c,$$

$$\mathrm{Re}\,\omega \lesssim \omega_{B_i}, \qquad \gamma \simeq (V/v_{Te})\,\omega_{B_i}. \tag{18.13}$$

The instability picture becomes more complicated if the ions have a non-Maxwellian velocity distribution (effects of this nature have been discussed by Rudakov).

Besides the purely beam instabilities, the presence of a current and inhomogeneity can lead to beam-gradient instabilities, of which the current-convective instability of Kadomtsev is a representative. An instability of this kind may be especially important if there are large transverse gradients of the current in the plasma (for example, in the case of a skin current). In this case the growth rates may exceed the ion cyclotron frequency (see §2.4) and the longitudinal wave number may be of the same order as the transverse wave number. Perturbations of this kind are insensitive to a shear and can develop even if $\Theta \simeq 1$.

§18.3. Slow Gradient Instabilities in Current Systems with a Large Shear

The magnetic field of systems with a current may have a fairly large shear in a region in which the plasma pressure decreases. The presence of the shear will mean that the majority of the gradient instabilities important in the case of a shear-free field will be suppressed.

The effect of a shear on the various different types of gradient instabilities was investigated in Chapters 8 and 9. It follows from the analysis of these chapters that the most difficult perturbations to stabilize are those with $\partial \ln T/\partial \ln n_0 \geq 1$ having a longitudinal phase velocity of the order of the ion thermal velocity, $\omega/k_{\|} \simeq v_{T_i}$ (ion-acoustic branch), or of the order of the thermal velocity of the electrons, $\omega/k_{\|} \simeq v_{T_e}$ (electron-acoustic branch). The first have transverse wave numbers, growth rates, and frequencies of order

$$k_{\perp} \simeq \Theta/\rho_i, \quad \gamma \simeq \mathrm{Re}\,\omega \simeq v_{T_i}\Theta/a, \tag{18.14}$$

and the second,

$$k_{\perp} \simeq \Theta/\rho_e, \qquad \gamma \simeq \mathrm{Re}\,\omega \simeq v_{T_e}\Theta/a. \tag{18.15}$$

Because of the longitudinal inhomogeneity of the magnetic field and the resulting trapped particles, the above instabilities may be augmented by an instability due to the unfavorable magnetic drift of the trapped particles (see §§11.9 and 16.3) and the dissipative instability due to collisions between the trapped particles and the untrapped particles. In both cases the transverse wavelength may be comparable with the gradient scale of the plasma, $k_{\perp}a \simeq 1$. The growth rate of the first of these instabilities is approximately [see (11.91)]

$$\gamma \simeq (a/R)^{3/4}\,\omega^{*}, \tag{18.16}$$

and the second (see the papers of Kadomtsev and Pogutse) is

$$\gamma = (a/R)^2\,\omega^{*2}/\nu_{ei}. \tag{18.17}$$

In contrast to the instabilities predicted for a plasma in a homogeneous field, these instabilities are not sensitive to a shear.

The first of them may be eliminated if a max $J_{||}$ system (cf. §16.3) can be created, and the second disappears if the plasma temperature is raised.

Bibliography

Aspects of magnetohydrodynamic stability

1. J. M. Greene and J. L. Johnson, Phys. Rev. Lett., 7:401 (1961).
2. C. Mercier, Nucl. Fusion, Suppl., 2:801 (1962).
3. C. Mercier, Nucl. Fusion, 4:213 (1964).
4. J. M. Greene and J. L. Johnson, Phys. Fluids, 5:510 (1962).
5. L. S. Solov'ev, Zh. Eksp. Teor. Fiz., 53:626 (1967) [Sov. Phys. – JETP, 26:400 (1968)].
6. L. S. Solov'ev, Zh. Eksp. Teor. Fiz., 53:2063 (1967) [Sov. Phys. – JETP, 26:1167 (1968)].
7. J. M. Greene and J. L. Johnson, Plasma Physics, 10:729 (1968).
8. V. D. Shafranov, Nucl. Fusion, 8:253 (1968). In [1-8], the reader can find the derivation, discussion, and different forms of expression of the general condition of magnetohydrodynamic stability of a plasma in closed systems with magnetic surfaces. On the basis of the general condition the magnetohydrodynamic stability of different specific systems with circular cross section is investigated in [8].
9. B. B. Kadomtsev and O. P. Pogutse, Dokl. Akad. Nauk SSSR, 170:811 (1966) [Sov. Phys. – Doklady 11:858 (1967)].
10. A. A. Ware and F. A. Haas, Phys. Fluids, 9:956 (1966).
11. V. D. Shafranov and É. I. Yurchenko, Zh. Eksp. Teor. Fiz., 53:1157 (1967) [Sov. Phys. – JETP, 26:682 (1968)]. In [9-11] a condition of magnetohydrodynamic stability is deduced for the case of a torus with circular cross section. In [11] this result is analyzed in more detail.
12. V. D. Shafranov, ZhETF Pis. Red., 6:975 (1967) [JETP Letters 6:387 (1967)]. Discussion of the correspondence between the results that follow from the general magnetohydrodynamic condition and the results of a model treatment of the stability of closed systems made in the papers of Furth and Rosenbluth and Kulsrud cited in Chapter 17. It is shown that simulation does not always give a correct representation of the boundary of magnetohydrodynamic stability.
13. V. D. Shafranov and E. I. Yurchenko, Nucl. Fusion, 8:329 (1968).
14. L. S. Solov'ev, V. D. Shafranov, and E. I. Yurchenko, in: Plasma Physics and Controlled Nuclear Fusion Research, Vol. 1, IAEA, Vienna (1969), p. 175. In [13, 14] an investigation is made of the limiting pressure of a plasma in various types of closed systems.

Beam instabilities. For the bibliography on this question see Chapter 12 of Volume 1;

Slow gradient instabilities

15. B. B. Kadomtsev and O. P. Pogutse, in: Reviews of Plasma Physics, Vol. 5, Consultants Bureau, New York (1970), p. 249. Detailed review of all the principal types of instability in toroidal systems with a longitudinal current and corresponding bibliography. We shall merely mention some of the later papers.

16. A. A. Galeev, R. Z. Sagdeev, and H. V. Wong, Phys. Fluids, 10:1535 (1967). Investigation of the trapped-particle instability with allowance for a radial electric field.
17. O. P. Pogutse, Zh. Eksp. Teor. Fiz., 52:1536 (1967) [Sov. Phys. – JETP, 25:1021 (1967)]. Analysis of the instability associated with the finite diameter of the orbits.
18. O. Pogutse, Phys. Lett., 27A:63 (1967). Study of the effect of finite diameter of the orbits on the trapped-particle instability.
19. R. Z. Sagdeev and A. A. Galeev, Dokl. Akad. Nauk SSSR, 180:839 (1968) [Sov. Phys. – Doklady, 13:562 (1968)]. Investigation of the dissipative instability of trapped particles with allowance for damping due to interaction with untrapped particles.